London Mathematical Society Lecture Note Series: 349

Model Theory with Applications to Algebra and Analysis

Volume 1

ZOÉ CHATZIDAKIS
CNRS – Université Paris 7

DUGALD MACPHERSON
University of Leeds

ANAND PILLAY
University of Leeds

ALEX WILKIE
University of Manchester

CAMBRIDGE
UNIVERSITY PRESS

CAMBRIDGE UNIVERSITY PRESS
Cambridge, New York, Melbourne, Madrid, Cape Town, Singapore, São Paulo, Delhi

Cambridge University Press
The Edinburgh Building, Cambridge CB2 8RU, UK

Published in the United States of America by Cambridge University Press, New York

www.cambridge.org
Information on this title: www.cambridge.org/9780521694841

First published 2008

Printed in the United Kingdom at the University Press, Cambridge

A catalogue record for this book is available from the British Library

Library of Congress Cataloguing in Publication data

ISBN 978-0-521-69484-1 paperback

Table of contents for Volume 1

Table of contents for Volume 2

Preface

These two volumes contain both expository and research papers in the general area of model theory and its applications to algebra and analysis. The volumes grew out of the semester on "Model Theory and Applications to Algebra and Analysis" which took place at the Isaac Newton Institute (INI), Cambridge, from January to July 2005. We, the editors, were also the organizers of the programme. The contributors have been selected from among the participants and their papers reflect many of the achievements and advances obtained during the programme. Also some of the expository papers are based on tutorials given at the March-April 2005 training workshop. We take this opportunity, both as editors of these volumes and organizers of the MAA programme, to thank the Isaac Newton Institute and its staff for supporting our programme and providing a perfect environment for mathematical research and collaboration.

The INI semester saw activity and progress in essentially all areas on the "applied" side of model theory: o-minimality, motivic integration, groups of finite Morley rank, and connections with number theory and geometry. With the exception of motivic integration and valued fields, these topics are well represented in the two volumes.

The collection of papers is more or less divided into (overlapping) themes, together with a few singularities. Aspects of the interaction between stability theory, differential and difference equations, and number theory, appear in the first six papers of volume I. The first paper, based on Pillay's workshop tutorial, can also serve as a fast introduction to model theory for the general reader, although it quickly moves to an account of Mordell-Lang for function fields in characteristic 0. The "arithmetic of differential equations" figures strongly in Pillay's paper on the Grothendieck-Katz conjecture and its nonlinear generalizations, as well as in Bertrand's paper which initiates the investigation of versions of Ax-Schanuel for nonisoconstant semiabelian varieties over function fields. The Galois theory of difference equations is rather a hot topic and the Chatzidakis-Hardouin-Singer paper compares definitions and concepts that have arisen in algebra, analysis, and model theory.

Interactions of complex analytic geometry with model theory and logic (in the form of stability, o-minimality, as well as decidability issues) appear in papers 7 to 10 of volume 1. The papers by Peterzil-Starchenko and Moosa-Pillay (on nonstandard complex analysis and compact Kähler manifolds respectively) are comprehensive accounts of important projects, which contain new results and set the stage for future research. In the first, o-minimality is the model-theoretic tool. In the second it is stability. Wilkie's paper characterizes the holomorphic functions locally definable from a given family of holomorphic functions, and Macintyre's paper is related to his work on the decidability of Weierstrass functions. They are both set in the o-minimal context.

The o-minimality theme is continued in papers 12 and 13 of volume 1 from a (real) geometric point of view. In particular Rolin's paper is a comprehensive account of the most modern techniques of finding o-minimal expansions of the real field.

In recent years Zilber has been exploring connections between model theory and noncommutative geometry, and in his paper in volume I he succeeds in interpreting certain "quantum algebras" as Zariski structures. Fesenko's short note contains a wealth of speculations and questions, including the use of nonstandard methods in noncommutative geometry.

Definable groups of "finite dimension" in various senses (finite Morley rank, finite SU-rank, o-minimal) figure strongly in papers 1 to 5 of volume II. Papers 1 and 2 contain new and striking general results on groups of finite Morley rank, coming out of techniques and results developed in work on the Cherlin-Zilber conjecture. Paper 3 gives an overview of a model-theoretic approach to asymptotics and measure stimulated by the analogous results and concepts for finite and pseudofinite fields. The article by Hrushovski and Wagner, on the size of the intersection of a finite subgroup of an algebraic group with a subvariety, generalizes a theorem of Pink and Larsen. Otero's paper gives a comprehensive description of work since the 1980's on groups definable in o-minimal structures. This includes an account of the positive solution to "Pillay's conjecture" on definably compact groups which was proved during the Newton semester.

Hilbert's 10th problem and its generalizations, as well as first order properties of function fields, appear in papers 6 to 8 of volume II. The Pheidas-Zahidi and Eistenträger papers are based on tutorials given at INI, and give a comprehensive account of work on Hilbert's 10th problem for the rational field and for various rings and fields of functions. Paper 8 proves among other things definability of the constant field in function fields whose constant field is "large". The three papers together give a good picture of an exciting and very active subject at the intersection of logic and number theory.

The volumes are rounded off by important papers on Hrushovski constructions, ordered abelian groups, and continuous logic. In particular the paper 10 in volume II (based again on a tutorial) is an elementary and self-contained presentation of "continuous logic" or the "model theory of metric structures" which is fast becoming an autonomous area of model theory with links to both stability and functional analysis.

Zoé Chatzidakis
Dugald Macpherson
Anand Pillay
Alex Wilkie

Contributors

Itaï Ben Yaacov
Institut Camille Jordan
Université Claude Bernard (Lyon-1)
43 boulevard du 11 novembre 1918
69622 Villeurbanne cédex
France

Alexander Berenstein
Departamento de Matematicas
Universidad de los Andes
Carrera 1 Nro 18A-10
Bogotá
Colombia

Daniel Bertrand
Institut de Mathématiques de Jussieu
Université Paris 6
Boite 247
4 place Jussieu
75252 Paris cedex 05
France

Alexandre Borovik
School of Mathematics
University of Manchester
Oxford Road
Manchester M13 9PL
UK

Zoé Chatzidakis
UFR de Mathématiques
Université Paris 7 - Case 7012
2 place Jussieu
75251 Paris cedex 05
France.

Gregory Cherlin
Department of Mathematics
Rutgers University
110 Frelinghuysen Rd
Piscataway, NJ 08854
USA

Kirsten Eisenträger
Department of Mathematics
The Pennsylvania State University
109 McAllister Building
University Park, PA 16802
USA

Richard Elwes
School of Mathematics
University of Leeds
Leeds LS2 9JT
UK

Ivan Fesenko
Department of Mathematics
University of Nottingham
Nottingham NG7 2RD
UK

Olivier Frécon
Laboratoire de Mathématiques et Applications
Université de Poitiers
Téléport 2 - BP 30179
Boulevard Marie et Pierre Curie
86962 Futuroscope Chasseneuil cedex
France

Charlotte Hardouin
IWR
Im Neuenheimer Feld 368
69120 Heidelberg
Germany

C. Ward Henson
Department of Mathematics
University of Illinois
1409 W. Green St.
Urbana, IL 61801
USA

Ehud Hrushovski
Einstein Institute of Mathematics
The Hebrew University of Jerusalem
Jerusalem 91904
Israel

Eric Jaligot
Institut Camille Jordan
Université Claude Bernard (Lyon-1)
43 boulevard du 11 novembre 1918
69622 Villeurbanne cédex
France

Angus Macintyre
School of Mathematics
Queen Mary, University of London
Mile End Road
London E1 4NS
UK

Dugald Macpherson
School of Mathematics
University of Leeds
Leeds LS2 9JT
UK

Rahim Moosa
Department of Pure Mathematics
200 University Avenue West
Waterloo, Ontario N2L 3G1
Canada

Margarita Otero
Departamento de Matemáticas
Universidad Autónoma de Madrid
28049 Madrid
Spain

Ya'acov Peterzil
Department of Mathematics
University of Haifa
Haifa
Israel

Thanases Pheidas
Department of Mathematics
University of Crete
Knossos Avenue
71409 Iraklio, Crete
Greece

Anand Pillay
School of Mathematics
University of Leeds
Leeds LS2 9JT
England

Bjorn Poonen
Department of Mathematics
University of California
Berkeley, CA 94720-3840
USA

Florian Pop
Department of Mathematics
University of Pennsylvania, DRL
209 S 33rd Street
Philadelphia, PA 19104
USA

Jean-Philippe Rolin
Université de Bourgogne
I.M.B.
9. Avenue Alain Savary
BP 47870
21078 Dijon Cedex
France

Damian Rössler
Institut de Mathématiques de Jussieu
Université Paris 7 Denis Diderot
Case Postale 7012
2, place Jussieu
F-75251 Paris Cedex 05
France

Thomas Scanlon
University of California, Berkeley
Department of Mathematics
Evans Hall
Berkeley, CA 94720-3840
USA

Philip Scowcroft
Department of Mathematics
and Computer Science
Wesleyan University
Middletown, CT 06459
USA

Michael F. Singer
North Carolina State University
Department of Mathematics
Box 8205
Raleigh, North Carolina 27695-8205
USA

Sergei Starchenko
Department of Mathematics
University of Notre Dame
Notre Dame, IN 46556
USA

Alexander Usvyatsov
UCLA Mathematics Department
Box 951555
Los Angeles, CA 90095-1555
USA

Frank Wagner
Institut Camille Jordan
Université Claude Bernard (Lyon-1)
43 boulevard du 11 novembre 1918
69622 Villeurbanne cédex
France

A.J. Wilkie
School of Mathematics
Alan Turing Building
The University of Manchester
Manchester M13 9PL
UK

Karim Zahidi
Dept of Mathematics, statistics and
actuarial science
University of Antwerp
Prinsenstraat 13
B-2000 Antwerpen
Belgium

Martin Ziegler
Mathematisches Institut
Albert-Ludwigs-Universität Freiburg
D79104 Freiburg
Germany

B. Zilber
Mathematical Institute
24 - 29 St. Giles
Oxford OX1 3LB
UK

Model theory and stability theory, with applications in differential algebra and algebraic geometry

Anand Pillay[†]

University of Leeds

This article is based around parts of the tutorial given by E. Bouscaren and A. Pillay at the training workshop at the Isaac Newton Institute, March 29 - April 8, 2005. The material is treated in an informal and free-ranging manner. We begin at an elementary level with an introduction to model theory for the non logician, but the level increases throughout, and towards the end of the article some familiarity with algebraic geometry is assumed. We will give some general references now rather than in the body of the article. For model theory, the beginnings of stability theory, and even material on differential fields, we recommend [5] and [8]. For more advanced stability theory, we recommend [6]. For the elements of algebraic geometry see [10], and for differential algebra see [2] and [9]. The material in section 5 is in the style of [7]. The volume [1] also has a self-contained exhaustive treatment of many of the topics discussed in the present article, such as stability, ω-stable groups, differential fields in all characteristics, algebraic geometry, and abelian varieties.

1 Model theory

From one point of view model theory operates at a somewhat naive level: that of point-sets, namely (definable) subsets X of a fixed universe M and its Cartesian powers $M \times \cdots \times M$. But some subtlety is introduced by the fact that the universe M is "movable", namely can be replaced by an elementary extension M', so a definable set should be thought of more as a functor.

† This work is supported by a Marie Curie Chair as well as NSF grants DMS-0300639 and the FRG DMS-0100979

Subtlety or no subtlety, model theory operates at a quite high level of generality.

A (1-*sorted*) structure M is simply a set (also called M) together with a fixed collection of distinguished relations (subsets of $M \times \cdots \times M$) and distinguished functions (from $M \times \cdots \times M$ to M). We always include the diagonal $\{(x,x) : x \in M\} \subset M \times M$ among the distinguished relations. (Example: Any group, ring, lattice,... is a structure under the natural choices for the distinguished relations/functions.) These distinguished relations/functions are sometimes called the *primitives* of the structure M. From the collection of primitives, one constructs using the operations composition, finite unions and intersections, complementation, Cartesian product, and projection, the class of ∅-*definable* sets and functions of the structure M. Let us call this class $Def_0(M)$, which should be seen as a natural "category" associated to the structure M: the objects of $Def_0(M)$ are the ∅-definable sets (certain subsets of $M \times \cdots \times M$) and the morphisms are ∅-definable functions $f : X \to Y$ (i.e. $graph(f)$ is ∅-definable). The category $Def(M)$ of definable (with parameters) sets in M is obtained from $Def_0(M)$ by allowing also fibres of ∅-definable functions as objects: if $f : X \to Y$ is in $Def_0(M)$ and $b \in Y$ then $f^{-1}(b)$ is a definable set (defined with parameter b). For A a subset of the (underlying set of M) $Def_A(M)$ denotes the category of definable sets in M which are *defined over A*, namely defined with parameter which is a tuple of elements of A. By convention, by a definable set we mean a set definably possibly with parameters. By a uniformly definable family of definable sets we mean the family of fibres of a definable map $f : X \to Y$.

We give a couple of examples.

The reals.
Consider the structure consisting of $\mathbb{R}$ with primitives $0, 1, +, -, \cdot$. Then the natural total ordering on $\mathbb{R}$ is a 0-definable set, being the projection on the first two coordinates of $\{(x, y, z) \in \mathbb{R}^3 : y - x = z^2 \text{ and } x \neq y\}$. Tarski's "quantifier elimination" theorem says that the definable sets in $(\mathbb{R}, 0, 1, +, -, \cdot)$ are precisely the *semialgebraic sets*, namely finite unions of subsets of $\mathbb{R}^n$ of the form

$$\{x \in \mathbb{R}^n : f(x) = 0 \text{ and } g_1(x) > 0 \text{ and } \ldots g_r(x) = 0\}$$

where f and the g_i are polynomials over $\mathbb{R}$.

Algebraically closed fields

Consider the field $\mathbb{C}$ of complex numbers. An (affine) algebraic variety is a subset $X \subseteq \mathbb{C}^n$ defined by a finite system of polynomial equations in n-variables and with coefficients from $\mathbb{C}$. If the equations have coefficients from $\mathbb{Q}$ we say that X is defined over $\mathbb{Q}$. A morphism between algebraic varieties X and Y is a map from X to Y given by a tuple of polynomial functions. Such a morphism is over $\mathbb{Q}$ if the polynomial functions have coefficients from $\mathbb{Q}$. View $\mathbb{C}$ as a structure with primitives $0, 1, +, -, \cdot$. Then it is a theorem (quantifier-elimination) that the category $Def_0(\mathbb{C})$ consists, up to Boolean combination, of the affine algebraic varieties defined over $\mathbb{Q}$ with morphisms defined over $\mathbb{Q}$. Likewise $Def(\mathbb{C})$ is (up to Boolean combination) just the category of algebraic varieties and morphisms. Everything we have said applies with any algebraically closed field K in place of $\mathbb{C}$ and with the prime field in place of $\mathbb{Q}$.

Given a structure M, the *language* or *signature* $L = L(M)$ of M is an indexing of the primitives, or rather a collection of (relation/ function) symbols corresponding to the primitives of M. We call M an L-structure or structure for the signature L. There is a natural notion of an L-structure M being a substructure or extension of an L-structure N (generalizing the notions subgroup, subring, ...). But somewhat more crucial notions for model theory are those of *elementary substructure* and *elementary extension*. We may take the Tarski-Vaught criterion as a definition: So assume that M, N are L-structures and M is a substructure of N (notationally $M \subseteq N$). Then M is an elementary substructure of N if whenever $X \subseteq N^n$, $X \in Def_M(N)$, and $X \neq \emptyset$, then $X \cap M^n \neq \emptyset$.

It is usual to begin by introducing first order formulas and sentences of L, define the notion of their satisfaction/truth in L-structures, and develop the rest of the theory afterwards. So the first order formulas of L are built up in a syntactically correct way, with the aid of parentheses, from primitive formulas $R(x_1, \ldots, x_n)$, $f(x_1, \ldots, x_n) = y$ (where R, f are relation/function symbols of L and $x_1, \ldots, x_n, y$ are "variables" or "indeterminates") using $\neg, \wedge, \vee$, and quantifiers $\exists x, \forall x$. Among the L-formulas are those with no unquantified variables. These are called L-sentences. An L-formula with unquantified variables $x_1, \ldots, x_n$ is often written as $\phi(x_1, \ldots, x_n)$. Given an L-structure M and L-sentence σ there is a natural notion of "σ is true in M" which is written $M \models \sigma$, and for $\phi(x_1, \ldots, x_n)$ an L-formula, and $b = (b_1, \ldots, b_n) \in M^n$, a natural notion of "$\phi$ is true of b in M", written $M \models \phi(b)$.

As the logical operations $\neg, \vee, \ldots$ correspond to complementation, union, $\ldots$ we see that for M an L-structure the 0-definable sets of M come from L-formulas: if $\phi(x_1, \ldots, x_n)$ is an L-formula, then $\{b \in M^n : M \models \phi(b)\}$ is a 0-definable set in M and all 0-definable sets of M occur this way. Depending on one's taste, the syntactic approach may be more easily understandable. For example if M is a group $(G, \cdot)$, then the centre of G is a 0-definable set, defined by the formula $\forall y(x \cdot y = y \cdot x)$.

Likewise the definable sets in M are given by L-formulas with parameters from M.

With this formalism M is an elementary substructure of N (N is an elementary extension of M) if $M \subseteq N$ are both L-structures and for each L-formula $\phi(x_1, \ldots, x_n)$ and tuple $b = (b_1, \ldots, b_n) \in M^n$, $M \models \phi(b)$ iff $N \models \phi(b)$.

The compactness theorem of first order logic gives rise to elementary extensions of M of arbitrarily large cardinality, as long as (the underlying set of) M is infinite. Such an elementary extension N could be considered as some kind of "nonstandard" extension of M, in which all things true in M remain true. If X is a definable set in M then X has a canonical extension, say $X(N)$ to a definable set in N (in fact $X(N)$ is just defined in the structure N by the same formula which defines X in M). The usefulness of passing to an elementary extension N of a structure M is that we can find such elementary extensions with lots of symmetries (automorphisms) and "homogeneity" properties. Such models play the role of Weil's universal domains in algebraic geometry (and Kolchin's universal differential fields in differential algebraic geometry). The relative unfashionability of such objects in modern algebraic geometry is sometimes an obstacle to the grasp of what is otherwise the considerably naive point of view of model theory. Another advantage of such nonstandard models is that uniformly definable families of definable sets have explicit "generic fibres".

Given a cardinal κ, a structure M is called κ-*compact* if whenever $\{X_i : i \in I\}$ is a collection of definable subsets of M with the finite intersection property, and $|I| < \kappa$, then $\cap_{i \in I} X_i \neq \emptyset$.

Under some mild set-theoretic assumptions, any structure M has κ-compact elementary extensions of cardinality κ for sufficiently large cardinals κ. There is a related notion, κ-*saturation*: M is said to be κ-saturated if for any subset of M of cardinality $< \kappa$ any collection of A-definable subsets of M, which has the finite intersection property, has nonempty intersection. For κ strictly greater than the cardinality of L (number of L-formulas), κ-saturation coincides with κ-compactness.

Let κ be any uncountable cardinal. Then any algebraically closed field K of cardinality κ is κ-compact. Moreover let $\lambda < \kappa$, let $K[x_i : i < \lambda]$ be the polynomial ring in λ unknowns over K, and let S be any proper ideal of this ring, there is a common zero $(a_i)_{i<\lambda}$ of S whose coordinates lie in K.

We finish this section with some additional notation, conventions, and examples (aimed at the nonlogician).

Fix a language L. An L-*theory* is a set Σ of L-sentences which has a model. If σ is an L-sentence we write $\Sigma \models \sigma$ to mean that every model of Σ is a model of σ. The L-theory Σ is said to be *complete* if for every L-sentence σ either $\Sigma \models \sigma$ or $\Sigma \models \neg\sigma$.

A complete theory is often denoted by T. If T is such then we are interested in models of T and definable sets in such models. It has been a convention to choose a κ-compact model $\bar{M}$ of T of cardinality κ for some large κ. Then every model of T of cardinality $< \kappa$ is (up to isomorphism) an elementary substructure of $\bar{M}$. So when we speak of a model of T we refer to a small (cardinality $< \kappa$) elementary substructure of $\bar{M}$. We use $A, B, \ldots$ to denote small subsets of (the underlying set of) $\bar{M}$.

Complete types play an important role in model theory, especially in stability theory. If a is a tuple (usually finite) from $\bar{M}$ and A a subset of $\bar{M}$ then $tp(a/A)$ denotes the set of formulas $\phi(x)$ with parameters from A which are true of a in $\bar{M}$. Working rather with definable sets, $tp(a/A)$ can be identified with the collection of A-definable subsets of $\bar{M}^n$ which contain the point a, and is an ultrafilter on the set of A-definable subsets of $\bar{M}^n$. Then tuples a and b have the same type over A if there is an automorphism of $\bar{M}$ fixing A pointwise and taking a to b.

A basic example of a complete theory is ACF_p the theory of algebraically closed fields of characteristic p (where p is a prime, or is 0). The language here is the language of rings $(0, 1, +, -, \cdot)$. (Note that it is easy to write down first order sentences giving the axioms.)

A theory Σ (complete or not) in language L is said to have *quantifier-elimination* if every L-formula $\phi(x_1, \ldots, x_n)$ is equivalent (in models of Σ) to a quantifier-free L-formula. For example the (incomplete) theory ACF has QE. For specific theories it is important to have some kind of quantifier-elimination or relative quantifier elimination theorem so as to understand to some extent definable sets. But as far as the general theory of definability goes one can always assume quantifier-elimination by expanding the language by new relation symbols $R_\phi(x_1, \ldots, x_n)$ for each formula $\phi(x_1, \ldots, x_n)$.

The origin of stability theory was the (abstract) study of theories T which are *uncountably categorical*. The convention here is that the language L is countable, and then T is said to be uncountably categorical if for every (equivalently some, by Morley's theorem) uncountable λ, T has exactly one model of cardinality λ up to isomorphism. ACF_p is uncountably categorical.

Differentially closed fields.

An important example in this series of talks will be *differentially closed fields*. The relevant complete theory is DCF_0. The language here is the language of differential rings, namely the language of rings together with a new unary function symbol ∂. The axioms are the axioms for fields of characteristic 0 with a derivation ∂ (∂ is an additive homomorphism, and $\partial(x \cdot y) = \partial(x) \cdot y + x \cdot \partial(y)$), together with axioms which state that any finite system of differential polynomial equations and inequations with parameters (in finitely many indeterminates) which has a solution in differential field extension already has a solution in the model in question. It is a nontrivial fact that one can find such axioms, and in fact there are much simpler axioms (referring just to single differential equations in one indeterminate) which suffice, as shown by Blum. The theorem is that DCF_0 is complete and has quantifier-elimination.

Other important complete theories with interest are the theory of separably closed fields of characteristic p with Ershov invariant e (meaning that the dimension of K over K^p is p^e), and the theory of nontrivially valued algebraically closed fields of a given pair of characteristics. The latter is an important first order context for dealing with "infinitesimals".

Many-sorted structures

It is natural to consider many-sorted structures and theories in place of one-sorted ones. In this more general context, a structure M will be a family $(M_s : s \in S)$ of universes. The primitive relations and functions will be be on and between Cartesian products of universes. The language L of the structure will include the set of sorts S. Moreover any variable x comes equipped with a specific sort, and thus quantifiers will range over designated sorts. The whole machinery of first order logic (elementary extensions, saturation, formulas, theories) generalizes without difficulty to the many-sorted context. In fact this fits in very well with the above-mentioned notion of the category of definable sets in a given structure. For example, given a one-sorted structure M, form a new many-sorted structure $Def_0(M)$ whose sorts are the $\emptyset$-definable sets of M and whose relations are those induced by $\emptyset$-definable sets in M. For example, if X and Y are $\emptyset$-definable sets of M, and $R \subseteq X \times Y$ is a $\emptyset$-definable relation of M, then we have a corresponding basic relation in $Def_0(M)$ between the sort of X and that sort of Y. (Note that $Def_0(M)$ will automatically have quantifier-elimination in this presentation.) In fact one can go further: we can consider not only $\emptyset$-definable sets in M but also quotients of such by $\emptyset$-definable equivalence relations. So we take as sorts all sets of the form X/E where X is $\emptyset$-definable in M and E is an $\emptyset$-definable equivalence relation on X. Again we take as relations things induced by $\emptyset$-definable relations on M. Note that among the new basic functions will be the canonical surjections $X \to X/E$. We call this new many-sorted structure M^{eq}. The point is that M^{eq} is the "same" as M. (The technical term for sameness here is "bi-interpretable")

A typical example is obtained when we start with ACF_0 say and form the category of algebraic varieties defined over $\mathbb{Q}$ (again with the induced structure). We call this many-sorted structure AG_0 (algebraic geometry in characteristic 0).

A somewhat richer structure is the many-sorted structure $\mathcal{A}$ whose sorts are compact complex analytic spaces (up to biholomorphism) and whose relations are analytic subvarieties of (finite) cartesian products of sorts. More details will be given in a subsequent paper in the volume. The structure $\mathcal{A}$ is NOT κ-saturated for any cardinal κ. This is because every element of every sort is essentially named by a constant. We let CCM denote the first order theory of $\mathcal{A}$. Among the sorts in $\mathcal{A}$ are the projective algebraic varieties, and in this way AG_0 can be seen as a "subcategory" of CCM. CCM has quantifier elimination.

In many cases, it is not necessary to pass to M^{eq} in that the quotient sets are already present in M. This is when M (or $Th(M)$) has so-called elimination of imaginaries. So the structure M is said to have *elimination of imaginaries* if whenever X and E are $\emptyset$-definable sets in M ($X \subset M^n$ say and E an equivalence relation on X) then there is another $\emptyset$-definable $Y \subset M^k$ and a $\emptyset$-definable surjective function from X to Y such that $f(x_1) = f(x_2)$ iff $E(x_1, x_2)$.

ACF_p, DCF_0, CCM and $SCF_{p,e}$ all eliminate imaginaries (the latter after naming a p-basis).

The notions of algebraic and definable closure are important in what follows: Given a possibly many-sorted structure which eliminates imaginaries (for example M^{eq}), and a subset A of M, $dcl(A) = \{f(a) : a$ a finite tuple from A and f a $\emptyset$-definable function$\}$. We let $acl(A)$ denote the union of all finite A-definable sets. (For a structure which does not necessarily eliminate imaginaries, we have described what are usually called $dcl^{eq}(A), acl^{eq}(A)$.)

2 Stability

We will fix a many-sorted structure $M = (M_s : s \in S)$, which we assume to be saturated (κ-saturated of cardinality κ for some large κ) for convenience. We also assume that M has 'elimination of imaginaries. Let $X, Y, \ldots$ denote definable sets, and $A, B, \ldots$ sets of parameters.

What kind of relationships between definable sets can be formulated at this general level?

Definition 2.1 (i) X and Y are *fully orthogonal* if every definable $Z \subseteq X^n \times Y^m$ is up to Boolean combination of the form $Z_1 \times Z_2$.
(ii) At the opposite extreme: X is *internal* to Y if there is a definable surjective map f from Y^n to X (for some n).

A naive example of full orthogonality is the case where M consists of two infinite sorts M_0, M_1 with no additional relations. Put $X = M_0$ and $Y = M_1$. A rather trivial example of X being internal to Y is when $X = Y^n$ for some n. A more interesting example is when Y is equipped with a definable group structure, and there is a definable strictly transitive action of Y on X. Then the choice of a point $x \in X$ yields a definable bijection between Y and X.

A slight weakening of internality is *almost internality* where the map f above is replaced by a definable relation $R \subset Y^n \times X$ such that for any $x \in X$, there are only finitely many, but at least one, $y \in Y^n$ such that $R(y, x)$. In any case internality is a fundamental model-theoretic notion. The subtlety is that X and Y may be $\emptyset$-definable, and X may be internal to Y but only witnessed by a definable function defined with additional parameters. In such a situation there will be an associated nontrivial Galois group arising: a definable group naturally isomorphic to the group of permutations of X induced by automorphisms of M which fix Y pointwise.

Note that if X is finite then X is fully orthogonal to any Y. In AG

orthogonality is vacuous. In fact in AG if X, Y are infinite definable sets, then each is almost internal in the other.

Stability is an assumption on M (or on $Th(M)$) which played a very large role in Shelah's classification theory program, but also has many consequences for the structure of definable sets. For stable structures the study of *complete types* plays an important role.

Definition 2.2 M is *stable* if there is no definable relation $R(x, y)$ and a_i, b_i in M for $i < \omega$ such that for all $i, j < \omega$, $R(a_i, b_j)$ iff $i < j$.

As the ordering on $\mathbb{R}$ in the structure $(\mathbb{R}, <, +, \cdot)$ is definable we see that this structure is unstable. On the other hand AG is stable.

Independence (also called nondividing, nonforking).
Under the assumption of stability, a notion of *freeness* can be developed, giving meaning to "a is independent (free) from B over A" where $A \subset B$ are sets of parameters and a is a finite tuple of elements of M. In the case of AG (as a 1-sorted structure), assuming $A \subset B$ are subfields $F_1 < F_2$ of K this will mean precisely that $tr.deg(F_1(a)/F_1) = tr.deg(F_2(a)/F_2)$.

The precise definition depends on the notion of *indiscernibility*: a sequence $(b_i : i \in \omega)$ of tuples b_i of the same length is said to be indiscernible over a set A if for all n, $tp(b_{i_1}, \ldots, b_{i_n}/A) = tp(b_{j_1}, \ldots, b_{j_n}/A)$ whenever $i_1 < \cdots < i_n$ and $j_1 < \cdots < j_n$.

Definition 2.3 Let a, b be possibly infinite tuples, and A a set of parameters. We say that $p(x, b) = tp(a/A, b)$ *divides* over A if there is an A-indiscernible sequence $(b_i : i < \omega)$ with $b_0 = b$ such that $\{p(x, b_i) : i < \omega\}$ is inconsistent (not realized in M).

For T stable (or more generally "simple") nondividing is our notion of freeness and it has good properties: so a is free from b over A if $tp(a/A, b)$ does not divide over A, and we have properties such as
symmetry: a is free from b over A iff b is free from a over A;
free extensions: for every a, A and b there is a' with $tp(a'/A) = tp(a/A)$ and a' is free from b over A;
small bases: for any finite tuple a and set A, there is $A_0 \subseteq A$ of cardinality $\leq |L|$ such that a is free from A over A_0.

Stationarity. A characteristic property of independence in stable theories is "uniqueness of generic types" or "uniqueness of free extensions"

Fact 2.4 *Assume M stable. Let $A \subseteq B \subset M$ be sets of parameters. Assume A is algebraically closed. Let a_1, a_2 be tuples such that $tp(a_1/A) = tp(a_2/A)$ and each of a_1, a_2 is independent from B over A. Then $tp(a_1/B) = tp(a_2/B)$.*

We express the above fact by saying that complete types over algebraically closed sets are *stationary*.

In the case of AG, any stationary type is the "generic" type of an (absolutely) irreducible variety: Fact 2.4 says that if V is an irreducible variety over F, and a_1, a_2 are generic points of V over F then there is an automorphism of K fixing F pointwise and taking a_1 to a_2.

The notion of a "general" or "generic" point of a definable set may make sense in many contexts, especially where there is a notion of "dimension" for definable sets. However in the case of stable theories Fact 2.4 leads to an independence-theoretic characterization of full orthogonality:

Lemma 2.5 *(T stable.) Let X, Y be $\emptyset$-definable sets. Then X is fully orthogonal to Y iff for any set A of parameters, and $a \in X$ and $b \in Y$, a is independent from b over A.*

There is a notion of orthogonality for stationary types: p and q are orthogonal iff for any set A of parameters including the domains of p and q, and a realizing p independent from A over $dom(p)$ and and b realizing q independent from A over $dom(q)$ then a is independent from b over A. So the lemma above can be restated as: X and Y are fully orthogonal if and only if for all complete stationary types $p(x)$ containing $x \in X$ and $q(y)$ containing $y \in Y$, p is orthogonal to q.

Generalizing the notion of smallest field of definition of an algebraic variety, is the notion of the canonical base of a stationary type:

Fact 2.6 *(T stable.) Assume $tp(a/A)$ is stationary. Then there is smallest $A_0 \subset A$ such that a is independent from A over A_0 and $tp(a/A_0)$ is stationary. A_0 is called the canonical base of p.*

Morley rank and t.t. theories.
The notions of stability theory are a little clearer for so-called totally transcendental (t.t.) theories. T is said to be *t.t* if every definable set has an ordinal valued Morley rank.

Again work in a possibly many sorted saturated structure M. Let X

be a definable set in M. Then we define $RM(X) \geq 0$ if X is nonempty, and $RM(X) \geq \alpha + 1$ if there are disjoint definable subsets $X_1, X_2, \ldots$ of X such that $RM(X_i) \geq \alpha$ for all $i < \omega$ (with the obvious definition at limit ordinals.) Note that $RM(X) = 0$ iff X is finite and nonempty. It is a fact that if $RM(X) = \alpha$ then there is a greatest positive integer d such that X can be partitioned into d definable sets of rank α. We call d the Morley degree (or multiplicity) of X. Note that X has Morley rank 1 and Morley degree 1 just if X is infinite but cannot be partitioned into two infinite definable sets. Such X is also called *strongly minimal*.

Any *t.t.* theory T is stable.

Assume T to be *t.t.*. Let X be a definable set of Morley rank α and degree 1, and let A be any set of parameters over which X is defined. Then a generic point of X over A is (by definition) some $c \in X$ such that $c \notin Y$ for any definable set Y defined over A such that $RM(Y) < \alpha$. Our assumptions on X imply that $tp(c/A)$ is unique, or equivalently that any two generic points of X over A are conjugate by an automorphism of M fixing A pointwise. In any case, this generic type of X over A is stationary, and its free extension over any $B \supseteq A$ is simply the generic type of X over B.

Algebraic examples.

(i) In AG any definable set X has finite Morley rank which coincides with the algebraic-geometric dimension of the Zariski closure of X. In particular if we consider AG as a 1-sorted structure then the universe is strongly minimal.

(ii) Assume T to be stable. Then any infinite field definable in M is both additively and multiplicatively *connected*, namely has no definable additive or multiplicative subgroup of finite index. Hence any infinite ultraproduct of finite fields is unstable.

(iii) If T is *t.t.* then any definable group G in M has the *DCC* on definable subgroups. So $SCF_{p,e}$ is not *t.t.*

(iv) If $(A, +)$ is a commutative group then $Th(A, +)$ is stable. $Th(A, +)$ is *t.t* iff A has the *DCC* on definable subgroups.

(v) If V is a vector space over a field F, then $Th((V, +, 0, \lambda)_{\lambda \in F})$ is strongly minimal.

(vi) CCM is *t.t.* The Morley rank of a compact complex manifold X

(as a definable set in $\mathcal{A}$) is bounded by the dimension of X as a complex manifold.

Differentially closed fields continued.

Fix a saturated model $(\mathcal{U}, 0, 1, +, -, \cdot, \partial)$ of DCF_0. Let $\mathcal{C}$ denote the field of constants. An (affine) differential-algebraic variety, or Kolchin closed (or differential Zariski closed), set is some $X \subset \mathcal{U}^n$ which is the common zero set of a finite system of differential polynomial equations $P(\bar{x}, \partial(\bar{x}), ., \partial^r(\bar{x})) = 0$. (So here $P(\cdots)$ denotes a polynomial over $\mathcal{U}$.) As DCF_0 has QE any definable set $Y \subseteq \mathcal{U}^n$ is a finite Boolean combination of Kolchin closed sets, namely is a finite union of locally differential Zariski closed sets.

Let X be a Kolchin closed set, defined over k. We call X *finite-dimensional* if for all $a \in X$, $tr.deg(k(a, \partial(a), \ldots, \partial^r(a))/k)$ is finite.

Fact 2.7 X *is finite dimensional if $RM(X)$ is finite.*

A typical example of a finite-dimensional set is $\mathcal{C}$ (which is in fact strongly minimal). The subsets of $\mathcal{C}^n$ definable in $\mathcal{U}$ are just those definable in the structure $(\mathcal{C}, +, \cdot)$. So $\mathcal{C}$ with the "induced structure" is AG_0.

Typical examples of infinite-dimensional sets are positive-dimensional algebraic varieties: if $X \subset \mathcal{U}^n$ is an algebraic variety, then $RM(X) = \omega \cdot d$ where d is the algebraic-geometric dimension of X (or equivalently $RM(X)$ in the structure $(\mathcal{U}, +, \cdot)$).

Independence in DCF_0 has the following description: a is independent from b over k if $k(a, \partial(a), \ldots)$ and $k(b, \partial(b), \ldots)$ are algebraically disjoint over k (equivalently independent over k in the sense of AG).

The 1-sorted structure $\mathcal{U}$ has infinite Morley rank, but nevertheless we can build from $\mathcal{U}$ a very rich many-sorted structure of finite Morley rank, the algebraic-geometric significance of which will examined in subsequent lectures.

So fix some small differential subfield k of $\mathcal{U}$. Consider the family of all finite-dimensional Kolchin closed sets, defined over k (or alternatively the family of all sets of finite Morley rank definable over k). Make this into a many-sorted structure, whose sorts are the Kolchin closed sets under consideration, and whose relations are those induced from k-definable sets in $\mathcal{U}$. Call this many-sorted structure $\mathcal{D}_k$, which inherits the saturation of $\mathcal{U}$.

As mentioned above $\mathcal{C}$ is one of the sorts of $\mathcal{D}_k$, and can be identified with AG_0. We will see that orthogonality is a nonvacuous notion for $\mathcal{D}_k$.

In subsequent sections $\mathcal{D}$ will denote $\mathcal{D}_k$ for some small k.

Theories/structures of finite Morley rank.

We consider a many-sorted (saturated) structure M such that every sort (as a definable set) has finite Morley rank. It follows that every definable set in M has finite Morley rank. The general picture is that the structure of definable sets in M is controlled by strongly minimal sets in M modulo "fibrations". What we mean by this is: (i) any infinite $\emptyset$-definable set X of Morley degree 1 (but Morley rank arbitrary) is (generically) nonorthogonal to some definable strongly minimal set Y, and moreover (ii) let Y be as in (i), then there is a $\emptyset$-definable map f from X onto a $\emptyset$-definable set Z, such that $RM(Z) > 0$ and Z is internal to Y.

This picture is not completely true, for some technical but important reasons. Strongly minimal sets should be replaced by definable sets whose generic type is "minimal". We will see later that the category of compact complex spaces has finite Morley rank, and there is a pre-existing notion of the definable sets we have in mind. In the CCM context these are called *simple* ccm's. So we could call a definable set Y of Morley rank $m > 0$ and Morley degree 1, *simple* if there is no definable subset Y' of Y of Morley rank $0 < m' < m$ passing through a generic point of Y. In any case, replacing strongly minimal by *simple*, the picture above is true.

Nonorthogonality is an equivalence relation between strongly minimal sets:

Fact 2.8 *(T stable.) Let X, Y be strongly minimal sets. Then the following are equivalent:*
(i) X is not fully orthogonal to Y.
(ii) The generic types of X and Y are nonorthogonal,
(iii) X and Y are mutually almost internal (that is there some definable $R \subseteq X \times Y$ such that R projects onto X and onto Y and R is finite-to-finite).

3 Modularity

The stability-theoretic notion "modularity" stems from matroid theory. However, it has a substantial "geometric" content.

Let us work with a stable theory T, and let X be a $\emptyset$-definable set in M. The most abstract definition of modularity is: " X is modular if for any tuple a from X and any b from M, a is independent from b over $acl(a) \cap acl(b)$".

This description may be rather inaccessible. So let us specialise to the case where T has finite Morley rank (or is just $t.t$). By a normalized definable family of definable sets, we mean a definable family $\{Y_z : z \in Z\}$ of definable sets such that for some fixed m, each Y_z has Morley rank m and Morley degree 1 and for $z \neq z'$, $RM(Y_z \cap Y_{z'}) < m$.

Then X is modular if for any n, whenever $\{Y_z : z \in Z\}$ is a normalised definable family of definable subsets of X^n and $a \in X^n$, then $\{z \in Z : a$ is a generic point of $Y_z\}$ is finite. (So intuitively only finitely many Y_z pass through a given point of X^n.)

Modularity is a robust notion. For example (i) if X is modular and Y is almost internal to X then Y is modular, (ii) If $f : X \to Y$ is definable, Y is modular, and every fibre is modular then X is modular.

It follows from the observations made above that if T has finite Morley rank and every strongly minimal set is modular then T is modular (namely every definable set is modular).

Notice that any positive-dimensional algebraic variety X (as a definable set in an ambient algebraically closed field) is nonmodular as we can find infinite definable families of curves on $X \times X$ all passing through the same (generic) point.

As a positive consequence of modularity we have:

Fact 3.1 *If X is $\emptyset$-definable and modular and has a $\emptyset$-definable group operation, then every definable subset of X is a Boolean combination of translates of $acl(\emptyset)$-definable subgroups.*

The *old* Zilber conjecture stated roughly that (in the context where T has finite Morley rank) any nonmodular definable set X is related in some fashion to algebraic geometry.

One formal version is:

Conjecture (T of finite Morley rank.) If X is a nonmodular definable set, then there is a strongly minimal algebraically closed field internal to X.

In the special case where X is strongly minimal, the conjecture is equivalent to: X is nonmodular iff X is nonorthogonal to some strongly minimal algebraically closed field.

The conjecture is false (counterexamples were given by Hrushovski), but in the algebraic examples to be studied in this tutorial, the conjecture is true in a very strong way.

Strong conjecture (T of finite Morley rank.)
(i) Let $(Y_z : z \in Z)$ be a normalized definable family of definable subsets of a definable set X. Let $a \in X$ and let $Z_a = \{z \in Z : a$ is generic on $Y_z\}$. Then Z_a is internal to some strongly minimal definable algebraically closed field K.
(ii) Let K be an algebraically closed field definable in $M \models T$. Then any subset of K^n definable in M is definable in $(K, +, \cdot)$.

The strong conjecture can be seen as an "algebraicity" conjecture: certain definable sets are "algebraic varieties". The strong conjecture will be true in both the structure $\mathcal{D}$ (and will be "proved" below), as well as in the structure $\mathcal{A}$ of compact complex spaces. In $\mathcal{D}$ there is (up to definable isomorphism) only one definable algebraically closed field, namely the field of constants. The truth of the strong conjecture is thus related to the issue of when an algebraic variety over $\mathcal{U}$ "descends" to the field of constants, and as such connects to diophantine geometry over function fields. Likewise in $\mathcal{A}$ the truth of the strong conjecture is bound up with the issue of when a compact complex manifold or analytic space is "algebraic", namely bimeromorphic to a complex algebraic variety.

Appropriate versions of the strong conjecture are also valid in separably closed fields and generic difference fields.

Moreover in all these examples there will exist infinite modular definable sets (as opposed to the situation in AG).

4 Algebraic varieties over differential fields

We study in more detail finite-dimensional definable sets in DCF_0, that is the many-sorted structure $\mathcal{D}$ introduced earlier. In section 5 we outline a proof of the strong algebraicity conjecture for $\mathcal{D}$ and deduce the "Mordell-Lang conjecture for function fields in characteristic zero".

We start by recalling a more algebraic-geometric description of definable sets of finite Morley rank in DCF_0.

Work in a saturated differentially closed field $(\mathcal{U}, 0, 1, +, \cdot, \partial)$. Let $K, L, \ldots$ denote small differential subfields of $\mathcal{U}$. If $a = (a_1, \ldots, a_n)$ is a n-tuple of elements of $\mathcal{U}$, $\partial(a)$ denotes $(\partial(a_1), \ldots, \partial(a_n))$.

If $P(x_1, \ldots, x_n)$ is a polynomial over K, then dP denotes the sequence of partial derivatives $(\partial P/\partial x_1, \ldots, \partial P/\partial x_n)$, and P^∂ denotes the polynomial obtained by applying ∂ to the coefficients of P.

The basic observation which follows from the definition of a derivation, is:

(∗) If $P(a) = 0$, then $dP(a) \cdot \partial(a) + P^\partial(a) = 0$.

Let $V \subseteq \mathcal{U}^n$ be an irreducible algebraic variety defined over K. Let $I_K(V)$ be the set of polynomials over K vanishing on V. Let $T_\partial(V)$ denote the algebraic subvariety of $\mathcal{U}^{2n}$ defined by the equations $P(x) = 0$ together with $dP(x) \cdot u + P^\partial(x) = 0$, for P ranging over a set of generators of $I_K(V)$. Here $u = (u_1, \ldots, u_n)$. The projection to the first n coordinates gives a surjective morphism $\pi : T_\partial(V) \to V$. From (∗) we see that if $a \in V$ then $(a, \partial(a)) \in T_\partial(V)$.

An important (and even defining) property of differentially closed fields is:

Fact 4.1 *Let $V \subseteq \mathcal{U}^n$ be an irreducible algebraic variety over K and W an algebraic subvariety (also defined over K) of $T_\partial(V)$ which projects onto V. Then there is a generic point a of V over K such that $(a, \partial(a)) \in W$.*

If the variety V above is defined over the constants of K (that is $I_K(V)$ is generated by polynomials over C_K) then the P^∂ terms in the defining polynomials of $T_\partial(V)$ disappear and one has precisely the (Zariski) tangent bundle $T(V)$ of V, a "classical" object.

Definition 4.2 (i) By a D-variety (V, s) defined over K, we mean an algebraic variety V defined over K, together with a morphism (polynomial map) $s : V \to T_\partial(V)$ also defined over K such that $\pi \circ s = id$.
(ii) Given a D-variety (V, s), $(V, s)^\sharp$ (or just $V^\sharp$ is s is understood) is $\{x \in V : s(x) = (x, \partial(x))\}$.

Again in the case that V is defined over C_K, s is what is called a vector field (but defined over K) of V. In any case $V^\sharp$ is obviously a definable set of finite Morley rank in $\mathcal{U}$, and in fact every finite-dimensional definable set is essentially of this form: Let X be a definable set over K of Morley rank n say and Morley degree 1. Let $p(x) = tp(a/K)$ be its "generic type". So $tr.deg(K(a, \partial(a), \ldots, \partial^r(a), \ldots)/K)$ is finite. It follows that for some tuple $b = (a, \partial(a), \ldots, \partial^r(a), c)$ where $c \in K(a, \ldots, \partial^r(a))$, we have $\partial(b) = s_1(b)$ where s_1 is a polynomial function over K. Let V be the (irreducible) algebraic variety over K whose generic point is b. Let $s(x) = (x, s_1(x))$. Then s is a morphism from V to $T_\partial(V)$ and a section of $\pi : T_\partial(V) \to V$, and b is a generic point of $(V, s)^\sharp$ over K. As a and b are interdefinable over K, we obtain a definable bijection between $X \setminus Y$ and $(V, s)^\sharp \setminus Z$ for Y, Z of Morley rank $< n$.

Given an (irreducible) algebraic D-variety (V, s) over K, we are usually interested in the (unique) generic type of $(V, s)^\sharp$. In that case all we really care about is the fact that s is a *rational* section of π (so not necessarily everywhere defined). So we might define a *rational D*-variety to be an algebraic variety X together with a *rational* section s of π.

Likewise given an arbitrary definable set of Morley rank n and degree 1 we are in general interested in whether X is *generically* internal to $\mathcal{C}$, rather than internal to $\mathcal{C}$. So generically internal to $\mathcal{C}$ just means that $X \setminus Y$ is internal to $\mathcal{C}$ for some Y of Morley rank $< RM(X)$, or equivalently the generic type of X is internal to $\mathcal{C}$.

There is a natural notion of a morphism (or rational map) f between algebraic D-varieties (V, s) and (Y, t). Namely f should be a morphism (or rational map) defined over $\mathcal{U}$ between the algebraic varieties V and W, and $(df + f^\partial) \circ s$ should equal $t \circ f$. (In general, $d_\partial(f) =_{def} df + f^\partial$ is to df as $T_\partial(V)$ is to $T(V)$.)

Definition 4.3 Let (V, s) be an irreducible algebraic D-variety. Then
(i) (V, s) is isotrivial if there is an algebraic variety V_0 over $\mathcal{C}$ such that (V, s) is isomorphic to $(V_0, 0)$ over $\mathcal{U}$.
(ii) (V, s) is generically isotrivial if there is a Zariski open set $U \subset V$ such that $(U, s|U)$ is isotrivial.

Lemma 4.4 (V, s) *is generically isotrivial if and only if* $(V, s)^\sharp$ *is generically internal to* $\mathcal{C}$.

It is not hard to see that any (irreducible) algebraic variety X over $\mathcal{U}$ can be equipped with the structure of a *rational D*-variety, as this is just a matter of extending the derivation ∂ of $\mathcal{U}$ to a derivation on the function field of X. It is a more delicate issue to determine when an algebraic variety over $\mathcal{U}$ admits the structure of a D-variety, and to classify the possible D-variety structures. Buium examined this issue in some detail. Among his results is:

Theorem 4.5 [3]. *Let (V, s) be a projective D-variety over $\mathcal{U}$ (namely the underlying algebraic variety V is projective). Then (V, s) is isotrivial.*

So far we have seen that (for any differential subfield k of $\mathcal{U}$) the many-sorted structure $\mathcal{D}_k$ "essentially" coincides with the category whose objects are of the form (V, s), V a variety over k and $s : V \to T_\partial(V)$ a regular (or rational) section defined over k, and with morphisms as described above. (In the case of definable groups, this relationship is very tight.)

However such objects (V, s) are still quite far from "geometry". In fact a small restriction will enable us to considerably geometrize the picture. Let us assume that our saturated model $\mathcal{U}$ has size the continuum, whereby so does $\mathcal{C}$ and so we may suppose $\mathcal{C}$ to be the field $\mathbb{C}$ of complex numbers. Let (V, s) be a D-variety, where V is defined over $\mathbb{C}(a)$ say where $tp(a/\mathbb{C})$ has finite Morley rank (equivalently $tp(a)$ has finite Morley rank). We may assume that $\partial(a) \in \mathbb{C}(a)$. Then a is the generic point of an (irreducible) variety B over $\mathbb{C}$, and if $\partial(a) = s_0(a)$ then s_0 is a rational section of the tangent bundle $T(B)$, defined over $\mathbb{C}$. Let b be a generic point of $(V, s)^\sharp$ over $\mathbb{C}(a)$. Then (a, b) is the generic point of an algebraic variety X say defined over $\mathbb{C}$, and we have a natural projection $q : X \to B$ defined over $\mathbb{C}$. As $\partial(a, b) \in \mathbb{C}(a, b)$, X is equipped with a (rational) vector field s_1 defined over $\mathbb{C}$ lifting s_0. Namely the image of s_1 under the differential of q is s_0. So to (V, s) we have associated a dominant morphism $q : (X, s_1) \to (B, s_0)$ of varieties over $\mathbb{C}$ with (rational) vector fields over $\mathbb{C}$. From the latter we can reconstruct $(V, s)^\sharp$ as the fibre of $(X, s_1)^\sharp \to (B, s_0)^\sharp$ over a. So in this context, (V, s) can be considered as the generic fibre of $q : (X, s_1) \to (B, s_0)$. The lifting of s_0 to s_1 is more or less a "connection" on the fibration $q : X \to B$.

So the general point is that the many-sorted structure $\mathcal{D}_\mathbb{C}$ DOES have a geometric interpretation as the category of varieties with vector field over $\mathbb{C}$, and by considering generic fibres of maps we obtain arbitrary D-varieties over $\mathcal{U}$ as long as they are defined over "finite-dimensional" parameters. We now have a reasonably geometric account of internality/isotriviality.

Lemma 4.6 *Let (V, s) be the generic fibre of $(X, s_1) \to (B, s_0)$ where (X, s_1) and (B, s_0) are over $\mathbb{C}$. Then $(V, s)^\sharp$ is generically internal to $\mathcal{C}$ (or (V, s) is generically isotrivial) if and only if there are complex varieties with vector fields (B', s_0') and $(F, 0)$, and a dominant map $(B', s_0') \to (B, s_0)$ over $\mathbb{C}$, such that $(X, s_1) \times_{(B, s_0)} (B', s_0')$ birationally embeds (over $\mathbb{C}$) into $(F, 0) \times (B', s_0')$.*

Linear differential equations

We work in $\mathcal{U}$. Let K be a differential subfield with algebraically closed field of constants C_K. An nth order (homogeneous) linear differential equation over K is something of the form:

$$(*). \qquad a_n \partial^n(y) + a_{n-1}\partial^{n-1}(y) + \cdots a_0 y = 0, \text{ where } a_i \in K$$

Fact 4.7 *The set of solutions of $(*)$ in $\mathcal{U}$ is an n-dimensional vector space over $\mathcal{C}$. (So in particular, the set of solutions is internal to $\mathcal{C}$.)*

We can easily rewrite $(*)$ as a first order *system*

$$(**) \qquad\qquad\qquad \partial(y) = Ay$$

where y is a column vector $(y_1, \ldots, y_n)^t$ and A is an $n \times n$ matrix over K. We redefine a linear differential equation of degree n over K to be something of the form $(**)$. A solution is a column vector with entries from $\mathcal{U}$. Now $\partial - A$ is a "derivation" on $\mathcal{U}^n$ with respect to ∂, namely is additive and obeys the Leibniz law with respect to scalar multiplication by elements of $\mathcal{U}$. This is also called a *connection* on the $\mathcal{U}$-vector space $\mathcal{U}^n$, or a ∂-module over $\mathcal{U}$. So this gives another "definition" of a linear DE.

Again the solution set V of $(**)$ is an n-dimensional vector space over $\mathcal{C}$. We are interested in finding a fundamental system of solutions to $(**)$, namely a $\mathcal{C}$-basis of V. In fact such a basis will form a nonsingular $n \times n$ matrix U over $\mathcal{U}$. So the columns of U will form a basis of $\mathcal{U}^n$ consisting of solutions of $(**)$. For such a matrix U we have $\partial(U) = AU$. So we can redefine a linear DE over K to be something of the form

$$(***) \qquad\qquad\qquad \partial(X) = AX,$$

where A is an $n \times n$ matrix over K and X is a unknown matrix ranging over GL_n. Note that the map $X \to AX$ is an *invariant* vector field on GL_n, defined over K, and any invariant vector field on GL_n has this form.

Now assume $K = C_K(t)$ where $\partial(t) = 1$. So K is the function field of the affine line (or the projective line) over C_K, which is equipped with the constant (rational) vector field 1. Write $A = A_t$ to reflect the dependence on t. So as in the previous analysis, we obtain a bundle $q : GL_n \times \mathbb{P}^1 \to \mathbf{P}^1$, with a lifting of the constant vector field to the (rational) vector field $(t, X) \to (1, A_t X)$, and this vector field is GL_n-invariant on each fibre. This is precisely what is called a connection on a principal GL_n-bundle over $\mathbb{P}^1$, which is yet another "definition" of a linear DE over $\mathbb{C}(t)$. This can be done for any curve in place of the projective line. If we want to work over a higher dimensional base and maintain the language of connections on a principal bundle, we would need to work with several commuting derivations in place of just the one.

5 The strong conjecture for $\mathcal{D}$

We give a brief sketch of a proof of the strong conjecture (from section 3) for the category $\mathcal{D}$, and show how this has algebraic-geometric consequences.

Theorem 5.1 *Let X be a definable set of finite Morley rank in $\mathcal{U}$, defined over an algebraically closed differential field K. Let $(Y_z : z \in Z)$ be a normalized definable family of definable subsets of X. Let $z_0 \in Z$ and $a \in Y_{z_0}$ be generic over z_0. Then $\{z \in Z : a \in Y_z\}$ is generically internal to $\mathcal{C}$.*

Sketch of proof.
We have definable sets of finite Morley rank, X, $(Y_z : z \in Z)$, $a \in X$, and more or less have to prove that $Z_a = \{z \in Z : a$ is (generic) on $Y_z\}$ is internal to $\mathcal{C}$.
Suppose X to be defined over the differential field K, and we may assume a to be a generic point of X over K.

By the discussion in section 4 we may assume that $X = (V, s)^\sharp$, where (V, s) is an irreducible D-variety defined over K. Likewise we may assume Y_z to be of the form $(W_z, s|W_z)^\sharp$. Here $(W_z)_z$ is an algebraic family of irreducible algebraic subvarieties of V. So $Y_{z_1} = Y_{z_2}$ iff $W_{z_1} = W_{z_2}$ (so we may assume that z is a generator of the field of definition of W_z).

The first remark concerns just the algebraic varieties V and W_z. Fix z and we have $a \in W_z \subseteq X$. We have the local ring $\mathcal{O}(V)_a$ of rational functions on V defined at a. Likewise we have $\mathcal{O}(W_z)_a$ as well as a surjective homomorphism (of $\mathcal{U}$-algebras) $f_z : \mathcal{O}(V)_a \to \mathcal{O}(W_z)_a$. The kernel of f_z is the set of rational functions on V defined at a which vanish on some Zariski open subset of W. It is rather easy to see that the subvariety W_z of V is determined by $ker(f_z)$.

But the ring $\mathcal{O}(V)_a$ has a filtration by powers of its maximal ideal $\mathcal{M}(V)_a$ (the set of rational functions on V vanishing at a). Namely $\cap_r (\mathcal{M}(V)_a)^r) = 0$. Likewise with $\mathcal{O}(W_z)_a$. For each r, f_z induces a surjective linear map between the finite-dimensional $\mathcal{U}$-vector spaces

$\mathcal{O}(V)_a/(\mathcal{M}(V)_a)^r$ and $\mathcal{O}(W_z)_a/(\mathcal{M}(W_z)_a)^r$. Let us call this map $f_{z,r}$. By compactness it follows that:

Claim. There is r such that the map taking $z \in Z_a$ to $ker(f_{z,r})$ is generically injective.

On the other hand as $a \in (V,s)^\sharp$, the local ring $\mathcal{O}(V)_a$ is equipped with a derivation extending ∂, and $(\mathcal{M}(V)_a)^r$ will be a differential ideal, so

$\mathcal{O}(V)_a/(\mathcal{M}(V)_a)^r$ is equipped with the structure of a ∂-module over $\mathcal{U}$. Thus we have a linear differential equation as in section 4. After fixing a fundamental system of solutions, this ∂-module can be identified with $(\mathcal{U}^m, 0)$. Now $Ker(f_{z,r})$ turns out to be a ∂-submodule of $(\mathcal{U}^m, 0)$, and thus a vector subspace which is defined over $\mathcal{C}$.

Hence the map given by the Claim will be a definable embedding of Z_a into $Gr(\mathcal{U}^m)(\mathcal{C})$. (Here $Gr(\mathcal{U}^m)$ is the Grassmanian, namely the algebraic variety whose elements are vector subspaces of $\mathcal{U}^m$.) This proves the theorem.

Corollary 5.2 *Let G be a connected group of finite Morley rank definable in $\mathcal{U}$. Suppose G has a definable subset X of Morley degree 1 such that $Stab(X)$ is trivial, and X, X^{-1} generate G in finitely many steps. Then G is internal to $\mathcal{C}$. Thus there is a connected algebraic group H defined over $\mathcal{C}$ such that G is definably isomorphic to $H(\mathcal{C})$.*

Sketch of proof.
We will assume that G is commutative and use additive notation. Let $RM(X) = m$. By $Stab(X)$ we mean by definition

$$\{c \in G : RM(X \cap (X+c)) = m\}$$

(equivalently the Morley rank of the symmetric difference of X and $X+c$ is $< m$, as X has Morley degree 1). The assumption that $Stab(X) = \{0\}$ means that the family $\{X - d : d \in G\}$ is normalized. Let $a \in G$ be generic. Let $Z = \{d \in G : a \in X - d\}$. Then $d \in Z$ iff $a + d \in X$, so Z is in definable bijection with X. On the other hand, a Morley rank computation shows that for d generic in Z, a is generic in $X - d$. Hence by Theorem 5.1, Z is generically internal to $\mathcal{C}$. Thus also X is generically internal to $\mathcal{C}$. Our finite generation assumption implies that G is generically internal, thus internal, to $\mathcal{C}$. Thus there is a definable (in $\mathcal{U}$) bijection between G and a definable group $G_1 \subseteq \mathcal{C}^n$. As $\mathcal{C}$ with structure induced from $\mathcal{U}$ is just an algebraically closed field without additional structure, G_1 is definably isomorphic (in $\mathcal{U}$) to an algebraic group in the sense of $\mathcal{C}$, namely a group $H(\mathcal{C})$ where H is a connected algebraic group defined over $\mathcal{C}$.

An abelian variety is a connected commutative algebraic group whose underlying variety is projective. A semiabelian variety is a connected commutative algebraic group A which has an algebraic subgroup T isomorphic to some power of the multiplicative group and such that A/T is an abelian variety. Semi-abelian varieties are divisible as abstract commutative groups. Buium [2] proves that any abelian variety A over our differentially closed field $\mathcal{U}$ has a definable (in $\mathcal{U}$) subgroup G of finite Morley rank such that A/G is a vector space over $\mathcal{C}$. The argument goes through for a semiabelian variety A. It follows that if Γ is a "finite rank" subgroup of $A(\mathcal{U})$, in the sense that Γ is contained in the divisible hull of a finitely generated subgroup Γ_0 of $A(\mathcal{U})$, then there is a finite Morley rank definable subgroup H of $A(\mathcal{U})$ which contains Γ. (The image of Γ in A/G will be contained in a finite-dimensional $\mathcal{C}$-vector space V, so we can take H to be the preimage of V in A.)

Corollary 5.3 *([4]) Let A be a semiabelian variety, and X an irreducible subvariety, all defined over an algebraically closed field K of characteristic 0. Let k be an algebraically closed subfield of K (for example take k to be the algebraic closure of $\mathbb{Q}$). Let Γ be a "finite rank" subgroup of $A(K)$ (in the sense above). Assume that*
(i) $0 \in X$ and X generates A (in finitely many steps),
(ii) $Stab_A(X) =_{def} \{a \in A : a + X = X\} = \{0\}$, and
(iii) $X \cap \Gamma$ is Zariski-dense in X.
Then there is an isomorphism f of A with a semiabelian variety A_0 defined over k such that moreover $f(X) = X_0$ is also defined over k.

Sketch of proof.
We may assume that $K = \mathcal{U}$ for a differentially closed field $\mathcal{U}$ such that $k = \mathcal{C}$. Identify A with its group $A(\mathcal{U})$ of $\mathcal{U}$-rational points. Work in the structure $\mathcal{U}$. By the above discussion, let G be a definable subgroup of A of finite Morley rank containing Γ. We will assume G to be connected. Let $Y = G \cap X$. Then Y is definable, in fact Kolchin closed, and Zariski-dense in X. We will assume Y to be of Morley degree 1. By Zariski denseness, $Stab_G(Y) \subseteq Stab_A(X)$. Hence $Stab_G(Y)$ is trivial. There is no harm in assuming that Y generates G. Note that G is Zariski dense in A. By Corollary 5.2, G is definably isomorphic to $B(\mathcal{C})$ for some (commutative, connected) algebraic group B defined over $\mathcal{C}$. By

quantifier-elimination in DCF_0 the isomorphism i of $B(\mathcal{C})$ with G is given by the restriction of functions quantifier-free definable over $\mathcal{U}$ in the ring language to $B(\mathcal{C})$. Hence, as G is Zariski-dense in A, i extends to a surjective homomorphism $j : B \to A$ of (commutative) algebraic groups. Note that the unipotent part of B is defined over $\mathcal{C}$ and in the kernel of j, hence is trivial (as $g|B(\mathcal{C})$ is injective). So B is a semi-abelian variety, whereby any algebraic subgroup of B is defined over $\mathcal{C}$. We conclude that $ker(j)$ is trivial, and $j : B \to A$ is an isomorphism. As Y is Zariski-dense in X, $j^{-1}(Y)$ is Zariski-dense in $j^{-1}(X)$. But $j^{-1}(Y) \subset B(\mathcal{C})$, so $j^{-1}(X)$ is defined over $\mathcal{C}$. The proof is complete.

References

[1] E. Bouscaren (editor), *Model Theory and Algebraic Geometry*, Lecture notes in Mathematics **1696**, Springer-Verlag 1999.

[2] A. Buium, *Differentially algebraic groups of finite dimension*, Lecture Notes in Mathematics **1506**, Springer 1992.

[3] A. Buium, *Differential function fields and moduli of algebraic varieties*, Lecture Notes in Mathematics **1226**, Springer-Verlag, 1986.

[4] E. Hrushovski, The Mordell-Lang conjecture for function fields, *J. Am. Math. Soc.* **9** (1996), 667-690.

[5] D. Marker, *Model theory - an introduction*, Springer-Verlag, 2002.

[6] A. Pillay, *Geometric stability theory*, Oxford University Press, 1996.

[7] A. Pillay and M. Ziegler, Jet spaces of varieties over differential and difference fields, *Selecta Math., New ser.* **9** (2003), 579-599.

[8] B. Poizat, *A Course in model theory*, Springer-Verlag, 2000.

[9] Van der Put and M. Singer, *Galois Theory of Linear Differential Equations*, Springer-Verlag, 2003.

[10] I. Shafarevich, *Basic Algebraic Geometry 1, 2*, Springer-Verlag 1994.

Differential algebra and generalizations of Grothendieck's conjecture on the arithmetic of linear differential equations

Anand Pillay[†]
University of Leeds

Summary

We prove that a nonlinear version of the Grothendieck-Katz conjecture (essentially in a form given by Ekedahl, Shepherd-Barron and Taylor) is equivalent to the original Grothendieck-Katz conjecture together with a certain differential algebraic geometric/model-theoretic statement: a type over $\mathbb{C}(t)$ with "p-curvature 0 for almost all p" is nonorthogonal to the constants.

1 Introduction

The Grothendieck-Katz conjecture [5], which we will call (**G**) says that if

$$(*) \qquad\qquad dY/dt = AY$$

is a linear differential equation (in vector form), where A is an $n \times n$ matrix over $\mathbb{Q}(t)$, and for almost all primes p the reduction

$$(*)_p \qquad\qquad dY/dt = A_p Y$$

of the equation mod p, has a fundamental matrix of solutions with coefficients from the separable closure $\mathbb{F}_p(t)^{sep}$ of $\mathbb{F}_p(t)$ then the original equation $(*)$ has a fundamental matrix of solutions with coefficients from $\mathbb{Q}(t)^{alg}$.

In [4] a somewhat different conjecture (but also relating reduction mod p and differential equations) was stated and studied. We will call this

† This work is supported by a Marie Curie Chair as well as NSF grants DMS-0300639 and the FRG DMS-0100979

Conjecture (**F**): If (X, F) is an irreducible smooth variety equipped with a rank 1 algebraic subbundle F of $T(X)$, all defined over $\mathbb{Q}$, and for almost all p the "p-curvature" of the reduction mod p, (X_p, F_p), of (X, F) is 0, then F is "algebraically integrable", namely there is a rational map f from X to a variety Y such that on some Zariski open $U \subseteq X$, $Ker(df) = F$.

In the paper [4] it was pointed out that (**F**) implies (**G**), and so (**F**) can be considered as some kind of nonlinear Grothendieck-Katz. On the other hand, Chatzidakis and Hrushovski, in [3] give an interesting translation of this nonlinear Grothendieck-Katz into a differential algebraic/model-theoretic framework, as well as giving an equivalent *vector field* version of (**F**). The current paper is very influenced by, and in a sense continues, this work of Chatzidakis and Hrushovski. We will essentially prove that Conjecture (**F**) is equivalent to Conjecture (**G**) together with:

(**D**) Working in a differentially closed field of characteristic 0, if $r(x)$ is a complete type over $\mathbb{Q}(t)$ and the reduction of r mod p has "p-curvature 0" for almost all primes p, then r is *nonorthogonal* to the constants.

The conclusion of (**D**) is a "geometric" statement in differential algebra, concerning nontrivial maps into the constants, possibly "after base change". Whereas by the translation in [3] the conclusion of (**F**) is related to the existence of such maps into the constants, but without base change. The gap is explained by a "differential Galois group" and this is roughly how (**G**) is used.

To obtain the precise equivalence of (**F**) with [(**G**) and (**D**)] we will work in a more general setting (as is actually done in [4]), with arbitrary fields of characteristic 0 in place of $\mathbb{Q}$ and interpret "reduction mod p" and "for almost all p" accordingly. Also our proofs will be using what is on the face of it a somewhat stronger statement/conjecture (**G'**) in place of (**G**): (**G'**) concerns logarithmic differential equations on arbitrary algebraic groups rather than just on GL_n. However J.-B. Bost in [1] proves (as a corollary of a result on the existence of integral subvarieties of foliations) the equivalence of (**G'**) and (**G**) over number fields, and he has informed us that the same is true over arbitrary fields of characteristic 0.

In sections 2 and 3 we give some differential algebraic and model-theoretic background, We translate between various languages, and we summarise and expand somewhat on [3]. In particular we explain the meaning of (**D**) in the category of algebraic varieties equipped with vector fields. The proof of the main result is then given in section 4. All we use about the "p-curvature 0 for almost all p" hypothesis is that it is preserved under various natural operations. But we nevertheless give and use the definitions involving p-curvature, for at least cultural reasons.

I would like to thank D. Bertrand, J.-B. Bost, and S. Suer for various helpful discussions on the issues treated in the current paper. Thanks also to the Isaac Newton Institute for its hospitality in the spring of 2005, when this work germinated.

2 Differential algebra

We treat differential algebra as close to synonymous with the model theory of differential fields. Our notation and conventions are largely as in [3]. The reader can also see [7] for more on the model theory of differential fields. Here we recall some notation.

Let us start with a differential field (K, ∂) of arbitrary characteristic, and an irreducible algebraic variety X defined over K. The "shifted tangent bundle" $T_\partial(X)$ of X is the variety over K defined locally by
(i) the equations defining X, and
(ii) the equations $\sum_{i=1}^{i=n} (\partial P(x_1, \ldots, x_n)/\partial x_i) u_i + P^\partial(x_1, \ldots, x_n)$ where $P(x_1, \ldots, x_n)$ ranges over polynomials over K generating the ideal of X and P^∂ denotes the result of hitting the coefficients of P with ∂.

By a "shifted vector field" on X (over K) we mean a morphism $s : X \to T_\partial(X)$ which is defined over K and also a section of the canonical surjection $T_\partial(X) \to X$.

Such a pair (X, s) is called an algebraic D-variety over (K, ∂). If $\partial = 0$ on K then $T_\partial(X)$ coincides with the (Zariski) tangent bundle $T(X)$ of X and (X, s) is simply a variety over K together with a vector field on X (over K).

We will mainly be working in the situation where s is just a rational section, namely defined on a Zariski open subset of X. We will call such objects *rational D-varieties* (X, s) over differential fields (K, ∂).

Note that a rational D-variety (X, s) over (K, ∂) corresponds to a derivation ∂_s from the function field $K(X)$ of X to itself which extends ∂: if f is a K-rational function on X then $\partial_s(f) = \sum_{i=1}^{i=n} (\partial f/\partial x_i) s_i(x) + f^\partial(x)$.

In characteristic $p > 0$ the pth iterate of a derivation is also a derivation and for (X, s) a rational D-variety over K we define $s^{(p)}$ to be the K-rational section $X \to T_{\partial^p}(X)$ corresponding to the derivation ∂_s^p of

$K(X)$. (Likewise if s is a regular section of $T_\partial(X) \to X$, $s^{(p)}$ is a regular section of $T_{\partial^p}(X) \to X$.)

There is a natural "category" of algebraic D-varieties over (K, ∂). Given a morphism $f : X \to Y$ of varieties over K we can form the shifted derivative $df_\partial : T_\partial(X) \to T_\partial(Y)$ $(df_\partial = df + f^\partial)$, and f is by definition a morphism from (X, s) to (Y, t) if $df_\partial \circ s = t \circ f$ on X. If (X, s), (Y, t) are rational D-varieties over K then this makes sense if f is rational and dominant. Again if ∂ is the 0-derivation on K, the category is a familiar one of varieties over K equipped with (rational) vector fields over K.

For any differential field extension (L, ∂') of (K, ∂) we define the set of (L, ∂')-valued points of (X, s), notationally $(X, s)^\sharp(L, \partial')$ as $\{a \in X(L) : \partial'(a) = s(a)\}$, where the latter is understood in local coordinates.

The main *complete theory* relevant to our considerations is DCF_0 the theory of differentially closed fields of characteristic 0 (in the language of differential rings $+, -, \cdot, 0, 1, \partial$). The characteristic property of differentially closed fields (K, ∂) of characteristic zero, is that if (X, s) is an algebraic D-variety over (K, ∂) then $(X, s)^\sharp(K, \partial)$ is Zariski dense in X. We will fix a "saturated" differentially closed field $(\mathcal{U}, \partial)$ (a universal domain of uncountable cardinality κ say, in the sense of Kolchin). Every differential field of characteristic zero we consider will be a differential subfield of $\mathcal{U}$ of cardinality $< \kappa$. $\mathcal{C}$ will denote the field of constants of $\mathcal{U}$. If we take κ to be the continuum then $\mathcal{C}$ can be identified it with the complex field $\mathbb{C}$. In any case the theory DCF_0 is complete, has quantifier-elimination, and is ω-stable. If K is a differential subfield of $\mathcal{U}$ and (X, s) an (irreducible, rational) D-variety over K then there is a "generic" (over K) point a of $(X, s)^\sharp(\mathcal{U})$, say a, namely a is a generic point over K of the algebraic variety X and $\partial(a) = s(a)$. Moreover $tp(a/K)$ is unique, namely all generic points of $(X, s)^\sharp$ over K have the same complete type over K in the sense of the differential field $\mathcal{U}$. If a is a finite tuple from $\mathcal{U}$, then $dim(a/K)$ denotes the transcendence degree over K of the differential field $K\langle a \rangle$ generated by K and a. For any a and algebraically closed K, $dim(a/K)$ is finite if and only if (after possibly replacing a by some $(a, \partial(a), \ldots, \partial^n(a))$), a is a generic point over K of $(X, s)^\sharp$ for some irreducible rational D-variety (X, s) over K. If $K < L$ are differential fields (contained in $\mathcal{U}$) and a a tuple from $\mathcal{U}$ we say that a is independent from L over K if $K\langle a \rangle$ is algebraically disjoint from L over K. We will also let t denote (when appropriate) an element of $\mathcal{U}$ such that $\partial(t) = 1$. Thus for example the differential subfield $\mathcal{C}(t)$ of $\mathcal{U}$ is precisely $(\mathcal{C}(t), d/dt)$.

Now we recall some key model-theoretic notions used in this paper.

Assume K to be an algebraically closed differential field (differential subfield of $\mathcal{U}$). We say that $tp(a/K)$ is *nonorthogonal* to $\mathcal{C}$ if for some $L > K$ such that a is independent from L over K, there is some c in $acl(\mathcal{C} \cap acl(L\langle a \rangle)) \backslash acl(L)$. We say that $tp(a/K)$ is (almost) *internal* to $\mathcal{C}$ if for some $L > K$ such that a is independent from L over K, a is contained in the (algebraic closure of) the differential field generated by K and $\mathcal{C}$. So internality implies almost internality implies nonorthogonality. On the other hand a basic result in stability theory (see [8]) says:

Fact 2.1 *Suppose $tp(a/K)$ is nonorthogonal to $\mathcal{C}$. Then there is $b \in K\langle a \rangle \setminus K^{alg}$ such that $tp(b/K)$ is internal to $\mathcal{C}$.*

We will be mainly considering algebraic D-varieties (X, s) which are over either (a) fields of constants such as a number field or $\mathbb{C}$, or (b) differential fields $(K\langle b \rangle, \partial)$, where $\partial = 0$ on K and $dim(b/K)$ is finite. In the latter situation we may assume that $K(b) = K\langle b \rangle$, so that $\partial(b) = s'(b)$ for some K-rational function t. Then we can interpret (X, s) as the "generic fibre" of a suitable dominant morphism of rational D-varieties $(W, s'') \to (Y, s')$ where Y is the locus of b over K and so both (W, s'') and (Y, s') are defined over the field of constants K, and s', s'' are rational vector fields. For example a D-variety over $(K(t), d/dt)$ is a generic fibre of a dominant morphism $(W, s'') \to (\mathbb{P}^1, 1)$. over K.

Here is a routine translation of internality/nonorthogonality to the constants, in the cases which will be relevant to us.

Fact 2.2 *(i) Let (X, s) be an irreducible variety together with (rational) vector field, defined over an algebraically closed field K. Consider K as a field of constants (namely a subfield of $\mathcal{C}$). Then the generic type of $(X, s)^{\sharp}$ is internal to $\mathcal{C}$ (in the structure $\mathcal{U}$) if and only if there are (Y, s') and $(Z, 0)$ over K such that $(X, s) \times (Y, s')$ is birationally isomorphic to $(Z, 0) \times (Y, s')$ over (Y, s'). Namely iff (X, s) is "trivializable" after base change. Likewise the generic type of $(X, s)^{\sharp}$ is almost internal to $\mathcal{C}$ if there are (Y, s') and $(Z, 0)$ as above and a generically finite-to-one rational dominant morphism over (Y, s') from $(X, s) \times (Y, s')$ to $(Z, 0) \times (Y, s')$.*

(ii) Let again K be an algebraically closed field (of constants). Let $\pi : (X, s) \to (Y, s')$ be a dominant rational morphism, defined over K, with irreducible generic fibres. Let a be a K-generic point of $(X, s)^{\sharp}$. Let $b = \pi(a)$. Then b is a K-generic point of $(Y, s')^{\sharp}$. Moreover $tp(a/K, b)$

is nonorthogonal to C *(in* $\mathcal{U}$*) if and only if there is some* (Z, s'') *over* K *and dominant rational* $\pi' : (Z, s'') \to (Y, s')$ *and a dominant rational* $f : (X, s) \times_{(Y, s')} (Z, s'') \to (\mathbb{A}^1, 0) \times (Z, s'')$ *over* (Z, s'').

Of course in the notions internality/almost internality, one could also "specify" the required base change For example, let us supposing $q(x)$ to be a complete type over $K = K^{alg}$ and $K < L$, say that q is (almost) internal to C over L, if for some (any) a realizing q with a independent from L over K, a is contained in (the algebraic closure of) the field generated by L and C.

Maybe the first order theory in characteristic p most relevant to our considerations is the theory of the differential field $((\mathbb{F}_p(t))^{sep}, d/dt)$, which we call $SCF_{1,\partial}$, and essentially coincides with the theory of separably closed fields K of characteristic p such that K has dimension p over its pth powers. (However Carol Wood's theory, the model companion of differential fields of positive characteristic is also in the background.) A characteristic property of $SCF_{1,\partial}$ is:

Fact 2.3 ([3]) *Let* (K, ∂) *be a model of* $SCF_{1,\partial}$, *and* (X, s) *an (irreducible, rational) algebraic D-variety defined over* (K, ∂). *Then* $s^{(p)} = 0$ *iff* $(X, s)^{\sharp}(K, \partial)$ *is Zariski-dense in* X *(iff there is an elementary extension* (K, ∂) *of* (K, ∂) *and a generic point* $a \in X(L)$ *of* X *over* K *such that* $\partial(a) = s(a)$*).*

We now discuss "reduction mod p" following [4]. Suppose we are in situation (a), an algebraic variety X over C say together with a vector field s. Then there is a finitely generated $\mathbb{Z}$-algebra R such that X and s are "defined over" R in the sense that definitions of X and s without denominators, and with coefficients from R can be given. For any maximal ideal $\mathbf{p}$ of R we can reduce (X, s) mod $\mathbf{p}$ to obtain $(X_{\mathbf{p}}, s_{\mathbf{p}})$ defined over the residue field $R/\mathbf{p}$ which will be a finite field. Abusing terminology somewhat we will say that "$s_p^{(p)} = 0$ for almost all p" if for all maximal ideals $\mathbf{p}$ of R outside some proper closed subscheme of $Spec(R)$, $s_{\mathbf{p}}^{(p)} = 0$ where p is the characteristic of the reside field $R/\mathbf{p}$. This does not depend on the choice of R. If (X, s) is defined over a number field, then "for almost all p" corresponds to "for all but finitely many prime ideals of the ring of integers of the number field".

Suppose now we are in situation (b). In fact we will consider just the case where (X, s) is over (L, ∂) where K is a finite extension of $C(t)$ and ∂ is the unique extension of d/dt to a derivation on L. For simplicity assume here that (L, ∂) is precisely $(C(t), d/dt)$. Then again (X, s) is over $R(t)$ for some finitely generated $\mathbb{Z}$-algebra R. We can reduce the coefficients from R modulo any maximal ideal $\mathbf{p}$ of R to obtain $(X_{\mathbf{p}}, s_{\mathbf{p}})$ over $(F(t), d/dt)$ for some finite field F. And "$s_p^{(p)} = 0$ for almost all p" will have the same meaning as before.

It is convenient to extend this notation to types as follows. Suppose K to be a field of constants, and let $L = K^{alg}$ or $K(t)^{alg}$. Let a be a finite tuple from $\mathcal{U}$ such that $L\langle a \rangle = L(a)$ and $tr.deg(L(a)/L)$ is finite (namely $dim(a/L)$ is finite. Then a is the generic point of an irreducible D-variety (X, s) defined over L and with abuse of notation we say that $tp(a/L)$ has "p-curvature 0 for almost all p" if "$s_p^{(p)} = 0$ for almost all p".

The following exercise will be used heavily. It is rather routine, using density of separable points, as in the proof of Proposition 5.11 in [3]. The proof in section 4 of the main theorem uses just these properties of "p-curvature 0", so one could actually ignore the actual definition of "p-curvature 0" if one wished.

Lemma 2.4 *(with above notation) Assume that $tp(a/L)$ has p-curvature 0 for almost all p. Then*
(i) if $b \in L(a)$ and $L\langle b \rangle = L(b)$ then $tp(b/L)$ has p-curvature 0 for almost all p,
(ii) If c is a tuple of constants and $L_1 = L(c)^{alg}$ then $tp(a/L_1)$ has p-curvature 0 for almost all p,
(iii) if $a_1, \ldots, a_n$ is an L-independent sequence of realizations of $tp(a/L)$ then $tp(a_1, \ldots, a_n/L)$ has p-curvature 0 for almost all p.

Finally we return to the Grothendieck-Katz conjecture and a "mild" generalization. Let K be a field of constants and $L = K(t)^{alg}$. Then a linear differential equation (in vector form) over L is something of the form

$$(*) \qquad\qquad \partial(Y) = AY,$$

where Y is a $n \times 1$ vector of unknowns and A an $n \times n$ matrix with coefficients from L. A so-called fundamental matrix of solutions to $(*)$ is an $n \times n$ matrix Z with coefficients from $\mathcal{U}$ whose columns form a basis of the (n-dimensional) $\mathcal{C}$-vector space of solutions to $(*)$. Equivalently (via the Wronskian argument) $Z \in GL_n(\mathcal{U})$ and $\partial(Z) = AZ$. Now the map $g \to Ag$ on GL_n is a vector field s_A say, defined over L. In fact s_A is an invariant vector field, and the matrix A over L is precisely the element of the Lie algebra gl_n of G corresponding to s_A. Moreover any invariant vector field on GL_n has this form. In any case (GL_n, s_A) is a very special kind of algebraic D-variety over L, and so the Grothendieck-Katz conjecture says (with above notation, and using Fact 2.3):

(**G**): Suppose A is an $n \times n$ matrix over L and s_A has p-curvature 0 for almost all p. Then there is $g \in GL_n(L)$ such that $\partial(g) = Ag$.

It is a more or less obvious thing to try to generalize (**G**) to arbitrary algebraic groups (over the constants). In fact the case of linear groups is already equivalent to (**G**). In any case we obtain

Conjecture (**G'**): Let L be as before (namely $K(t)^{alg}$ for some algebraically closed field K of constants). Let G be a connected algebraic group over K and s an invariant vector field on G defined over L. Assume that s has p-curvature 0 for almost all p. Then there is $g \in G(L)$ such that $\partial(g) = s(g)$.

The invariant vector field s on G corresponds to some L-rational element a of the Lie algebra $Lie(G)$ of G. The equation $\partial(x) = s(x)$ on G coincides with the Kolchin's "logarithmic differential equation" $l\partial(x) = a$ (on G). (This is because Kolchin's logarithmic derivative $l\partial$ is precisely the composition of $\partial : G \to T(G)$ with the differential of translation to the identity.)

Logarithmic differential equations on algebraic groups over the constants are bound up with Kolchin's theory of *strongly normal* extensions of differential fields and their Galois theory, as we will see in section 4.

3 Integrability, connections, principal bundles

Let X be smooth and irreducible variety over an algebraically closed field of any characteristic. Let F be an "involutive distribution" on X, that is an algebraic vector subbundle of the tangent bundle $T(X)$. Involutive means closed under Lie brackets. We say that F or (X, F) is *algebraically integrable* if there is a rational map f from X to a variety Y such that $Ker(df) = F$ on a nonempty Zariski open subset U of X. We will be considering the case of rank 1-subbundles of $T(X)$, namely $dim(F_x) = 1$ for all $x \in X$, which are automatically involutive. Moreover in this case there is a rational vector field $s : X \to T(X)$ generating (or spanning) F. Note that a rational vector field on X corresponds to a derivation (of the function field of X). In characteristic $p > 0$, the pth iterate of a derivation is also a derivation. So one can ask whether or not F is closed under the pth power operation. If it is we say that F (or (X, F)) has

p-curvature 0. It is worth noting than in this case (where F is a rank 1 subbundle of $T(X)$) whether or not the pth iterate of a generating vector field of F is a scalar multiple of the original, does not depend on the choice of the generating vector field.

Now suppose that (X, F) is defined over a field K of characteristic 0. Then as in section 2, (X, F) is defined over a finitely generated $\mathbb{Z}$-algebra R and for any maximal ideal $\mathbf{p}$ one can reduce mod $\mathbf{p}$ to obtain $(X_{\mathbf{p}}, F_{\mathbf{p}})$. And we say that F_p (or (X_p, F_p) has p-curvature 0 for almost all p if for all maximal ideals $\mathbf{p}$ of R in an open dense subscheme, $(X_{\mathbf{p}}, F_{\mathbf{p}})$ has p-curvature 0 where p is the characteristic of the finite field $R/\mathbf{p}$.

The Ekedahl, Shepherd-Barron, Taylor conjecture [4], in the framework of rank 1 bundles (X, F) is then:

Conjecture (**F**): Let (X, F) be defined over a field K of characteristic 0. Then (X, F) is algebraically integrable if and only if for almost all p, (X_p, F_p) has p-curvature 0.

We now state the key result from the second part of [3] (omitting the notion "parametric integrability"). Although stated in [3] over number fields the proof remains valid in our more general context. The conclusions of (**F'**) and (**F''**) refer to definability in the differentially closed field $\mathcal{U}$ with field of constants $\mathcal{U}$, and as before t denotes an element of $\mathcal{U}$ with $\partial(t) = 1$.

Proposition 3.1 (5.16, [3]) *The following are equivalent:*
(i) Conjecture **F**.
(ii) Conjecture **F'**: *If (X, s) is an irreducible algebraic D-variety defined over algebraically closed $K < \mathcal{C}$ and, $s_p^{(p)} = 0$ for almost all p, then the generic type over K, q of $(X, s)^\sharp$ is almost internal to $\mathcal{C}$ over $K(t)^{alg}$.*
(iii) Conjecture **F''** : *If $K < \mathcal{C}$ and (X, s) is an irreducible D-variety over $K(t)^{alg}$ with $s_p^{(p)} = 0$ for almost all p, then the generic type q of $(X, s)^\sharp$ is almost internal to $\mathcal{C}$ over $K(t)^{alg}$ (so without base change).*

It is worthwhile giving the following relationship between the conclusions of (**F**) and (**F'**), which does not appear explicitly in [3] (although it should have):

Remark 3.2 *Let (X, F) be a variety X with rank 1 distribution F, over an algebraically closed field K of characteristic 0. Assume $K < \mathcal{C}$. Then (X, F) is algebraically integrable if and only if there is some rational vector field s representing F such that in $\mathcal{U}$ the generic type of $(X, s)^\sharp$ is almost internal to $\mathcal{C}$ over $K(t)^{alg}$.*

Proof. Right to left is proved in [3].

Left to right: Let $g : X \to Y$ witness the algebraic integrability of F. We may assume that the generic fibre of g is irreducible. For $a \in X$ generic, let $C_a = g^{-1}(g(a))$, an integral curve of F. Some projection of C_a on one of the coordinates gives a dominant rational (étale) map f_a from C_a to $\mathbb{P}^1$. So df_a is an isomorphism between the tangent space of C_a at a and the tangent space of $\mathbb{P}^1$ at $f_a(a)$. But the tangent space to C_a at a is precisely F_a. Define $s(a)$ to be the tangent vector to $C(a)$ such that $df_a(s(a)) = 1$. s is then a rational vector field on X which represents F and such that for general $a \in X$, $s|C(a)$ maps to the constant (rational) vector field 1 on $\mathbb{P}^1$ under df_a. Note that the map taking $x \in X$ to $f_x(x) \in \mathbb{P}^1$ is defined over the same parameters as X, s and g are defined. Let us call this (dominant, rational) map from X to $\mathbb{P}^1$, h. So $dh(s) = 1$.

Note that everything defined so far is defined over K. Now let us work again in the differentially closed field $(\mathcal{U}, \partial)$ whose field of constants is precisely $\mathcal{C}$. Let a be a generic point (over K) of $(X, s)^\sharp$ (So $a \in X(\mathcal{U})$.) As f is defined over the constants, we see that $\partial(h(a)) = dh(\partial(a)) = dh(s(a)) = 1$. Likewise $\partial(g(a)) = dg(\partial(a)) = dg(s(a))$. But $s(a) \in F_a \in ker(dg)$ so $\partial(g(a)) = 0$. Let $t_1 = h(a)$ and $c = g(a)$. So as the curve $C(a)$ is defined over c, and the projection from $C(a)$ to $\mathbb{P}^1$ is dominant and also defined over $K(c)$, we see that $a \in acl(K, t_1, c)$. But as $\partial(t_1) = 1$, $t_1 = t + d$ for some constant d. Thus $a \in acl(K, t, d, c)$, where d, c are constants. We have shown that the generic type of $(X, s)^\sharp$ is almost internal to $\mathcal{C}$ over $K(t)$, completing the proof.

Let us note quickly that Conjecture (**G'**) follows from Conjecture (**F**), using Proposition 3.1. So let G be a connected algebraic group over algebraically closed $K < \mathcal{C}$, and let s_A be an invariant vector field on G defined over $K(t)^{alg}$ with p-curvature 0 for almost all p. (**F"**) gives us some $g \in G(\mathcal{U})$ such that $g \in acl(K(t)^{alg}, \mathcal{C})$ and $\partial(g) = s_A(g)$. But there is an elementary substructure L of $\mathcal{U}$ such that $K(t)^{alg} < L$ and K is the constants of L (take for L the prime model over $K(t)$ for example). So we obtain $g \in G(L)$ with $\partial(g) = s_A(g)$ and g in the algebraic closure of $K(t)$ together with the constants of L, namely $g \in G(K(t)^{alg})$ as required.

Finally we discuss "connections on principal bundles" to relate to the language of Bost's work [1]. The theory of connections on a principal fibre bundle is well-established in differential geometry. Analogous notions in the algebraic geometric context seem to be less well-documented although well-known. In the appendix to [1] this algebraic-geometric theory is explained, and the reader is referred there for more details. In fact the notion of a connection on a principal bundle corresponds to Kolchin's "logarithmic derivations on algebraic groups" which play a fundamental role in his Galois theory of "strongly normal extensions". As we are working with ordinary differential fields (rather than fields with several commuting derivations), we will consider G-bundles over curves rather than over higher dimensional varieties.

Work over a fixed algebraically closed field K. Fix a connected algebraic group G over K and a smooth (not necessarily complete) curve B over K. By a principal G-bundle over B, we mean roughly a smooth irreducible variety X, together with an action of G on X, and a smooth surjective G-invariant morphism from X to B, such that $X \times G$ is naturally isomorphic to $X \times_B X$. A special case of a principal G-bundle over B is $X = G \times B$ with the action $g(h, b) = (gh, b)$. Such a thing is called a trivial G-bundle. It is stated as a basic fact in [1] that a principal G-bundle over B becomes trivial after base change by some surjective étale morphism $B' \to B$. As we will be free to make such base changes, we will assume that our principal G-bundles are trivial.

So let us fix a (trivial) G-bundle $X = G \times B$, with $q : X \to B$ the natural projection. Identify $T(G)$ with the "kernel" T_q of dq. A connection on the G-bundle $q : X \to B$ is then defined to be a splitting $T(G) \oplus H$ of $T(X)$ where H is a G-equivariant algebraic subbundle of $T(X)$. Namely for any $(x, u) \in H$ and $g \in G$, the image of (x, u) under the action by g and its derivative, is also in H. We let (X, H) denote the corresponding "principal G-bundle with connection". The connection is said to be isotrivial (or even trivial) if the splitting $T(X) = T(G) \oplus H$ is isomorphic to the trivial splitting $T(X) = T(G) \oplus T(B)$. The connection is said to be "trivial on some finite étale covering" if there is a finite étale morphism $\nu : B' \to B$ such that the unique lifting of the connection (X, H) over B to a connection over B' is isotrivial. Such a property is not guaranteed. On the other hand a principal G-bundle with connection (X, H) is also a rank 1 subbundle of $T(X)$, and in characteristic p one can ask whether its p-curvature is 0. Likewise in characteristic 0 we have the notion "p-curvature of (X_p, H_p) being 0 for almost all p".

A straightforward translation yields that Conjecture (**G'**) is precisely:

(∗∗) If G is a connected algebraic group, (X, H) is a principal G-bundle with connection over a smooth curve B, all defined over a field K of characteristic 0, and H has p-curvature 0 for almost all p, then (X, H) becomes trivial after a finite étale extension $B' \to B$.

For example let us show just what we will need below: that $(**)$ implies $(\mathbf{G'})$ (group-by-group). So let G be a connected algebraic group over an algebraically closed field of constants K, and let s be an invariant vector field on G defined over $K(t)^{alg}$, so defined over a differential subfield $K(b)$ of $K(t)^{alg}$ containing $K(t)$. Then $\partial(b) = s'(b)$ for some K-rational function s' and b is the generic point of $(B, s')^\sharp$, where B is a curve over K. Let g be a generic point over $K(t)^{alg}$ of $(G, s_A)^\sharp$. Then (g, b) is a K-generic point of $G \times B$, and after cutting down B suitably the map taking (x, y) to $(s(x, y), s'(y))$ is a G- invariant vector field on $G \times B$, say $s''(x, y)$. As $s^{(p)} = 0$ for almost all p, also $s''^{(p)} = 0$ for almost all p. Let H be the subbundle of $T(G \times B)$ spanned by s''. So H has p-curvature 0 for almost all p. The statement says that H becomes trivial after a finite extension $\pi : B' \to B$ of curves. Let $\pi(b') = b$. So $b' \in K(t)^{alg}$ and $\partial(b') = s_2(b')$. The triviality statement implies that the D-variety $(G \times B', s''(x, \pi(y')))$ is isomorphic (over K) to $(G \times B', (0, s_2(y')))$ (even via an isomorphism which is the identity on B). So some K-generic (g, b') of $(G \times B)$ with $\partial(g, b'') = s''(g, b'')$ is birational over K to some generic (g', b'') of $G \times B$ such that $\partial(g') = 0$ and $s_2(b'') = \partial(b'')$. Without loss $b'' = b'$. As above we can find $g \in G(K(t)^{alg})$ with $\partial(g, b') = s''(g, b')$ namely $\partial(g) = s(g)$.

We will make key use of the following result of J.-B. Bost:

Theorem 3.3 *Let (X, H) be a principal G-bundle over curve B with connection, over a field K of characteristic 0. Assume G is commutative and H has p-curvature 0 for almost all p. Then (X, H) is "trivial on some finite étale covering of B".*

The result is Theorem 2.9 of [1] in the case where K is a number field, and Bost has explained [2] how the proof generalizes to arbitrary K. It follows from the equivalence of $(**)$ and $(\mathbf{G'})$ that $(\mathbf{G'})$ is true for commutative G. On the other hand, as in [1] if $H \to G \to G/H$ is a short exact sequence of connected algebraic groups and $(\mathbf{G'})$ is true for H and G/H then it is true for G. Hence, Chevalley's structure theorem for algebraic groups implies:

Proposition 3.4 *Conjecture $(\mathbf{G'})$ is equivalent to the Grothendieck-Katz conjecture $(\mathbf{G})$.*

4 Proof of main result

We complete the paper with the proof of the equivalence of (**F**) with [(**G**) and (**D**)]. We will use in a serious way the "Galois theory" (existence of definable automorphism groups) associated with internality. This general theory (due to Zilber, Hrushovski, and others) is expounded in some detail in section 4 of Chapter 7 of [8] which the reader is referred to. Kolchin's differential Galois theory (the theory of strongly normal extensions which has been already referred to) is a special case of the general model-theoretic set-up, but of course has characteristics which are specific to the differential setting. One of these is the so-called "primitive element theorem" of Kolchin, which can be found in the last chapter of [6] but which we summarise briefly, using model-theoretic language. So fix a differential field F (subfield of $\mathcal{U}$) with algebraically closed field $\mathcal{C}_F$ of constants. A finitely generated differential extension $E = F\langle a \rangle$ of F is defined to be a *strongly normal* extension of F, if (a) any realization b of $tp(a/F)$ is in $dcl(F, a, \mathcal{C})$ (so in particular $tp(a/F)$ is internal to $\mathcal{C}$ in the sense of the ambient differentially closed field $\mathcal{U}$), and (b) $\mathcal{C}_E = \mathcal{C}_F$ (no new constants). The model-theoretic content of (b) is that $tp(a/F)$ is "almost orthogonal" to $\mathcal{C}$. This actually implies that $tp(a/F)$ is isolated. The general theory tells us that the group of elementary permutations of the set of realizations of $tp(a/F)$ which fix $\mathcal{C}$ pointwise is definable, as well as its action. This definable Galois group has various existences, one of which is as a group internal to $\mathcal{C}$, defined over $\mathcal{C}_F$, and thus of the form $G(\mathcal{C})$ for some algebraic group G defined over $\mathcal{C}_F$. If F is algebraically closed, then G is a connected algebraic group. Moreover in this case (F algebraically closed), the primitive element theorem states that E can be a generated (even as a field) by an element $g \in G(E)$ such that $l\partial(g) = a$ for some F-rational point a of the Lie algebra $Lie(G)$ of G. Namely E is generated by a solution of a logarithmic differential equation on G over F. This will be used below.

Recall the conventions: $\mathcal{U}$ is a saturated differential closed field of characteristic 0 with field of constants $\mathcal{C}$. K denotes an algebraically closed subfield of $\mathcal{C}$, and $t \in \mathcal{U}$ is an element with $\partial(t) = 1$.
We repeat:

Conjecture (**F**): if (X, F) is a smooth irreducible variety together with an algebraic subbundle of $T(X)$, all defined over K, with p-curvature 0 for almost all p, then (X, F) is algebraically integrable.

Conjecture (**G**): If $A \in gl_n(K(t)^{alg})$ is such that $(s_A)_p^{(p)} = 0$ for almost all p, then there is $g \in GL_n(K(t)^{alg})$ such that $g^{-1}\partial(g) = A$.

Conjecture (**D**): if (X, s) is an irreducible algebraic D-variety defined over $K(t)^{alg}$ and $s_p^{(p)} = 0$ for almost all p, then the generic type of $(X, s)^\sharp$ is nonorthogonal to $\mathcal{C}$.

Proposition 4.1 *Assume the truth of (**G**). Then the following are equivalent:*
*(i) Conjecture (**F**),*
*(ii) Conjecture (**D**),*
(iii) whenever (X, s) is an irreducible algebraic D variety over K and $s_p^{(p)} = 0$ for almost all p, then the generic type of $(X, s)^\sharp$ is almost internal to $\mathcal{C}$.

Proof. We know from Proposition 3.1 that (i) implies each of (ii) and (iii).

(ii) implies (i): Let (X, s) be an irreducible algebraic D-variety over $K(t)^{alg}$ such that $s_p^{(p)} = 0$ for almost all p. By Proposition 3.1 we must show that the generic type of $(X, s)^\sharp$ is almost internal to $\mathcal{C}$ over $K(t)^{alg}$. Namely, if a is a generic point of $(X, s)^\sharp$ then $a \in acl(K(t)^{alg} \cup \mathcal{C})$. We will prove this by induction on $dim(a/K(t)^{alg})$ (which in this case equals $tr.deg(K(t)^{alg}(a)/K(t)^{alg})$). By our assumptions (including the truth of (**D**)), $tp(a/K(t)^{alg})$ is nonorthogonal to $\mathcal{C}$. By Fact 2.1 there is $b \in dcl(K(t)^{alg}(a)) \setminus K(t)^{alg}$ such that $tp(b/K(t)^{alg})$ is internal to $\mathcal{C}$. We may assume that $K(t)^{alg}\langle b \rangle = K(t)^{alg}(b)$, hence by 2.4, $tp(b/K(t)^{alg})$ also has p-curvature 0 for almost all p. By Lemma 7.4.2(ii) of [8] we obtain some $K(t)^{alg}$-independent tuple $b' = (b_1, \ldots, b_n)$ of realizations of $tp(b/K(t)^{alg})$ with $b_1 = b$, such that for any realization b'' of $tp(b'/K(t)^{alg})$, b'' is in the (differential) field generated by $K(t)^{alg}(b')$ together with the constants $\mathcal{C}$. Note $K(t)^{alg}\langle b' \rangle = K(t)^{alg}(b')$. Also note by 2.4 that $tp(b'/K(t)^{alg})$ has p-curvature 0 for almost all p. Let $L = \mathcal{C} \cap K(t)^{alg}(b')$. Then
(i) L^{alg} is the field of constants of $L(t)^{alg}(b')$.
Moreover it remains true that
(ii) for every realization b'' of $tp(b'/L(t)^{alg})$, b'' is in the (differential) field generated by $L(t)^{alg}(b')$ together with $\mathcal{C}$.
By (i) and (ii), $L(t)^{alg}(b')$ is a strongly normal extension of $L(t)^{alg}$, in the sense of Kolchin [6].

By the primitive element theorem discussed above, there is a connected algebraic group G over L^{alg}, the differential Galois group of the strongly normal extension, and an element $g \in G(\mathcal{U})$ such that $l\partial(g) = d \in Lie(G)(L(t)^{alg})$ and moreover $L(t)^{alg}(b') = L(t)^{alg}(g)$. By 2.4, $tp(b'/L(t)^{alg})$ has p-curvature 0 for almost all p, hence by 2.4 again $tp(g/L(t)^{alg})$ does also. On the other hand, as $dim(G) = dim(b'/L(t)^{alg}) = dim(g/L(t)^{alg})$ it follows that g is a generic point over $L(t)^{alg}$ of the D-variety (G, s_d). Hence s_d has p-curvature 0 for almost all p. By assumption (**G**) and Proposition 3.4, there is $g' \in G(L(t)^{alg})$ such that $l\partial(g') = d$. But then $g = g'h$ for some $h \in G(\mathcal{C} \cap L(t)^{alg}(g)) = G(L^{alg})$, whence $g \in G(L(t)^{alg})$ too. Hence $b' \in L(t)^{alg}$. So the conclusion is that there is some tuple c of constants such that $b \in acl(K(t), c)$. But $b \in dcl(K(t)(a))$ and $b \notin K(t)^{alg}$. It follows that $dim(a/K(c)(t)^{alg}) < dim(a/K(t)^{alg})$. By 2.4 $tp(a/K(c)(t)^{alg})$ has p-curvature 0 for almost all p, hence we can apply the induction hypothesis to conclude that $a \in acl(K(c)(t)^{alg} \cup \mathcal{C})$. But then $a \in acl(K(t)^{alg} \cup \mathcal{C})$ and we have proved what we wanted.

(iii) implies (i). We use the equivalence of (**F'**) with (**F**) from 3.1. Namely we take (X, s) an irreducible D-variety over K and we want to conclude that if a is a generic point over $K(t)^{alg}$ of $(X, s)^\sharp$ then $a \in acl(K(t) \cup \mathcal{C})$. The assumption (ii) tells us that $tp(a/K)$ is almost internal to $\mathcal{C}$ and so on general grounds there is $b \in dcl(K, a)$ such that $a \in acl(K, b)$ and $tp(b/K)$ is internal to $\mathcal{C}$. The argument in the proof of (ii) implies (i) now kicks in to yield that $b \in acl(K(t)^{alg} \cup \mathcal{C})$, so as $a \in acl(K, b)$ also $a \in acl(K(t)^{alg} \cup \mathcal{C})$ and we finish. $\qquad\square$

References

[1] J.-B. Bost, Algebraic leaves of algebraic foliations over number fields, *Publ. Math. IHES*, no. **93** (2001), 161-221.

[2] J.-B. Bost, Personal communication, Oct. 2005.

[3] Z. Chatzidakis and E. Hrushovski, Asymptotic theories of differential fields, *Illinois Journal of Math.*, **47** (2003), 593-618.

[4] T. Ekedahl, N.I. Shepherd-Barron and R. L. Taylor, A conjecture on the existence of compact leaves of algebraic foliations, preprint 1999.

[5] N. Katz, A conjecture in the arithmetic theory of differential equations, *Bull. Soc. Math. de France*, **110** (1982), 203-239.

[6] E. R. Kolchin, *Differential Algebra and Algebraic groups*, Academic Press, 1973.

[7] D. Marker, Model theory of differential fields, in *Model Theory of Fields, revised edition*, D. Marker, M. Messmer, and A. Pillay, Lecture Notes in Logic 5, A.K. Peters, 2005.

[8] A. Pillay, *Geometric Stability Theory*, Oxford University Press, 1996.

Schanuel's conjecture for non-isoconstant elliptic curves over function fields

Daniel Bertrand

Institut de Mathématiques de Jussieu

Summary

We discuss functional and number theoretic extensions of Schanuel's conjecture, with special emphasis on the study of elliptic integrals of the third kind.

Introduction

Schanuel's conjecture [La] on the layman's exponential function can be viewed as a measure of the defect between an algebraic and a linear dimension. Its functional analogue, be it in Ax's original setting [Ax1], Coleman's [Co], or Zilber's geometric interpretation [Zi], certainly gives ground to this view-point.

The same remark applies to the elliptic version of the conjecture, and to its functional analogue, as studied by Brownawell and Kubota [BK], and by J. Kirby [K1]. Here, the elliptic curve under consideration is constant. In the same spirit, we discuss in the first section of this note Ax's general theorem [Ax2] on the exponential map on a constant semi-abelian variety G, where transcendence degrees are controlled by the (linear) dimension of a certain "hull". We obtain a similar statement for the universal vectorial extension of G, and refer to the recent work of J. Kirby [K2, K3] for further generalizations of Ax's theorem, involving arbitrary differential fields, multiplicative parametrizations, and uniformity questions.

The naïve number-theoretic analogues of these functional results, however, are clearly false. The first counterexample which comes to mind is provided by periods: Riemann-Legendre relations are quadratic, and cannot be tracked back to hulls of the above type. Furthermore, the

41

theory of mixed motives shows that path integrals have as many reasons to be called periods as closed circuit ones, and we shall show in §2 that they too may obey non-linear constraints. The hopefully correct generalization of Schanuel's conjecture in this context, which is due to André [A2, A3], requires the introduction of a (motivic) Galois group.

In this theory, duality plays a crucial role. Going back to function fields, this makes the hypothesis of constancy of the ambient group sound rather unnatural. For instance, the dual of the one-motive attached to a non constant point on a constant elliptic curve is a non constant semi-abelian surface. And as soon as we allow for such variations, the functional statements cease to hold. Actually, the picture becomes closer to the number theoretic one, at least if we restrict to the "logarithmic" side of the conjecture: transcendence degrees are then controlled by a (linear differential) Galois group, which - not a surprise to model theorists - Manin's kernel theorem can help to compute. This was already noticed in [A1] and [B3] for pencils of abelian varieties (i.e., families of abelian integrals of the second kind), and the third part of the paper extends this approach to pencils of semi-abelian surfaces (elliptic integrals of the 3rd kind).

In fact, the hulls of §1 too can be interpreted as differential Galois groups (now in Kolchin's sense), if we restrict to the "exponential" side of the conjecture. But the specificity of Schanuel's conjecture lies precisely in its blending of exponentials and logarithms, and although we do not investigate this further here, it is likely that a similar blend of Galois groups, possibly with D-structures as in Pillay's theory [Pi], is required. The author can only thank (resp. apologize to) the organisers of the Newton conference for helping him to realize (resp. becoming aware so late of) the relevance of model theory to this circle of problems. He also thanks D. Masser, J. Kirby and Z. Chatzidakis for their comments on an earlier version of the paper.

1 Constant semi-abelian varieties

Let (F, ∂) be a differential field of characteristic 0. To give a common framework to the first and third parts of this study, we assume that (F, ∂) is differentially embedded in the field of meromorphic functions over a non empty domain U of the complex plane, and set $\mathcal{O}_F = F \cap \mathcal{O}_U$ (see below for a more algebraic presentation). We further assume that F

contains a not necessarily differential subfield K of transcendence degree 1 over $F^{\partial} = \mathbb{C}$.

Let G be a commutative algebraic group defined over $\mathbb{C}$, and let $\exp_G :$ $TG(\mathbb{C}) \to G(\mathbb{C})$ be the exponential map on its Lie algebra, identified with its tangent space TG at the origin. Since $\exp_G$ is analytic, it extends to a homomorphism from $TG(\mathcal{O}_F)$ to $G(\mathcal{O}_F)$, whose kernel is easily checked to coincide with that of $\exp_G$. Passing to quotients, we derive an injective homomorphism: $\overline{\exp}_G : TG(\mathcal{O}_F)/TG(\mathbb{C}) \to G(\mathcal{O}_F)/G(\mathbb{C})$. Notice that the periods of $\exp_G$ are lost in the process.

Suppose now that G is a semi-abelian variety. By rigidity, any algebraic subgroup H of G/F is then defined over $\mathbb{C}$, and $H(F)/H(\mathbb{C})$ embeds into $G(F)/G(\mathbb{C})$. To a given point $y \in G(F)$, we can therefore attach, without specifying fields of definitions, the smallest algebraic subgroup H of G such that $y \in H(F)$ *mod* $G(\mathbb{C})$. Its connected component though the origin is a semi-abelian subvariety G_y of G, which may be called the *relative hull* of the point y. For instance, the relative hull of a constant point $y \in G(\mathbb{C})$ is trivial; if $G = \mathbb{G}_m^s$ is a torus, a point $y = (y_1, \ldots, y_s) \in G(F)$ admits G as relative hull iff the classes of $y_1, \ldots, y_s$ in $F^*/\mathbb{C}^*$ are multiplicatively independent.

The following statement is a direct consequence of Ax's Theorem 3 in [Ax2]. I thank D. Masser for having drawn my attention to this reference. In fact, J. Kirby has recently reproved and extended this theorem in the setting of general differential fields F and general uniformizations. Furthermore, his results involve uniformity statements, a subject we shall not touch upon here. We refer to [K2], [K3] for more comments on these points.

Proposition 1.a ([Ax2], [K3]) *Let G be a semi-abelian variety defined over $\mathbb{C}$, let x be a point in $TG(\mathcal{O}_F)$, let $y = \exp_G(x)$, and let G_y be the relative hull of y. Then, $tr.deg.(K(x,y)/K) \geq dim(G_y)$.*

Proof of 1.a. If x is constant, the lower bound is trivial; otherwise, we deduce from [Ax2], Theorem 3, that $tr.deg.(\mathbb{C}(x,y)/\mathbb{C}) \geq dim(G_y) + 1$, where 1 stands for the rank of a jacobian matrix. Since $K/\mathbb{C}$ has transcendence degree 1, the claim easily follows.

In [Ax2], Ax assumes that the ambient group G admits no non trivial vectorial subgroup, but his argument readily extends to all G's admitting no non trivial vectorial *quotient*. For later applications, it seems more convenient to state the corresponding result in terms of universal extensions, as follows.

Let $\tilde{G}$ be the universal extension of G. If G is an extension of an abelian variety A by a torus T, this is the pull-back to G of the universal (vectorial) extension $\tilde{A}$ of A, which in turn is an extension of A by the dual $V \simeq \mathbb{G}_a^{dim(A)}$ of $H^1(A, \mathcal{O}_A)$, viewed as a vector group [NB: This should not be confused with the prolongation $\tau(A)$ of the standard D-group structure attached to A, which is an extension by TA, here split since A descends to $\mathbb{C}$, cf. [Bu], III, and [Ma]. Recall that $H^1(A, \mathcal{O}_A)$ is the tangent space of the dual of A, cf. [Mu], p. 130)]. In particular, the dimension of $\tilde{G}$ is equal to $2 dim(A) + dim(T)$. Following a suggestion of Z. Chatzidakis, we may also describe $\tilde{G}$ as the "largest" vectorial extension of G admitting no epimorphism to the additive group $\mathbb{G}_a$. The above Proposition can then be sharpened into

Proposition 1.b *Let G be a semi-abelian variety defined over $\mathbb{C}$, let x be a point in $TG(\mathcal{O}_F)$, let $y = \exp_G(x)$, let G_y be the relative hull of y and let $\tilde{G}_y$ be the universal vectorial extension of G_y. Furthermore, let $\tilde{x}$ be a lift of x to $T\tilde{G}(\mathcal{O}_F)$ and let $\tilde{y} = \exp_{\tilde{G}}(\tilde{x})$. Then,*

$$tr.deg.(K(\tilde{x}, \tilde{y})/K) \geq dim(\tilde{G}_y).$$

In particular, the equality holds true in either of the following situations:
  i) the logarithmic case, *where $\tilde{y}$ is defined over K;*
  ii) the exponential case, *where $\tilde{x}$ is defined over K.*

The denomination for these cases come from the classical Schanuel conjecture, where they respectively concern (i) Schneider's problem on the algebraic independence of $\mathbb{Q}$-linearly independent logarithms of algebraic numbers, and (ii) the Lindemann-Weierstrass theorem on the algebraic independence of the exponentials of $\mathbb{Q}$-linearly independent algebraic numbers. That the general inequality implies equalities in these special cases can be seen as follows: if $\tilde{x} \in T\tilde{G}(K)$, then, up to translations by a period and by a K-rational point of TV, it lies in the vector space $T\tilde{G}_y$. Since $\exp_{\tilde{G}}$ induces the identity on the vector group $TV \simeq V$, $\tilde{y} = \exp_{\tilde{G}}(\tilde{x})$ then differs from an element of $G_y(F)$ by a K-rational point. Its coordinates therefore generate over K a field of transcendence degree at most (and hence equal to) $dim(\tilde{G}_y)$. The argument can be reversed when we start with a point $\tilde{y} \in \tilde{G}(K)$. See Remark 1 below for a more intrinsic reformulation of these equalities.

Proof of 1.b. Once again, we may assume that x is not constant, and must then prove that $tr.deg.(\mathbb{C}(\tilde{x}, \tilde{y})/\mathbb{C}) \geq dim(\tilde{G}_y) + 1$. Since $\overline{\exp}_{G_y}$ is the restriction of $\overline{\exp}_G$ to TG_y, and since two lifts of x to $T\tilde{G}$ differ by an element of $TV \simeq V$, where $\exp_{\tilde{G}}$ reduces to the identity, we may also assume that $\tilde{x}$ lies in $T\tilde{G}_y$, and eventually, that $G_y = G$. In this case, any algebraic subgroup $G'/\mathbb{C}$ of $\tilde{G}$ projecting onto G_y coincides with $\tilde{G}$ (in other words, $\tilde{G}$ is an *essential extension* of G): indeed, the quotient $\tilde{G}/G'$ of the universal extension $\tilde{G}$ would otherwise be a non trivial vector group. In particular, $\tilde{G}_y = \tilde{G}$.

Let then $\mathbf{X}$ be the $\mathbb{C}$-algebraic group $T\tilde{G} \times \tilde{G}$, let $\mathbf{A}$ be the analytic subgroup of $\mathbf{X}$ made up by the graph of $\exp_{\tilde{G}}$, let $\mathbf{K}$ be the analytic curve defined by the image of $\{\tilde{x}, \tilde{y}\}$, viewed as a map from the complex domain U to $\mathbf{X}(\mathbb{C})$. Up to translation by a constant point, we may assume that $\mathbf{K}$ passes through the origin, and denote by $\mathbf{V}$ its Zariski closure in $\mathbf{X}$ over $\mathbb{C}$, so that $tr.deg.(\mathbb{C}(\tilde{x}, \tilde{y})/\mathbb{C}) = dim\mathbf{V}$. According to [Ax2], Theorem 1, there exists an analytic subgroup $\mathbf{B}$ of $\mathbf{X}$ containing both $\mathbf{A}$ and $\mathbf{V}$ such that $dim\mathbf{B} - dim\mathbf{V} \leq dim\mathbf{A} - dim\mathbf{K}$. We shall prove that $\mathbf{B} = \mathbf{X}$, and consequently, that

$$tr.deg.(\mathbb{C}(\tilde{x}, \tilde{y})/\mathbb{C}) = dim\mathbf{V} \geq dim\mathbf{X} - dim\mathbf{A} + dim\mathbf{K},$$

which is equal to $2dim\tilde{G} - dim\tilde{G} + 1 = dim\tilde{G} + 1 = dim\tilde{G}_y + 1$, as required.

Since $\mathbf{V}$ is a connected algebraic variety passing through the origin, the abstract group it generates in $\mathbf{X}$ is an algebraic subgroup $g(\mathbf{V})$ of $\mathbf{X} = T\tilde{G} \times \tilde{G}$. Since $\mathbf{V}$ contains $\mathbf{K}$, and since $G_y = G$, the image $G' \subset \tilde{G}$ of $g(\mathbf{V})$ under the second projection projects onto G, and therefore coincides with $\tilde{G}$. Let $T' \subset T\tilde{G}$ be the image of $g(\mathbf{V})$ under the first projection. We can now view $g(\mathbf{V})$ as an algebraic subgroup of $T' \times \tilde{G}$ with surjective images under the two projections. As is well known, any such subgroup of the product $T' \times \tilde{G}$ induces an isomorphism from a quotient of $\tilde{G}$ to a quotient of T'. More precisely, on setting $H = g(\mathbf{V}) \cap (0 \times \tilde{G})$, and $H' = g(\mathbf{V}) \cap (T' \times 0)$, we get an algebraic group isomorphism $\tilde{G}/H \simeq T'/H'$. But if these quotients are not trivial, the second one will admit $\mathbb{G}_a$ among its quotients, and the first one, hence $\tilde{G}$ itself, will share the same property. Again, since $\tilde{G}$ is a universal extension of a semi-abelian variety, this is impossible. Consequently, $\tilde{G}/H = 0$, and $g(\mathbf{V})$, hence $\mathbf{B}$, contains $0 \times \tilde{G}$. Finally, $\mathbf{B}$, which contains $\mathbf{A}$, projects onto $T\tilde{G}$ by the first projection. Hence, $\mathbf{B}$ does coincide with $T\tilde{G} \times \tilde{G} = \mathbf{X}$. (Notice that contrary to Kirby's general setting [K3], the algebraic groups $T\tilde{G}$ and $\tilde{G}$ do not play a symmetric role in this proof; it is likely, however, that Proposition 1.b could be reached by the method of [K3], §5.1.)

Propositions 1.a and b are better expressed in terms of the *logarithmic derivative map* $\partial \mathrm{Log}_G : G(F) \to TG(F)$ of the standard D-group structure attached to $G/\mathbb{C}$, cf. [Bu], [Pi], [Ma] - and [BC] for a historical perspective. To make the translation, view y as a section of the constant group scheme $G_U = G \times U$ over U, and the $\mathbb{C}$-vector space $\Omega^1 G$ of invariant differentials on G as a subspace of $H^0(G_U, \underline{\Omega}^1_{G_U})$. Since $(\exp_G^*)_0$ is the identity, the requirement $y = \exp_G(x)$ becomes: for any $\omega \in \Omega^1 G$, there exists an exact differential $dx_\omega \in d\mathcal{O}_F$ such the differential form $y^*(\omega) - dx_\omega$ on U kills the vector field ∂:

$$(x,y) \in (TG \times G)(\mathcal{O}_F) \text{ and } y^*(\omega)(\partial) = \partial x_\omega,$$

or more generally, denoting by $\partial \mathrm{Log}_G$ the standard logarithmic derivative on the constant group G: $(x,y) \in (TG \times G)(F)$ and $\partial x = \partial \mathrm{Log}_G(y)$. Indeed, the assignment $\omega \mapsto x_\omega$ is a linear form on $\Omega^1 G$ with values in F, defined up to linear forms on $\Omega^1 G$ with values in $F^\partial = \mathbb{C}$, i.e., as an element $x = x(y)$ of $TG(F)/TG(\mathbb{C})$, and the assignment $y \mapsto x(y) : G(F)/G(\mathbb{C}) \to TG(F)/TG(\mathbb{C})$ inverts on its image the map $\overline{\exp}_G$ defined above. Keeping in mind that these quotients do not affect fields of definitions over $\mathbb{C}$, and that all these notations should be indexed by ∂, we may then write $x = \mathrm{Log}_G(y)$, or more graphically

$$x_\omega(y) = \int^y \omega.$$

These notations remain meaningful for any closed, possibly singular, differential form ω on G, and can be extended to $\tilde{G}$. Proposition 1.b then reads: *let $y \in G(F)$, let $\tilde{y}$ be a lift of y to $\tilde{G}(F)$, and let $\tilde{x} = \mathrm{Log}_{\tilde{G}}(\tilde{y})$. Then $tr.deg.(K(\tilde{y}, \tilde{x})/K) \geq dim(\tilde{G}_y)$.*

Remark 1 We here assume that K is an algebraically closed *differential* subfield of (F, ∂), and consider the two special cases of Proposition 1.b.

i) In the "exponential" one, $\tilde{x}$ is a K-rational point of $T\tilde{G}$, and as explained above, we may assume wlog that it lies in $T\tilde{G}_y$. Set $\tilde{a} = \partial \tilde{x}$. Up to constants, $\tilde{y} = \exp_{\tilde{G}}(\tilde{x})$ is then a solution of the differential equation $\partial \mathrm{Log}_{\tilde{G}}(\tilde{y}) = \tilde{a}$, $\tilde{a} \in T\tilde{G}_y(K)$, to which Kolchin's differential Galois theory can be applied: indeed, $\tilde{G}_y$ being here constant, the differential extension $K(\tilde{y})/K$ is a strongly normal one, cf. [Pi], 3.2 and 3.8. In particular, its differential Galois group is an algebraic subgroup of $\tilde{G}_y$. Since its dimension is given by $tr.deg.K(\tilde{y})/K$, the proposition reduces in this case to the relation

$$\mathrm{Aut}_\partial(K(\tilde{y})/K) = \tilde{G}_y(\mathbb{C}).$$

ii) In the "logarithmic" one, $\tilde{y}$ is a K-rational point of G, which may be assumed wlog to lie in $\tilde{G}_y$. Set $\tilde{b} = \partial\mathrm{Log}_{\tilde{G}}(\tilde{y})$. Up to constants, $\tilde{x} = \mathrm{Log}_{\tilde{G}}(\tilde{y})$ is then a solution of the inhomogeneous linear equation $\partial\tilde{x} = \tilde{b}$, $\tilde{b} \in T\tilde{G}_y(K)$, to which the standard Picard-Vessiot theory can be applied. In particular, its differential Galois group is a vectorial subgroup of $T\tilde{G}_y$. Since its dimension is given by $tr.deg.K(\tilde{x})/K$, the proposition now reduces to the relation

$$\mathrm{Aut}_{\partial}(K(\tilde{x})/K) = T\tilde{G}_y(\mathbb{C}).$$

When G is a split product $A \times T$, this can be checked directly, as a slight amendment of the proof of Thm 3 of [A1] shows[1]. For a general study of split products, see [K3].

We now come back to the mixed case, and give a concrete translation of Proposition 1 (see Prop. 5 below for an even more concrete one). Let A be an abelian variety over $\mathbb{C}$, of dimension g, the elements of whose dual $\hat{A} = \mathrm{Pic}_0(A) \simeq \mathrm{Ext}^1(A, \mathbb{G}_m)$ we identify with the linear equivalence classes of residue divisors of differentials of the third kind on A. Let $\omega_1, \ldots, \omega_g$ be a basis of Ω^1_A over $\mathbb{C}$, let $\eta_1, \ldots, \eta_g$ be differential of the second kind on $A/\mathbb{C}$ whose cohomology classes generate a complement of Ω^1_A in $H^{dR}(A/\mathbb{C}) := H^1_{dR}(A/\mathbb{C})$, and let $\xi_{q_1}, \ldots, \xi_{q_r}$ be differentials of the third kind on $A/\mathbb{C}$, with residue divisors equivalent to $q_1, \ldots, q_r$ in $\hat{A}(\mathbb{C})$. Denote by G the extension of A by the torus $T = \mathbb{G}_m^r$ parametrized in $\mathrm{Ext}(A, T) \simeq \hat{A}^r$ by $q_1, \ldots, q_r$. Also, consider another torus $T' = \mathbb{G}_m^{r'}$.

Proposition 2 *In the above notation, assume that $q_1, \ldots, q_r \in \hat{A}(\mathbb{C})$ are linearly independent over $\mathbb{Z}$. Let y be a point of $A(F)$ whose relative hull A_y fills up A, and let $y' = (y'_1, \ldots, y'_{r'})$ be a point in $T'(F)$, whose relative hull $T'_{y'}$ fills up T'. Then,*

$$tr.deg._K K\left(y, y', \int^y \omega_i, \int^y \eta_i, \int^y \xi_{q_j}, \int^{y'_k} \frac{dt}{t}\right)_{\substack{1 \le i \le g, \ 1 \le j \le r \\ 1 \le k \le r'}} \ge 2g + r + r'.$$

Proof. Let us first deal with the case $r' = 0$. By Hilbert's Theorem 90 (see [Se]), there exists a $\mathbb{C}$-rational section s of the projection $p : G \to A$

---

1 In its appeal to Manin's theorem, the only property requested on the point $y \in G(K)$ is that its class modulo the *constant* sections of (the constant part of) G generate G_y; but this is precisely the definition of our relative hull. See also Footnote 4 below.

and elements $\Xi_1, \ldots, \Xi_r$ complementing $\{p^*(\omega_1), \ldots, p^*(\omega_g)\}$ into a basis of Ω^1_G such that $s^*(\Xi_j) = \xi_{q_j}$ for all j. Then $\mathbf{y} := s(y)$ lies in $G(F)$, projects to $y = p(\mathbf{y})$ in $A(F)$, and satisfies $\int^{\mathbf{y}} \Xi_j = \int^y \xi_{q_j}, \int^{\mathbf{y}} p^*(\omega_i) = \int^y \omega_i$, so that the field of definition over K of $\{\mathbf{y} : \mathbf{x} = \mathrm{Log}_G(\mathbf{y})\}$ coincides with $K(y, \int^y \omega_i, \int^y \xi_{q_j})$. Similarly (now by the very definition of the universal extension), the η_i's are pull-backs under a rational section of invariant forms $\tilde{\eta}_i$ on $\tilde{A}$, and the same argument provides a lift $\tilde{\mathbf{y}}$ of $\mathbf{y}$ to $\tilde{G}$ such that $K(\tilde{\mathbf{y}}, \mathrm{Log}_{\tilde{G}}\tilde{\mathbf{y}}) = K(y, \int^y \omega_i, \int^y \xi_{q_j}, \int^y \eta_i)$. According to Proposition 1.b, its transcendence degree over K is bounded from below by $dim(\tilde{G}_{\mathbf{y}})$. Now, the semi-abelian subvariety $G_{\mathbf{y}}$ of G projects onto A_y, which fills up A by hypothesis, and is thus an extension of A by a torus $\mathbb{G}_m^s$, parametrized by some points $w_1, \ldots, w_s$ in $\hat{A}(\mathbb{C})$. But (say by [B1], Prop. 1), such a semi-abelian variety can embed in G iff there exists an isogeny $\alpha \in \mathrm{End}(A)$ such that $\alpha^*(q_1), \ldots, \alpha^*(q_r)$ lie in the subgroup of $\hat{A}(\mathbb{C})$ generated by $w_1, \ldots, w_s$. Since the former are linearly independent over $\mathbb{Z}$, this forces $s = r$, so that the relative hull of $\mathbf{y}$ fills up G, whose universal extension has dimension $2g + r$. (In other words, the hypothesis on the points q_i means that G is an essential extension of A.)

For the general case, we introduce the semi-abelian variety $G \times T'$. A similar argument, combined with the hypothesis on y', shows that the relative hull of the point $(\mathbf{y}, y')$ is $G \times T'$, whence the required lower bound.

The point we made in the introduction about the limits of the functional setting is best illustrated by the following "counterexample" to Proposition 2, with $r = 1$ (and $r' = 0$). We say that an isogeny $f : A \to \hat{A}$ is antisymmetric if its transpose $\hat{f} : A \to \hat{A}$ satisfies $f + \hat{f} = 0$. Instead of the expected lower bound $2g + r + r' = 2g + 1$, we have:

Proposition 3 *Assume that the abelian variety $A/\mathbb{C}$ admits an antisymmetric isogeny f to $\hat{A}$, and let $y \in A(K)$ be such that $A_y = A$. There exists a differential of the third kind ξ_q on A/K, whose residue divisor lies in the equivalence class of the point $q = f(y) \in \hat{A}(K)$ such that $tr.deg._K K(\int^y \omega_i, \int^y \eta_i, \int^y \xi_q, i = 1, \ldots, g) = 2g$.*

Proof. In view of Proposition 1.b, the first $2g$ integrals generate over $K = K(y)$ a field of transcendence degree $2g$. As shown by the computational proof given in §2 in the case $g = 1$, the last one can be made to lie in this field (for general g, use the fact that the restriction of the Poincaré bundle to the graph of f is isotrivial).

Remark 2 In Proposition 3, q is non-torsion, but also *non constant*. In the setting of Proposition 1, this would correspond to a "semi-constant" semi-abelian variety, i.e., a non isoconstant extension G of the constant abelian variety A by the (constant) torus $\mathbb{G}_m$. It would be interesting to construct the corresponding differential equation $\partial \mathrm{Log}_{\tilde{G}} \tilde{y} = \partial \tilde{x}$ with the help of a D-group structure on $\tilde{G}$, viewed as an extension, in the category of D-groups, of the standard D-group structure of A by that of $\mathbb{G}_m \times \mathbb{G}_a^g$; see [Pi], [Ma] - and Remark 3.ii below for a slightly different suggestion.

2 Arithmetic interlude

We now turn to the number theoretic (i.e., honest) extension of Schanuel's conjecture to the semi-abelian variety $G/\mathbb{C}$. In this case, x lies in $TG(\mathbb{C})$, $y = \exp_G(x)$ in $G(\mathbb{C})$, and we want to bound from below the transcendence degree over $\mathbb{Q}$ of the field $k(\tilde{x}, \exp_{\tilde{G}}(\tilde{x}))$, where $k = \mathbb{Q}(G)$ denotes the field of definition of G (hence of its universal extension $\tilde{G}$). We shall give a pedestrian approach to the strategy proposed by Y. André in [A3], §23, and take advantage of this walk to write down the *full* period matrix of the "simplest interesting" one-motive. (For a general introduction to one-motives, see [De1].)

The conjecture should cover Schanuel's, and in particular imply the transcendency of π, so that we cannot mod out by the periods of $\exp_G$. Therefore, the lower bound must depend on x, rather than on y^2. The multiplicative and elliptic cases of the conjecture, as well as Wüstholz's theorem on linear forms in abelian integrals, suggest the introduction of the *Lie hull* of x, denoted by $\mathcal{G}_x$ and defined as smallest algebraic subgroup H of G such that $x \in TH(\mathbb{C})$. Again, there is no need to specify fields of definitions, since all algebraic subgroups of the semi-abelian variety G are defined on a finite extension of k. However, a statement of the type

$$tr.deg.(\mathbb{Q}(G, \tilde{x}, \tilde{y})/\mathbb{Q}) \geq dim(\tilde{\mathcal{G}}_x) \quad (??)$$

---

2 An alternative solution consists in replacing the base field $\mathbb{Q}$ by the field of all periods of $\tilde{G}$, as in [B2], Conjecture 2, and [Bn1]. This makes specific cases of the conjecture more difficult to check, but the Lie hull $\mathcal{G}_x$ can then be replaced by the *hull* of y, defined as the connected component $\mathcal{G}_y$ of the Zariski closure of $\mathbb{Z}y$ in G. Notice that the inclusion $\mathcal{G}_y \subset \mathcal{G}_x$ is often strict. No distinction between the two hulls needed to be made in the relative situation of §1, where we modded out by the (constant) periods of $\exp_G$.

is usually false. For instance, let $G = \mathbb{G}_m \times E$, where $E/\overline{\mathbb{Q}}$ is an elliptic curve with complex multiplications, and let ω_1, η_1 be a period and corresponding quasiperiod of E. The point $x = (2\pi i, \omega_1) \in TG(\mathbb{C})$ lifts in $T\tilde{G}$ to a point $\tilde{x}$ of the kernel of $\exp_{\tilde{G}}$, which may be represented by the vector $(2\pi i, \eta_1, \omega_1)$. By the CM hypothesis and Legendre relation (or Γ-function identities), $\eta_1\omega_1/\pi$ is an algebraic number, so that $\overline{\mathbb{Q}}(\tilde{x}, \tilde{y}) = \overline{\mathbb{Q}}(\omega_1, \pi)$ has transcendence degree (at most) 2 . On the other hand, the Lie hull of x is G itself, and its universal vectorial extension $\tilde{G}_x$ has dimension 3.

Counterexamples not involving vectorial extensions also abound. For instance, consider an abelian 4-fold G of primitive CM type, whose periods satisfy a Shimura relation (cf. [A3], §24.4), and let $0 \neq x = (\omega_1, \ldots, \omega_4)$ be such a period. Then, $tr.deg.(\mathbb{Q}(x, y)/\mathbb{Q}) \leq 3$, although $\mathcal{G}_x = G$ has dimension 4. But more to the point for our study, we shall now construct a counterexample involving a point y of infinite order on G (and for which $\mathcal{G}_x = \mathcal{G}_y = G$).

In the next paragraphs until Conjecture 1, we restrict to the logarithmic case of Schanuel's conjecture, i.e., assume that *G and $\tilde{y}$ are defined over a subfield k of $\overline{\mathbb{Q}}$*. Let thus E be an elliptic curve defined over the number field k by a Weierstrass equation $Y^2 = 4X^3 - g_2X - g_3$. Let $\wp, \zeta, \sigma$ be the standard Weierstrass functions attached to this model, and let $\omega_1, \omega_2, \eta_1, \eta_2$ be the periods and quasi-periods of $\wp$ and ζ. In particular, $\exp_E$ is represented by $(\wp, \wp')$, *quae* functions of the variable z defined by $dz = \exp_E^*(dX/Y)$, and $d\zeta = -\exp_E^*(XdX/Y)$. We also fix two complex numbers u, v, and assume that their images p, q under $\exp_E$ are *non torsion* points of $E(k)$. We do *not* require that p and q be linearly independent over $\text{End}(E)$. Denote by G the extension of E by $\mathbb{G}_m$ parametrized by $(-q) - (0)$.

Let us now puncture the curve E at the two points 0 and $-q$, and pinch it at two other k-rational points p_1, p_0 whose difference in the group E is p. The one-motive $M_0 = M(E, -q, p, p_0)$ attached by [De1] to the resulting open singular curve can be described as follows: there is a unique function $f_0 \in k(E)$ with value 1 at p_0 and divisor $(0) + (-q + p) - (p) - (-q)$, and by a well-known description of the set $G(k)$ ([Mu], p. 227), this defines a point y_0 in $G(k)$ lying above p, hence a one-motive $M_0 : \mathbb{Z} \to G : 1 \mapsto y_0$.

The de Rham realization $H^{dR}(M_0/k)$ of M_0 is the k-vector space generated by the differential of a rational function f on E such that $f(p_1)$ differs from $f(p_0)$ (say, by 1), the dfk $\omega = dX/Y$, the cohomology class of the dsk $\eta = XdX/Y$, and the dtk $\xi = \frac{1}{2}\frac{Y-Y(q)}{X-X(q)}\frac{dX}{Y}$. The residue divisor of ξ is $-(0)+(-q)$, and its pullback under $\exp_E$ is the logarithmic differential of the function

$$f_v(z) = \frac{\sigma(v+z)}{\sigma(v)\sigma(z)}e^{-\zeta(v)z},$$

whose quasi-periods are given by $e^{\lambda_i(v)}$, with

$$\lambda_i(v) = \eta_i v - \zeta(v)\omega_i, \text{ for } i = 1, 2.$$

The Betti realization of M_0 is the dual of the $\mathbb{Q}$-vector space $H_B(M_0, \mathbb{Q})$ generated by a small loop around the hole $-q$, the two standard loops on the elliptic curve E, and a "loop" from p_0 to p_1 on the pinched curve. Integrating the above differential forms along these loops, we obtain the period matrix of M_0. Not warranting signs, it may be written as

$$\Pi(u, v, \ell_0) := \begin{pmatrix} 2\pi i & \lambda_1(v) & \lambda_2(v) & g(u,v) - \zeta(v)u + \ell_0 \\ 0 & \eta_1 & \eta_2 & \zeta(u) \\ 0 & \omega_1 & \omega_2 & u \\ 0 & 0 & 0 & 1 \end{pmatrix},$$

where

$$g(u, v) = \ell n \frac{\sigma(u+v)}{\sigma(u)\sigma(v)},$$

and $e^{\ell_0} = \gamma_0 \in k^*$ can easily be computed in terms of p_0, p_1, q, using the triple addition formula for the σ-function ([WW], XX, ex. 20; NB: we modified η in its cohomology class so as to delete an additive k-rational factor from its last period $\int_{p_0}^{p_1} \eta$). It is fun to compute the matrix of cofactors of $\Pi(u, v, \ell_0)$, although the result is not a surprise: dividing by $2\pi i$, we get

$$\Pi'(v, u, \ell_0) = \begin{pmatrix} 1 & 0 & 0 & 0 \\ v & \omega_2 & \omega_1 & 0 \\ \zeta(v) & \eta_2 & \eta_1 & 0 \\ g(v,u) - \zeta(u)v + \ell_0 & \lambda_2(u) & \lambda_1(u) & 2\pi i \end{pmatrix}$$

which after some rearrangement, is the period matrix of the Cartier dual of M_0, given by a point y_0', lying above $-q$, on the extension of $\hat{E}$ by $\mathbb{G}_m$ parametrized by $p \in \mathrm{Pic}_0\hat{E} \simeq E$.

Let now y be an arbitrary point on $G(k)$ projecting onto p, i.e., of the form $y_0\gamma$ for some $\gamma \in \mathbb{G}_m$, let $M(y)$ be the corresponding one-motive, and let $\tilde{y}$ be a lift of y to $\tilde{G}(k)$. A logarithm $\tilde{x}$ of $\tilde{y}$ in $T\tilde{G}(\mathbb{C})$ is given by the last column (without its bottom entry) of the period matrix $\Pi(u, v, \ell)$, where $e^\ell = \gamma_0\gamma \in k^*$, and where $\zeta(u)$ should be replaced by $\zeta(u) + \beta$ for some $\beta \in k$ depending on $\tilde{y}$, so that $k(\tilde{x}, \tilde{y})$ coincides with the field

$$k(\tilde{x}) = k(u, \zeta(u), g(u, v) - \zeta(v)u + \ell).$$

Since q has infinite order, the extension G is not isotrivial, and since p too is non torsion, the Lie hull $\mathcal{G}_x$ of the projection x of $\tilde{x}$ to TG fills up G. Therefore, $\tilde{\mathcal{G}}_x$ has dimension 3, and if the statement (??) was correct, the three numbers $u, \zeta(u), g(u, v) - \zeta(v)u + \ell$ would be algebraically independent for any logarithm ℓ of an algebraic number, and elliptic logarithms u, v of non-torsion points on $E(\overline{\mathbb{Q}})$.

We can at last describe our counterexample. Assume that $g_3 = 0$, i.e., that E has complex multiplication by i. (Any CM field would work, modulo a finer choice of the dsk η.) Then, for any $\alpha \in \text{End}(E) = \mathbb{Z} \oplus \mathbb{Z}i$ with norm $N(\alpha) = \alpha\overline{\alpha}$, the functions $\zeta(\alpha z) - \overline{\alpha}\zeta(z)$ and the square of $\sigma(\alpha z)/\sigma(z)^{N(\alpha)}$ lie in the field $k(\wp(z), \wp'(z))$. Suppose now that $v = iu$, the important point being that $\alpha = i$ is totally imaginary. Then, $\zeta(v) = -i\zeta(u)$, and $\sigma(u+v)/\sigma(u)\sigma(v) = -i\sigma((1+i)u)/\sigma(u)^2$ is the square root γ' of an element of k^*, since $N(1 + i) = 2$. Choosing $\gamma = (\gamma_0\gamma')^{-1}$ and $\ell = -g(u, iu) = -\ell n(\gamma')$, and slightly extending k, we get a point $\tilde{x} \in T\tilde{G}(\mathbb{C})$ with $\tilde{y} = \exp_{\tilde{G}}(\tilde{x}) \in \tilde{G}(k)$ and $k(\tilde{x}) = k(u, \zeta(u))$. But this has transcendence degree at most 2 (in fact 2, according to a theorem of Chudnovsky), not 3!

This example, which translates word for word to the semi-constant situation of Proposition 3, is not mysterious. The one-motive $M = M(y)$ it corresponds to was discovered by Ribet in his study of Galois representations (cf [JR]), and is known to have a degenerate Mumford-Tate group. In general, this group $MT(M)$ is the semi-stabilizer in $GL_{\mathbb{Q}}(H_B(M, \mathbb{Q}))$ of all Hodge cycles occurring in the tensor constructions on $H_B(M, \mathbb{Q})$ and its dual (cf. [De2], p. 43, and [Br]). In the present case (cf. [B4], or more generally [Bn2]), its unipotent radical has

(i) dimension 5 if the points p and q are linearly independent over $\text{End}(E)$;

(ii) dimension 3 if there exists $\alpha \in \text{End}(E) \otimes \mathbb{Q}$ such that $p = \alpha q$ and $\alpha \neq -\overline{\alpha}$ is not antisymmetric (an automatic condition if E has no CM);

(iii) dimension 3 if $p = \alpha q$ with $\overline{\alpha} = -\alpha$, and y is not a Ribet point;

(iv) dimension 2 in the remaining case.

According to a conjecture of Grothendieck[3], the transcendence degree of the full field of periods of M should be equal to the dimension of $MT(M)$, and an elementary dimension count as in [A3], 23.2.1 (see also the proof of Prop. 1.i above) then implies, *still assuming that k is a number field*:

Proposition 4.a *In the above notation, assume that $dim MT(M) = tr.deg.(k(\Pi(u, v, \ell))/k)$. Then, the field $k(u, \zeta(u), g(u, v) - \zeta(v)u + \ell)$ has transcendence degree 3 in Cases (i,ii, iii). In Case (iv), it coincides with the field $k(u, \zeta(u), 2c\pi i)$ for some rational number c, and has transcendence degree 3 if $c \neq 0$, and 2 otherwise.*

Proof. Let us only treat the last two CM cases, again with $v = iu$. Then, the maximal reductive quotient of $MT(M)$ has dimension 2, while $k(\Pi(u, v, \ell)) = k(2\pi i, \omega_1, u, \zeta(u), \tilde{\ell})$, where $\tilde{\ell} := g(u, iu) + \ell$ is a logarithm of an algebraic number $\tilde{\gamma}$. In Case (iii), $dim(MT(M)) = 5$; by the Grothendieck conjecture, we have 5 algebraically independent numbers, any 3 of which must be algebraically independent. In Case (iv), $dim(MT(M)) = 4$, but $\tilde{\gamma}$ is a root of unity and $k(\Pi(u, v, \ell))$ reduces to $k(2\pi i, \omega_1, u, \zeta(u))$. We then have 4 algebraically independent numbers, any 3, or 2, of which must be algebraically independent. Note that in this last case, the transcendence degree of $k(\tilde{x})$ *depends* on the choice of the logarithm x of the point y, although the Lie hull of x always fills up G.

We now drop the assumption that $k \subset \mathbb{C}$ is a number field. The dimension count becomes hopeless, but as suggested in [A3], 23.2.2, a finer approach to the study of any specific period is provided by the $MT(M)$-torsor of all isomorphisms between $H_B(M)^*$ and $H^{dR}(M)$ which, up to homotheties, preserve the cohomology classes of Hodge cycles. The period matrix represents such an isomorphism. For $y \in G(\mathbb{C})$, the choice of a logarithm $x = \mathrm{Log}_G(y)$ of y determines a loop γ_x in $H_B(M)$, which projects to a generator of $H_B(M)/H_B(G)$, and which satisfies $g \cdot \gamma_x - \gamma_x \in H_B(G)$ for all $g \in MT(M)$. Define the *Mumford-Tate orbit* MT_x of x as the the Zariski closure in $H_B(G)$ of the orbit of γ_x under this affine action of $MT(M)$.

Conjecture 1 (following André, [A3], 23.4.1) *Let G be a semi-abelian variety defined over $\mathbb{C}$, let $\tilde{G}$ be its universal extension, let x be a point*

---

3 cf. [A3], 23.1.4, 23.3.2. This conjecture actually relates transcendence degrees to motivic Galois groups; in view of [De2] and [Br], Mumford-Tate groups are an acceptable substitute in the case of one-motives.

in $TG(\mathbb{C})$, *let* $y = \exp_G(x)$, *and let* MT_x *be the Mumford-Tate orbit of* x. *Let further* $\tilde{x}$ *be a lift of* x *to* $T\tilde{G}(\mathbb{C})$ *and let* $\tilde{y} = \exp_{\tilde{G}}(\tilde{x})$. *Then,*

$$tr.deg.(\mathbb{Q}(G, \tilde{x}, \tilde{y})/\mathbb{Q}) \geq dim(MT_x).$$

A "justification" of the conjecture is given in the proof of Theorem 1 below. Notice that for any semi-abelian variety G, $H^{dR}(G)$ is canonically isomorphic to $\Omega^1_{\tilde{G}}$, so that $dim H_B(G) = dim\tilde{G}$, and that for any $x \in TG(\mathbb{C})$, MT_x is necessarily contained in the Betti homology of the Lie hull $\mathcal{G}_x$ of x. In particular, $dim MT_x \leq dim\tilde{\mathcal{G}}_x$. As shown by the last case of Prop. 4.a), the inequality may be strict. However, if G is isogenous to a *split* product $A \times T$ as in [K2], and if y generates a Zariski dense subgroup of $\mathcal{G}_x$ (i.e., if $G_y = \mathcal{G}_x$), we deduce from [A1], Prop. 1, that the Mumford-Tate orbit MT_x coincides with $H_B(\mathcal{G}_x)$, and Conjecture 1 does imply that $tr.deg._{\mathbb{Q}}\mathbb{Q}(G, \tilde{x}, \tilde{y}) \geq dim(\tilde{\mathcal{G}}_x)$ in this special case.

As a companion to Prop. 4.a, now restricted to Cases (i) and (ii) of its discussion, here is another consequence of Conjecture 1.

Proposition 4.b *Let* E *be an elliptic curve with complex invariants* g_2, g_3, *and let* u, v, ℓ *be complex numbers such that* $\exp_E(u), \exp_E(v)$ *are not related by an antisymmetric relation over* $\text{End}(E)$. *Assume that Conjecture 1 holds true. Then,*

$$tr.deg.(\mathbb{Q}(g_2, g_3, u, \zeta(u), g(u,v) - \zeta(v)u + \ell, \wp(u), e^\ell)/\mathbb{Q}) \geq 3;$$

in particular, if g_2, g_3 *and* $\wp(u)$ *are algebraic, the numbers* $u, \zeta(u)$ *and* $\ell n\,(\sigma(u))$ *are algebraically independent; so are the numbers* $u, \zeta(u)$ *and* $\sigma(u)$.

Proof. Let γ_x be a loop complementing $H_B(G)$ in $H_B(M)$. In all cases except (iv) (and even in Case (iv), if we avoid a specific line in the choice of γ_x), the orbit of γ_x under the affine action of the unipotent radical of $MT(M)$ already fills up $H_B(G)$, so that the general inequality is clear. The other assertions, which could be checked by dimension count, concern Case (ii), with $v = u$. For the first one, recall that $\sigma(2u)/\sigma(u)^4 = -\wp'(u)$, and choose $-\ell$ as a logarithm of this algebraic number. For the last one, choose $\ell = -g(u, u)$. Notice that in order to reach the values of the σ function, we must here consider a *transcendental* point y on a semiabelian variety G defined over $\overline{\mathbb{Q}}$.

In the same spirit, but back into the functional context of Section 1, here is an application of Prop. 2. We recall that F is a differential field with constant field $\mathbb{C}$, and that K is a subfield of F of transcendence degree 1 over $\mathbb{C}$.

Proposition 5 *Let E be an elliptic curve with complex invariants g_2, g_3 and period lattice Ω, let $v_1, \ldots, v_n$ be complex numbers not lying $\Omega \otimes \mathbb{Q}$, and let $x_1, \ldots, x_n$ (resp. $x'_1, \ldots, x'_{r'}$) be elements of F linearly independent over $\mathrm{End}(E)$ (resp. $\mathbb{Z}$) modulo $\mathbb{C}$. Then,*

$$tr.deg_K K\left(x_i, x'_j, \zeta(x_i), \wp(x_i), \frac{\sigma(v_i + x_i)}{\sigma(x_i)\sigma(v_i)}, e^{x'_j}\right)_{\substack{1 \le i \le n \\ 1 \le j \le r'}} \ge 3n + r'.$$

Proof. For each $i = 1, \ldots, n$, let G_i be the extension of E by $\mathbb{G}_m$ parametrized by the divisor $(\exp_E(-v_i)) - (0)$. Then, $G = G_1 \times \cdots \times G_n$ is an extension of E^n by $\mathbb{G}_m^r$, with $r = n$, parametrized by $\mathbb{Z}$-linearly independent points $q_1, \ldots, q_n$ of $(\hat{E})^n(\mathbb{C})$: indeed, their collection can be represented by a diagonal matrix, none of whose diagonal entry is torsion. By hypothesis, the relative hull of the point $y = (\exp_E(x_1), \ldots, \exp_E(x_n))$ (resp. $y' = (e^{x'_1}, \ldots, e^{x'_{r'}})$) fills up E^n (resp. $\mathbb{G}_m^{r'}$). The result follows from Proposition 2, combined with the above computations, on choosing $\ell_i = -g(x_i, v_i)$ for $i = 1, \ldots, n$.

Remark 3 i) The "reason" for the validity of Proposition 1 is that the situation it concerns is akin to Case (i) above: in the notations of Prop. 2, the points q_i which parametrize the extension G are constant (since G is constant), while the relative hull G_y of y takes into account only the non-constant "parts" p_j of the points $\exp_E(x_j)$. No linear relation over $\mathrm{End}(A)$, antisymmetric or not, can then relate the p_j's to the q_i's.

ii) But for the very same reason, we can no longer take $v_i = x_i$ in Prop. 5, and contrary to [BK], the result falls short of the study of the elements $\sigma(x_i)$ themselves. To reach them, "semi-constant" semi-abelian varieties as in Remark 2 seem required. Here, though, is another suggestion: since $\frac{\sigma'}{\sigma} = \zeta$, the couples $(x = \mathrm{Log}_E(y), z) \in F \times F^*$ such that $\sigma(x) = z$ are solutions of the system $\frac{\partial z}{z} = t y^* \omega(\partial), \partial t = y^*(\eta)(\partial)$. This may be related to the Manin kernel of the split product of $\hat{E}$ by $\mathbb{G}_m$, its subgroup $\mathbb{G}_a \times \mathbb{G}_m$ being now endowed with a non standard D-group structure, as in [Pi], end of §2.

3 Non-isoconstant semi-abelian surfaces

From now on, $K = \mathbb{C}(S)$ is the field of rational functions on a smooth projective curve S over $\mathbb{C}$, t is a non-constant element of K, and ∂ is the rational vector field d/dt on S. More seriously, we only consider the "logarithmic case" of Schanuel's conjecture, i.e., assume that G and $\tilde{y}$ are defined over K. But we now allow G to be non constant. We start by recalling from [De1] and [A1] the general setting of smooth one-motives attached to such data. This reduces transcendence problems to the computation of an orbit under a Picard-Vessiot group. We then restrict to an elliptic pencil (punctured and pinched as in §2, now along rational sections), describe, in the style of Manin's paper [Mn], the corresponding extensions of its Picard-Fuchs equation, compute their Galois groups with the help of [B5], and apply the result to Schanuel's conjecture.

Let thus $\mathcal{A}$ be an abelian scheme over a non empty Zariski open subset $\mathcal{U}$ of S, let $\mathcal{G}$ be an extension of $\mathcal{A}$ by a constant torus $T_{\mathcal{U}}$ of relative dimension r over $\mathcal{U}$, let $\mathbf{y}$ be a section of $\mathcal{G}$ over $\mathcal{U}$, and let $f : \mathcal{M} \to \mathcal{U}$ be the smooth one-motive over $\mathcal{U}$ attached to the morphism $1 \mapsto \mathbf{y}$ from the constant group scheme $\mathbb{Z}_{\mathcal{U}}$ to $\mathcal{G}$. We denote by $A/K, G/K, y \in G(K), M/K$ the abelian and semi-abelian varieties, point and one-motive over K these data define at the generic point of S.

The first relative de Rham cohomology sheaf of $\mathcal{M}/\mathcal{U}$ is a locally free $\mathcal{O}_{\mathcal{U}}$-module equipped with a connexion ∇, which, restricted to the generic point and contracted with d/dt, defines a differential operator D on the K-vector space $H^{dR}(M/K)$. The quotient $H^{dR}(G/K)$ of $H^{dR}(M/K)$ by its (trivial rank one) D-submodule $H^{dR}(\mathbb{Z}/K) = (K, \partial)$ is itself an extension of the (trivial) D-module $H^{dR}(T/K) \simeq (K, \partial)^r$ by $H^{dR}(A/K)$.

The first relative Betti homology $R_1 f_* \mathbb{Z} := H_B(\mathcal{M}/\mathcal{U})$ is a constant sheaf over $\mathcal{U}$, whose dual generates over $\mathbb{C}$ the local system of horizontal vectors of ∇. In an analytic neighbourhood U of a point u_0 of $\mathcal{U}$, and relatively to a basis of $H^{dR}(M)$ respecting the above filtrations, its local sections provide a fundamental matrix of solutions for D of the shape

$$\begin{pmatrix} \mathbf{I}_r & \Lambda_1(t) & \Lambda_2(t) & \Gamma(t) \\ 0 & H_1(t) & H_2(t) & Z(t) \\ 0 & \Omega_1(t) & \Omega_2(t) & U(t) \\ 0 & 0 & 0 & 1 \end{pmatrix},$$

whose entries generate over K a Picard-Vessiot extension $F = F_{u_0}$, for which we set $\mathcal{O}_F = F \cap \mathcal{O}_U$. Its last column (without its bottom entry) represents a logarithm $\tilde{x} = \mathrm{Log}_{\tilde{G}}(\tilde{y}) \in T\tilde{G}(\mathcal{O}_F)$ of a K-rational point $\tilde{y}$ lifting y to the universal extension $\tilde{G}$ of G. The field of definition $K(\tilde{x}) = K(\tilde{x}, \tilde{y})$ of $\tilde{x}$ depends only on the image x of $\tilde{x}$ in TG.

Let now $PV(M) = \mathrm{Aut}_{\partial}(F/K)$ be the differential Galois group of the D-module $H^{dR}(M/K)$. For each g in $PV(M)$, $g \cdot \tilde{x} - \tilde{x}$ lies in $H_B(\mathcal{G}/\mathcal{U}) \otimes \mathbb{C}$, and depends only on x. We may therefore define the *Picard-Vessiot orbit* PV_x of x as the Zariski closure of the orbit of $\tilde{x}$ in $H_B(\mathcal{G}/\mathcal{U})$ under this affine action of $PV(M)$.

Theorem 1 *Let G be a semi-abelian variety defined over K, let $\tilde{G}$ be its universal extension, let x be a point in $TG(\mathcal{O}_U)$ such that $y = \exp_G(x)$ lies in $G(K)$, and let PV_x be the Picard-Vessiot orbit of x. Let further $\tilde{x}$ be a lift of x to $T\tilde{G}(\mathcal{O}_U)$ such that $\tilde{y} = \exp_{\tilde{G}}(\tilde{x})$ lies in $\tilde{G}(K)$. Then,*

$$tr.deg.(K(\tilde{x})/K) = dim(PV_x).$$

Proof. As explained in [Ka], Prop. 2.3.1 and Remark 2.3.3, this is a tautology once one is reminded that a fundamental matrix for D is a generic point of a K-torsor under $PV(M)$. We should point out that by exactly the same argument, the Grothendieck conjecture implies the "logarithmic case" of Conjecture 1, in the form: if G and $\tilde{y}$ are defined over $\overline{\mathbb{Q}}$, then $tr.deg.(\mathbb{Q}(\tilde{x})/\mathbb{Q}) = dim(MT_x)$.

In [A1], Y. André shows that $PV(M)$ is a normal subgroup of the derived group $\mathcal{D}MT(M_{u_0})$ of the Mumford-Tate group of the fiber M_{u_0} of $\mathcal{M}$ above a sufficiently general point $u_0 \in \mathcal{U}$, and gives non-obvious examples where *the inclusion $PV(M) \subset \mathcal{D}MT(M_{u_0})$ is strict*. On the other hand, as soon as $\mathcal{M}$ admits a special fiber M_{u_1} with an abelian Mumford-Tate group, Prop. 2 of [A1] shows that the two groups coincide. Now, at least theoretically, the main theorem of [Bn 2] provides a complete description of $MT(M_{u_0})$. Combined with Theorem 1, this gives a satisfactory answer to the logarithmic case of Schanuel's problem over function fields, under the proviso that $\mathcal{M}$ varies enough in its pencil to ensure both very small and rather large Mumford-Tate groups above various points of the base.

To dispense with this hypothesis, a more direct approach consists in computing the Picard-Vessiot group itself. Manin's kernel theorem provides such a possibility when the abelian variety A is not isoconstant; for abelian integrals of the second kind[4], this was already noticed in [A1],

---

4 i.e., when $G = A$. Actually, in this case, even the hypothesis on non-isoconstancy can be dispensed with. See [Ch], bottom of p. 388 and Footnote 1 above - as well as [AV] for an early application to transcendence! However, we do use it, at least formally, in the proof which follows.

Theorem 3 (and in less generality in [B3], Thm 5). We now extend this method to the study of elliptic integrals of the 3rd kind, where in parallel with §2, the description of the D-module $H^{dR}(M/K)$ can be made quite concrete, as follows.

Let E be an elliptic curve defined over the function field $K = \mathbb{C}(S)$ by the Weierstrass equation $Y^2 = 4X^3 - g_2(t)X - g_3(t)$. In the standard non-canonical way (cf. [Mn]), extend the derivation ∂ to the field $K(E)$ and to its space of differentials by setting $\partial x = 0, \partial(f(x,y)dx) = \partial f(x,y)dx$. Let $p(t), q(t)$ be two non torsion points on $E(K))$, possibly linearly dependent over $\mathrm{End}(E)$, and consider the differential of the third kind $\xi(t) = \frac{1}{4\pi i} \frac{Y - Y(q)}{X - X(q)} \frac{dX}{Y}$. Since the residues of ξ are the *constant* functions $\pm 1/2\pi i$ of K, the classical formula

$$\forall\, s(t) \in E(K),\ \partial(Res_{s(t)}\xi(t)) = Res_{s(t)}\partial\xi(t)$$

implies that $\partial\xi(t)$ is a differential of the *second* kind on E/K. By Gauss, its cohomology class is killed by a 2nd order fuchsian differential operator L_ξ, and a basis of local solutions of $(L_\xi \circ \partial)y = 0$ in an analytic neighbourhood U of a point u_0 of $\mathcal{U}$ is given by the periods $\lambda_1(t) = \int_{\gamma_1} \xi(t), \lambda_2(t)$ of $\xi(t)$ over loops γ_1, γ_2 of the fiber E_t, and the constant function 1, corresponding to the integral of ξ on a loop around $-q$. Now, the integral $\int_{p_0}^{p_1} \xi(t)$ of ξ between two sections p_0, p_1 differing by p in $E(K)$ is a locally analytic function $\Gamma(t)$, well defined up to the addition of a $\mathbb{Z}$-linear combination of the previous periods, so that $(L_\xi \circ \partial)(\Gamma(t)) := f_{p;p_0}(t)$ is a uniform function on a Zariski open subset of S, with moderate growth at infinity, hence a rational function on S. In brief, $\int_{p_0}^{p_1} \xi(t)$ provides the fourth solution to the 4th order linear differential operator $(\partial - \frac{\partial f_{p;p_0}}{f_{p;p_0}}) \circ L_\xi \circ \partial \in K[\partial]$. (Manin's paper dealt with the adjoint of the 3rd order operator $L_\xi \circ \partial$.)

From now on, we assume that the j-invariant of E is not constant. Then, L_ξ is equivalent to the standard irreducible Picard-Fuchs equation $L_{E/K}$ attached to the differential dX/Y on E, and the 4th order operator can be written in the form

$$\mathcal{L}_{p,q;p_0} = \partial_{p,p_0} \circ L_{E/K} \circ \partial_q,$$

where ∂_{p,p_0} and ∂_q are equivalent to ∂. In other words, the section p (respectively, q) provides an element N^p (respectively, N_q) of the group $\mathrm{Ext}(H^{dR}(E/K), \mathbf{1})$ of extensions of the D-module $H^{dR}(E/K) \simeq K[\partial]/K[\partial]L_{E/K}$ by the trivial D-module $\mathbf{1} = K[\partial]/K[\partial]\partial$ (respectively, of the group $\mathrm{Ext}(\mathbf{1}, H^{dR}(E/K))$) and the choice of p_0 then provides a blended extension (in the sense of [B5], Remark 6, and [Ha]) of N^p by

N_q. Now comes the main point: since p and q are non-torsion points on the non isoconstant curve E, Manin's theorem (cf. [Mn], [Ch], or [B3], Lemma 8) implies that both extensions N^p and N_q are unsplit; moreover, N_q and the adjoint N_p of N^p are linearly independent over $\mathbb{C}$ in $\mathrm{Ext}(\mathbf{1}, H^{dR}(E/K))$ if p and q are linearly independent over $\mathrm{End}(E)$. We can then appeal to the purely group theoretic arguments of [B5] to compute the Picard-Vessiot group of $\mathcal{L}_{p,q;p_0}$, as follows.

This PV group is an extension of that of $L_{E/K}$, which is $SL_2(\mathbb{C})$ since $L_{E/K}$ is irreducible and (antisymmetrically) self-adjoint, by its unipotent radical, which, on denoting the solution space of $L_{E/K}$ by $\mathcal{V} \simeq H_B(E_{u_0}) \otimes \mathbb{C}$ and in view of [B5], Thm 3, is isomorphic to

(i) an extension of $\mathcal{V} \times \mathcal{V}$ by $\mathbb{C}$ if p and q are linearly independent over $\mathrm{End}E$;

(ii) the Heisenberg group $\mathcal{H}$ on $\mathcal{V}$ otherwise, i.e., the extension of $\mathcal{V}$ by $\mathbb{C}$ given by the law $(c,v) \cdot (c',v') = (c + c' + \langle v \mid v' \rangle, v + v')$, where $\langle \mid \rangle$ denotes the canonical antisymmetric bilinear form induced on $\mathcal{V}$ by the intersection product.

Indeed, the other possibilities mentioned in [B5] (viz. that it becomes abelian, and reduces either to $\mathcal{V} \times \mathbb{C}$ or to $\mathcal{V}$, as in Cases (iii) and (iv) of §2 above) can occur only if the middle operator is symetrically self-adjoint, and the irreducibility of $L_{E/K}$ prevents this. Now, in both Cases (i) and (ii), the Picard-Vessiot orbit of the 4th solution $\int_{p_0}^{p_1} \xi(t)$ has dimension 3. We therefore deduce from Theorem 1 and the computations of §2 that for any choice $\ell(t), u(t), v(t)$ of analytic functions such that $e^\ell \in K^*$, and $\exp_E(u)$, $\exp_E(v)$ are non-torsion points on $E(K)$, the functions $u(t), \zeta(u(t)), g(u(t), v(t)) - \zeta(v(t))u(t) + \ell(t)$ are algebraically independent over K; in particular, $u(t), \zeta(u(t))$ and $\ln \sigma(u(t))$ are algebraically independent. More generally, we obtain the following theorem, which extends Thm 5 of [B3] to the study of $\ln \sigma$ (but still misses the σ-function itself).

Theorem 2 *Let $g_2(t), g_3(t)$ be algebraic functions such that g_2^3/g_3^2 is not constant, let $\wp_t$ be the Weierstrass function with invariants $g_2(t), g_3(t)$ and period lattice $\Omega(t)$, let $\{u_i : i = 1, \ldots, n\}$ be holomorphic functions on an open subset of $\mathbb{C}$, linearly independent over $\mathbb{Z}$ modulo $\Omega(t)$, and such that the functions $\wp_t(u_i(t))$ are algebraic. Then, the $3n$ functions $u_i(t), \zeta(u_i(t)), \ln \sigma(u_i(t))$ $(i = 1, \ldots, n)$ are algebraically independent over $\mathbb{C}(t)$.*

Proof. Let E be the corresponding elliptic curve; for $i = 1, \ldots, n$, set

$p_i = q_i = \exp_E(u_i)$ and denote by $\mathcal{L}_i$ the differential operator $\mathcal{L}_{p_i,p_i;p_{0,i}}$, for some choice of section $p_{0,i}$. All these are defined over an algebraic extension K of $\mathbb{C}(t)$. The unipotent radical R_u of the differential Galois group over K of the direct sum of the $\mathcal{L}_i$'s naturally embeds in $\mathcal{H}^n$, via the isomorphisms $\phi_i : R_u(PV(\mathcal{L}_i)) \simeq \mathcal{H}$ of Case (ii) above. We claim that the image of R_u coincides with $\mathcal{H}^n$ (which has dimension $2n + n = 3n$). The proposition then easily follows from Thm 1.

For each i, let ψ_i be the composition of ϕ_i with the projection from $\mathcal{H}$ to $\mathcal{V}$. Since the points q_i are linearly independent over $\text{End}(E)$, the extensions N_{q_i} which we defined above, are $\mathbb{C}$-linearly independent in $\text{Ext}(\mathbf{1}, H^{dR}(E/K))$, and Thm 2 of [B3] (or more generally, Thm 2.2.14 of [Ha]) implies that the image of R_u under $(\psi_1, \ldots, \psi_n)$ fills up $\mathcal{V}^n$. But since $\langle \,\mid\, \rangle$ is non degenerate, the derived group of any subgroup of $\mathcal{H}^n$ projecting onto $\mathcal{V}^n$ fills up $\mathbb{C}^n$, so that the only subgroup of $\mathcal{H}^n$ projecting onto $\mathcal{V}^n$ is $\mathcal{H}^n$ itself, and indeed, $R_u = \mathcal{H}^n$.

I shall not attempt here to formulate a Schanuel conjecture for smooth one-motives over an arbitrary base over $\mathbb{C}$, which would extend Proposition 1 and Theorem 1, and parallel Conjecture 1 of §2. The question is of course to find the correct analogue of the Mumford-Tate group. Let me only point out to the probable relevance of the algebraic D-group structure which Pillay's Galois groups [Pi] are endowed with. Already when G is constant, the two sides of Remark 1 show that the expected group should lie in the tangent bundle $T\tilde{G} \times \tilde{G}$ of $\tilde{G}$. In the non constant case, it is not difficult to guess that the prolongation $\tau(\tilde{G})$ of $\tilde{G}$, which in a sense is the natural habitat of Manin kernels (cf. [Ma]), will play a role. The ∂-Hodge structures of [Bu] may also have some bearing on these questions.

We finally mention two further directions of study:

i) the Fourier expansions of the functions $\zeta(z) - \frac{\eta_1}{\omega_1}z$, $\sigma(z)e^{-\frac{\eta_1}{2\omega_1}z^2}$, $\frac{\sigma(z+v)}{\sigma(z)\sigma(v)}e^{-\frac{\eta_1}{\omega_1}vz}$, are building blocks in the theory of q-difference equations. Can q-difference Galois groups and their higher dimensional analogues provide a new insight on Schanuel's conjecture?

ii) what about characteristic p analogues? We shall here merely refer to [AMP] for the algebraic independence of the $\mathbb{Z}_p$-powers $f^{\lambda_1}, \ldots, f^{\lambda_r}$ of a given power series f in $\mathbb{F}_p[[t]]$, and in closer relation to the present study, to [Pa] for a Galois theoretic solution of the logarithmic case of Schanuel's conjecture on powers of the Carlitz module.

References

[AMP] A. Allouche, M. Mendès-France, A. van der Poorten, Indépendance algébrique de certaines séries formelles, *Bull. Soc. Math. France* 116 (1988), no. 4, 449–454

[A1] Y. André, Mumford-Tate groups of mixed Hodge structures and the theorem of the fixed part, *Compo. Math.* 82 (1992) no. 1, 1–24.

[A2] Y. André, Quelques conjectures de transcendance issues de la géométrie algébrique, *Prép. Inst. Math. Jussieu* 121, 1997 (unpublished).

[A3] Y. André, *Une introduction aux motifs (motifs purs, motifs mixtes, périodes)*, Panoramas et Synthèses, No 17, Soc. Math. de France, Paris, 2004.

[AV] V.I. Arnol'd, V.A. Vasil'ev, Newton's Principia read 300 years later, *Notices Amer. Math. Soc.* 36 (1989), no. 9, 1148–1154.

[Ax1] J. Ax, On Schanuel's conjecture, *Ann. of Math.* (2) 93 1971 252–268.

[Ax2] J. Ax, Some topics in differential algebraic geometry I: Analytic subgroups of algebraic groups, *Amer. J. Math.* 94 (1972), 1195–1204.

[Bn1] C. Bertolin, Périodes de 1-motifs et transcendance, *J. Number Theory* 97 (2002), no. 2, 204–221.

[Bn2] C. Bertolin, The Mumford-Tate group of 1-motives, *Ann. Inst. Fourier* 52 (2002), no. 4, 1041–1059.

[B1] D. Bertrand, Endomorphismes de groupes algébriques : applications arithmétiques, in *Diophantine approximations and transcendental numbers (Luminy, 1982)*, Progr. Math., 31, Birkhäuser Boston, Boston, MA, 1983, 1–45.

[B2] D. Bertrand, Galois representations and transcendental numbers, in *New Advances in Transcendence Theory*, ed. A. Baker, Cambridge Univ. Press, Cambridge, 1988, 37–53.

[B3] D. Bertrand, Extensions de *D*-modules et groupes de Galois différentiels, in *p-adic analysis (Trento, 1989)*, Lecture Notes in Math., 1454, Springer, Berlin, 1990, 125–141.

[B4] D. Bertrand, Relative splitting of one-motives, in *Number theory (Tiruchirapalli, 1996)*, Contemp. Math., 210, Amer. Math. Soc., Providence, RI, 1998, 3–17.

[B5] D. Bertrand, Unipotent radicals of differential Galois groups and integrals of solutions of inhomogeneous equations, *Math. Ann.* 321, 2001, 645–666.

[BK] D. Brownawell, K. Kubota, The algebraic independence of Weierstrass functions and some related numbers, *Acta Arith.* 33 (1977), no. 2, 111–149.

[Br] J-L. Brylinski, 1-motifs et formes automorphes (théorie arithmétique des domaines de Siegel), in *Conference on automorphic theory (Dijon, 1981)*, Publ. Math. Univ. Paris VII, 15, Univ. Paris VII, Paris, 1983, 43–106.

[Bu] A. Buium, *Differential algebra and diophantine geometry*, Hermann, Paris, 1994.

[BC] A. Buium, P. Cassidy, Differential algebraic geometry and differential algebraic groups, in *Selected works of Ellis Kolchin with commentary*, eds. H. Bass, A. Buium and P. Cassidy, Amer. Math. Soc., Providence, RI, 1999, 567–636.

[Ch]   C-L. Chai, A note on Manin's theorem of the kernel, *Amer. J. Math.* 113 (1991), no. 3, 387–389.

[Co]   Robert F. Coleman, On a stronger version of the Schanuel-Ax theorem, *Amer. J. Math.* 102 (1980), no. 4, 595–624.

[De1]   P. Deligne, Théorie de Hodge III, *Publ. Math. IHES,* 44 (1974), 5–77.

[De2]   P. Deligne, Hodge cycles on abelian varieties (notes by J. Milne), in *Hodge Cycles, Motives, and Shimura Varieties,* P. Deligne, J. S. Milne, A. Ogus and K.-y. Shih eds., Lecture Notes in Math. 900, Springer-Verlag, 1982, Chapter I, 9–100.

[Ha]   C. Hardouin, *Structure galoisienne des extensions itérées de modules différentiels,* Thèse Univ. Paris 6, 2005.

[JR]   O. Jacquinot, K. Ribet, *Deficient points on extensions of abelian varieties by* $\mathbb{G}_m$, *J. Number Theory* 25 (1987), no. 2, 133–151.

[Ka]   N. Katz, *Exponential sums and differential equations,* Princeton Univ. Press, Princeton, NJ, 1990.

[K1]   J. Kirby, Exponential and Weierstrass equations, Preprint Oxford, 7/9/05, 27 p.

[K2]   J. Kirby, Dimension theory and differential equations; Talk at Inst. Math. Jussieu, Jan. 2006, slides available on author's home page http://www.maths.ox.ac.uk/~kirby/.

[K3]   J. Kirby, *The theory of exponential differential equations,* Ph. D. thesis, Oxford, 2006.

[La]   S. Lang, *Introduction to transcendental numbers,* Addison-Wesley, Reading, Mass., 1966.

[Ma]   D. Marker, Manin kernels, *Quaderni Mat.,* 6, Napoli, 2000, 1-21.

[Mn]   Ju. Manin, Rational points on algebraic curves over function fields, *Izv. Akad. Nauk SSSR Ser. Mat.* 27 1963 1395–1440.

[Mu]   D. Mumford, *Abelian varieties,* Oxford Univ. Press, London, 1974.

[Pa]   M. Papanikolas, Tannakian duality for Anderson-Drinfeld modules and algebraic independence of Carlitz logarithms, arXiv:math.NT/0506078 v1, June 2005.

[Pi]   A. Pillay, Algebraic *D*-groups and differential Galois theory, *Pacific J. Math.* 216 (2004), no. 2, 343–360.

[Se]   J-P. Serre, *Groupes algébriques et corps de classes,* Hermann, Paris 1959.

[WW]   E. Whittaker, G. Watson, *A Course of Modern Analysis. An Introduction to the General Theory of Infinite Processes and of Analytic Functions; with An Account of the Principal Transcendental Functions,* Cambridge Univ. Press, Cambridge 1978.

[Zi]   B. Zilber, Exponential sums and the Schanuel conjecture, *J. London Math. Soc.* (2) 65 (2002), no. 1, 27–44.

An afterthought on the generalized Mordell-Lang conjecture

Damian Rössler

CNRS - Université Paris 7 Denis Diderot

Summary

The generalized Mordell-Lang conjecture (GML) is the statement that the irreducible components of the Zariski closure of a subset of a group of finite rank inside a semi-abelian variety are translates of closed algebraic subgroups. In [6], M. McQuillan gave a proof of this statement. We revisit his proof, indicating some simplifications. This text contains a complete elementary proof of the fact that (GML) for groups of torsion points (= generalized Manin-Mumford conjecture), together with (GML) for finitely generated groups imply the full generalized Mordell-Lang conjecture.

1 Introduction

Let A be a semi-abelian variety over $\overline{\mathbb{Q}}$ (cf. beginning of Section 2 for the definition of a semi-abelian variety). We shall call a closed reduced subscheme of A *linear* if its irreducible components are translates of closed subgroup schemes of A by points of $A(\overline{\mathbb{Q}})$. Let Γ be a finitely generated subgroup of $A(\overline{\mathbb{Q}})$ and define

$$\mathrm{Div}(\Gamma) := \{a \in A(\overline{\mathbb{Q}}) \mid \exists n \in \mathbb{Z}_{\geqslant 1}, \ n \cdot a \in \Gamma\}.$$

Let X be a closed reduced subscheme of A. Consider the following statement:

   the variety $\mathrm{Zar}(X \cap \mathrm{Div}(\Gamma))$ *is linear*   (∗)

The generalized Mordell-Lang conjecture (GML) is the statement that (∗) holds for any data A, Γ, X as above. The statement (GML) with the supplementary requirement that $\Gamma = 0$ shall be referred to as (MM).

63

The statement (GML) with Div(Γ) replaced by Γ in ($*$) shall be referred to as (ML).

The statement (ML) was first proven by Vojta (who built on Faltings work) in [11]. The statement (MM) was first proven by Hindry (who built on work of Serre and Ribet) in [3]. Finally, McQuillan (who built on the work of the previous) proved (GML) in [6].

The structure of McQuillan's proof of (GML) has three key inputs: (1) the statement (ML), (2) an extension of (MM) to families of varieties and (3) the Kummer theory of abelian varieties.

In this text, we shall indicate some simplifications of this proof. More precisely, we show the following. First, that once (MM) is granted, a variation of (2) sufficient for the purposes of the proof is contained in an automatic uniformity principle proved by Hrushovski. See Lemma 2.3 for the statement of this automatic uniformity principle and a reference for the (short) proof. Secondly, we show that one can replace the Kummer theory of abelian varieties (3) by an elementary geometrical argument. The core of the simplified proof is thus an elementary proof of the following statement:

if (ML) *and* (MM) *hold then* (GML) *holds* ($**$)

and the proof of (GML) is then obtained by combining ($**$) with the statements (MM) and (ML), which are known to be true by the work of Hindry and Vojta.

We stress that our proof of ($**$) is independent of the truth or techniques of proof of either (ML) or (MM).

Notice that yet another proof of (GML) was given by Hrushovski in [4, Par. 6.5]. His proof builds on a generalisation of his model-theoretic proof of (MM) (which is based on the dichotomy theorem of the theory of difference fields) and on (ML). It also avoids the Kummer theory of abelian varieties but it apparently doesn't lead to a proof of ($**$). Finally, we want to remark that a deep Galois-theoretic result of Serre (which makes an earlier statement of Bogomolov uniform), which is used in McQuillan's proof of (GML) as well as in Hindry's proof of (MM) (see [3, Lemme 12]), was never published. Now different proofs of (MM), which do not rely on Serre's result, were given by Hrushovski in [4] and by Pink-Rössler in [8]. Our proof of ($**$) thus leads to another proof of (GML) which is independent of Serre's result.

The structure of the article is as follows. Section 3 contains the proof of (∗∗) and section 2 recalls the various facts from the theory of semi-abelian varieties that we shall need in Section 3. The reader is encouraged to start with Section 3 and refer to Section 2 as necessary.

Basic notational conventions. A (closed) subvariety of a scheme S is a (closed) reduced subscheme of S. If X is a closed subvariety of a $\overline{\mathbb{Q}}$-group scheme A, we shall write $\mathrm{Stab}(X)$ for the stabilizer of X in A, which is a closed group subscheme of A such that

$$\mathrm{Stab}(X)(\overline{\mathbb{Q}}) = \{a \in A(\overline{\mathbb{Q}}) \mid X + a = X\}.$$

If H is a commutative group, we write $\mathrm{Tor}(H) \subseteq H$ for the subgroup consisting of the elements of finite order in H. If T is a noetherian topological space, denote by $\mathrm{Irr}(T)$ the set of its irreducible components.

Acknowledgments. Many thanks to P. Vojta for his careful reading of a first version of this text and his detailed comments. My thanks also go to R. Pink, for his comments and suggestions of improvement and for many interesting conversations on matters related to this article. Finally, I am grateful to the referee for his work and for his suggestions.

2 Preliminaries

A *semi-abelian variety* A over an algebraically closed field k is by definition a commutative group scheme over k with the following properties: it has a closed subgroup scheme G which is isomorphic to a product of finitely many multiplicative groups over k and there exists an abelian variety B over k and a surjective morphism $\pi : A \to B$, which is a morphism of group schemes over k and whose kernel is G.

In the next lemma, let A be a semi-abelian variety as in the previous definition.

Lemma 2.1 *Let* $n \in \mathbb{Z}_{\geqslant 1}$. *The multiplication by* n *morphism* $[n]_A :$ $A \to A$ *is quasi-finite.*

Proof. We must prove that the fibers of $[n]_A$ have finitely many points or equivalently that they are of dimension 0. Moreover, since the function $\dim([n]_{A,a})$ (= dimension of the fiber of $[n]_A$ over a) is a constructible function of $a \in A$ (see [2, Ex. 3.22, chap. II]), it is sufficient to prove that the fibers of $[n]_A$ over closed points are finite. A fiber of $[n]_A$ over a closed point can be identified with the fiber $\ker [n]_A$ of $[n]_A$ over $0 \in A$. The scheme $\ker [n]_A$ is naturally fibered over $\ker [n]_B$. The scheme $\ker [n]_B$ consists of finitely many closed points because multiplication by n in B is a finite morphism, as B is an abelian variety (see [7, Prop. 8.1 (d)]. It will thus be sufficient to prove that the fibers of the morphism $\ker [n]_A \to \ker [n]_B$ are finite and furthermore each of these fibers can identified with the fiber of $\ker [n]_A \to \ker [n]_B$ over 0. By construction this fiber is the closed subscheme $\ker [n]_A \times_A G = \ker [n]_G$ of A. To prove that $\ker [n]_G$ has finitely many closed points, choose an identification $G \simeq G_m^\rho$ of G with a product of ρ multiplicative groups over k. The closed points of $\ker [n]_G$ then correspond to ρ-tuples of n-th roots of unity in k. This set is finite and this concludes the proof of the lemma. $\square$

Let now A be a semi-abelian variety over $\overline{\mathbb{Q}}$.

Theorem 2.2 (Kawamata-Abramovich) *Let X be a closed irreducible subvariety of A. The union $Z(X)$ of the irreducible linear subvarieties of positive dimension of X is Zariski closed. The stabilizer $\mathrm{Stab}(X)$ of X is finite if and only if the complement of $Z(X)$ in X is not empty.*

For the proof see [1, Th. 1 & 2] .

Let Y be a variety over $\overline{\mathbb{Q}}$ and let $W \hookrightarrow A \times_{\overline{\mathbb{Q}}} Y$ be a closed subvariety.

Lemma 2.3 (Hrushovski) *If* (MM) *holds then the quantity*

$$\mathrm{Sup}\{\#\mathrm{Irr}\big(\mathrm{Zar}(W_y \cap \mathrm{Tor}(A(\overline{\mathbb{Q}})))\big)\}_{y \in Y(\overline{\mathbb{Q}})}$$

is finite.

Notice that (MM) predicts that the irreducible components of the set $\mathrm{Zar}(W_y \cap \mathrm{Tor}(A(\overline{\mathbb{Q}})))$ are linear for each W_y, $y \in Y(\overline{\mathbb{Q}})$. In words, the content of the lemma is that if this is the case, then the number of these irreducible components can be bounded independently of $y \in Y(\overline{\mathbb{Q}})$. A self-contained proof of Lemma 2.3 can be found in [4, Postscript, Lemma 1.3.2, p. 52-53]. For an alternative presentation and an extension of Lemma 2.3, see the paper [10]. The proof is based on what logicians call the uniform definability of types in algebraically closed fields (see e.g. [10, Lemma 3.2]), which can be established using Chevalley's constructibility theorem and the compactness theorem of first order logic.

Suppose now that A has a model A_0 over a number field K.

In [9, Par. 1.2 and Prop. 2] it is shown that there exists a variety $\overline{A}_0$ projective over K and an open immersion $A_0 \hookrightarrow \overline{A}_0$ such that for all $n \in \mathbb{Z}_{\geqslant 1}$ the multiplication by n morphism $[n]_{A_0} : A_0 \to A_0$ extends to a K-morphism $[n]_{\overline{A}_0} : \overline{A}_0 \to \overline{A}_0$. Furthermore, it is shown in [9, Prop. 3] that the corresponding diagram

$$
\begin{array}{ccc}
A_0 & \lhook\joinrel\longrightarrow & \overline{A}_0 \\
\Big\downarrow{\scriptstyle [n]_{A_0}} & & \Big\downarrow{\scriptstyle [n]_{\overline{A}_0}} \\
A_0 & \lhook\joinrel\longrightarrow & \overline{A}_0
\end{array}
$$

is then cartesian.

Let Γ be a finitely generated subgroup of $A_0(K)$.

Lemma 2.4 (McQuillan) *The group generated by* $\mathrm{Div}(\Gamma) \cap A_0(K)$ *is finitely generated.*

Lemma 2.4 is McQuillan's Lemma 3.1.3 in [6]. As the proof given there is somewhat sketchy, we shall provide a proof of Lemma 2.4.

Proof. Let B be an abelian variety over $\overline{\mathbb{Q}}$ and $\pi : A \to B$ be a $\overline{\mathbb{Q}}$-morphism whose kernel G is isomorphic to a product of tori over $\overline{\mathbb{Q}}$. These data exist since A is a semi-abelian variety. Notice that for the purposes of the proof we may enlarge the field of definition K of A if necessary, since that operation will also enlarge the set $\mathrm{Div}(\Gamma) \cap A_0(K)$. Hence we may assume that B (resp. G) has a model B_0 (resp G_0) over K and that π has a model π_0 over K. Furthermore, we may assume that the isomorphism of G with a product of tori descends to a K-isomorphism of G_0 with a product of split tori over K. Now fix a compactification $\overline{A}_0$ of A_0 over K as above. We consider the following situation. The symbol V refers to an open subscheme of the spectrum $\mathrm{Spec}\,\mathcal{O}_K$ of the ring of integers of K and $\mathcal{A}_0$ is a semi-abelian scheme over V, which is a model of A_0. The symbol $\overline{\mathcal{A}}_0$ refers to a projective model over V of $\overline{A}_0$ and we suppose given an open immersion $\mathcal{A}_0 \hookrightarrow \overline{\mathcal{A}}_0$, which is a model of the open immersion $A_0 \hookrightarrow \overline{A}_0$. We also suppose that the multiplication by n morphism $[n]_{\mathcal{A}_0}$ on $\mathcal{A}_0$ extends to a V-morphism $[n]_{\overline{\mathcal{A}}_0} : \overline{\mathcal{A}}_0 \to \overline{\mathcal{A}}_0$

and we suppose that the corresponding diagram

$$
\begin{array}{ccc}
A_0 & \lhook\joinrel\longrightarrow & \overline{A}_0 \\
\Big\downarrow{\scriptstyle [n]_{A_0}} & & \Big\downarrow{\scriptstyle [n]_{\overline{A}_0}} \\
A_0 & \lhook\joinrel\longrightarrow & \overline{A}_0
\end{array}
$$

is cartesian. Let $\mathcal{B}_0$ (resp. $\mathcal{G}_0$) be a model of B_0 (resp. G_0) over V and let $\widetilde{\pi}_0$ be a model of π_0 over V. Furthermore, we assume that the K-isomorphism of G_0 with a product of split tori extends to a V-isomorphism of $\mathcal{G}_0$ with a product of split tori over V. We leave it to the reader to show that there are objects V, $\mathcal{A}_0$ etc. satisfying the described conditions.

The morphism $[n]_{\mathcal{A}_0}$ is then proper, because $[n]_{\overline{\mathcal{A}}_0}$ is proper (as $\overline{\mathcal{A}}_0$ is proper over V) and properness is invariant under base change. The morphism $[n]_{\mathcal{A}_0}$ is therefore finite, as it is quasi-finite by 2.1 (applied to each fiber of $\mathcal{A}_0$ over V). We may suppose without restriction of generality that Γ lies in the image of $\mathcal{A}_0(V)$ in $A_0(K)$; indeed this condition will always be fulfilled after possibly removing a finite number of closed points from V. Let $a \in \mathrm{Div}(\Gamma) \cap A_0(K)$. Choose an $n \in \mathbb{Z}_{\geqslant 1}$ such that $n \cdot a \in \Gamma$. Let E be the image of the section $V \to \mathcal{A}_0$ corresponding to $n \cdot a$. Consider the reduced irreducible component C of $[n]^*_{\mathcal{A}_0} E$ containing the image of a. The image of a is the generic point of C and by assumption the natural morphism $C \to E$ identifies the function fields of C and E. Furthermore, the morphism $C \to E$ is finite. Now let R_0 be the ring underlying the affine scheme V. In view of the above, we can write $C = \mathrm{Spec}\ R$, where R is a domain and the morphism $C \to E$ identifies R with an integral extension of R_0 inside the integral closure of R_0 in its own field of fractions. As R_0 is integrally closed (it is even a Dedekind ring) $C \to E$ is an isomorphism. Hence $a \in \mathcal{A}_0(V)$. Thus we only have to show that $\mathcal{A}_0(V)$ is finitely generated. This follows from the fact that $\mathcal{B}_0(V)$ is finitely generated by the Mordell-Weil theorem applied to B_0 and the fact that $\mathcal{G}_0(V)$ is finitely generated by the generalized Dirichlet unit theorem (see [5, chap. V, par. 1]). $\square$

Lemma 2.5 *Let $C > 0$. The set $\{a \in \mathrm{Tor}(A(\overline{\mathbb{Q}})) \mid [K(a) : K] < C\}$ is finite.*

Proof. If A_0 is an abelian variety over K then this follows from the fact that the Néron-Tate height of torsion points vanishes and from Northcott's theorem. We leave the general case as an exercise for the reader. □

3 Proof of ($**$)

In this section we shall prove (GML) using the results listed in Section 2 as well as (ML) and (MM).

First notice that to prove ($*$), one may assume without loss of generality that X is irreducible and that $\text{Zar}(X \cap \text{Div}(\Gamma)) = X$. The statement ($*$) then becomes

> *the variety X is a translate of a connected*   ($*$)′
> *closed group subscheme of A*

In the remaining of this section, we shall prove that (ML) and (MM) imply ($*$)′. The overall structure of our proof will be similar to McQuillan's proof of (GML), with some simplifications that we shall point out along the way.

First, we may suppose without restriction of generality that $\text{Stab}(X)$ is trivial.

To see the latter, consider the closed subvariety $X/\text{Stab}(X)$ of the quotient variety $A/\text{Stab}(X)$. The image of $\text{Div}(\Gamma)$ in $(A/\text{Stab}(X))(\overline{\mathbb{Q}})$ lies inside the group $\text{Div}(\Gamma_1)$, where Γ_1 is the image of Γ and the image of $\text{Div}(\Gamma)$ is dense in $X/\text{Stab}(X)$. So the assumptions of ($*$) hold for $A/\text{Stab}(X)$, Γ_1 and $X/\text{Stab}(X)$. Furthermore, by construction $\text{Stab}(X/\text{Stab}(X)) = 0$. Now if ($*$)′ holds in this situation, $X/\text{Stab}(X)$ is the translate of a connected closed group subscheme of $A/\text{Stab}(X)$. Hence $X/\text{Stab}(X)$ is a closed point. This implies that X is a translate of $\text{Stab}(X)$, thus proving ($*$)′ for A, Γ and X. It is thus sufficient to prove ($*$)′ for $A/\text{Stab}(X)$, Γ_1 and $X/\text{Stab}(X)$ and we may thus replace A by $A/\text{Stab}(X)$, Γ by Γ_1, X by $X/\text{Stab}(X)$. We then have $\text{Stab}(X) = 0$.

We may also assume without loss of generality that A (resp. X) has a model A_0 (resp. X_0) over a number field K such that $\Gamma \subseteq A_0(K)$ and such that the immersion $X \to A$ has a model over K as an immersion $X_0 \to A_0$.

Let U be the complement in X of the union of the irreducible linear subvarieties of positive dimension of X. By Theorem 2.2 the set U is non-empty and open in X and thus $U \cap \mathrm{Div}(\Gamma)$ is dense in X.

Let $a \in U \cap \mathrm{Div}(\Gamma)$ and let $\sigma \in \mathrm{Gal}(\overline{\mathbb{Q}}|K)$. By construction $\sigma(U) \subseteq U$ and we thus have $\sigma(a) - a \in U - a$. Now by definition, there exists $n = n(a) \in \mathbb{Z}_{\geqslant 1}$ such that $n \cdot a \in \Gamma \subseteq A_0(K)$. We calculate

$$n \cdot (\sigma(a) - a) = \sigma(n \cdot a) - n \cdot a = n \cdot a - n \cdot a = 0.$$

Thus $\sigma(a) - a \in \mathrm{Tor}(A(\overline{\mathbb{Q}})) \cap (U - a)$. The statement (MM) implies that $\mathrm{Tor}(A(\overline{\mathbb{Q}})) \cap (U - a)$ is finite and using Theorem 2.3, we see that $\#(\mathrm{Tor}(A(\overline{\mathbb{Q}})) \cap (U - a)) < C$ for some constant $C \in \mathbb{Z}_{\geqslant 1}$, which is independent of a. Now this implies that $\#\{\tau(a)|\tau \in \mathrm{Gal}(\overline{\mathbb{Q}}|K)\} = \#\{\tau(a) - a \mid \tau \in \mathrm{Gal}(\overline{\mathbb{Q}}|K)\} < C$.

A consequence of this conclusion is reached by McQuillan at the beginning of Par. 3.3 (p. 157) of [6] using Theorem 3.2.2 of that article. This last theorem is replaced by Theorem 2.3 in our context.

By Galois theory, we thus have $[K(a) : K] < C$. We deduce from this last inequality that

$$[K(\sigma(a) - a) : K] \leqslant [K(a, \sigma(a)) : K] < C^2.$$

By Lemma 2.5, we see that this implies that $\sigma(a) - a \in T$, where $T \subseteq A(\overline{\mathbb{Q}})$ is a finite set, which is dependent on C but independent of either a or σ. For each $b \in A(\overline{\mathbb{Q}}) \setminus A_0(K)$, choose $\sigma_b \in \mathrm{Gal}(\overline{\mathbb{Q}}|K)$ such that $\sigma_b(b) \neq b$. We know that either the set

$$\{b \in U \cap \mathrm{Div}(\Gamma) \mid b \in A(\overline{\mathbb{Q}}) \setminus A_0(K)\}$$

or the set

$$\{b \in U \cap \mathrm{Div}(\Gamma) \mid b \in A_0(K)\}$$

is dense in X. First suppose the former. In that case, there exists $t_0 \in T$, $t_0 \neq 0$ such that the set

$$\{b \in U \cap \mathrm{Div}(\Gamma) \mid b \in A(\overline{\mathbb{Q}}) \setminus A_0(K), \ \sigma_b(b) - b = t_0\}$$

is dense in X. Since $\sigma_b(b) \in U$ for all $b \in U$ such that $b \in A(\overline{\mathbb{Q}}) \setminus A_0(K)$, we see that this implies that $t_0 \in \mathrm{Stab}(X)(\overline{\mathbb{Q}})$. But $t_0 \neq 0$ so this contradicts our hypothesis that $\mathrm{Stab}(X) = 0$. Thus we deduce that the set $\{b \in U \cap \mathrm{Div}(\Gamma) \mid b \in A_0(K)\}$ is dense in X. By Lemma 2.4, the elements of this set are contained in a finitely generated group, so using (ML) we deduce $(*)'$.

The last part of our proof of $(*)'$ is similar to McQuillan's final argument in Par. 3.3 of [6]. His argument involves the Kummer theories of abelian varieties and tori; this is replaced in our context by the above elementary geometrical construction, based on Lemma 2.5.

References

[1] D. Abramovich, Subvarieties of semiabelian varieties, *Compositio Math.* **90**, no. 1, 37–52 (1994).

[2] R. Hartshorne, *Algebraic geometry*, Graduate Texts in Mathematics, No. 52. Springer-Verlag, New York-Heidelberg 1977.

[3] M. Hindry, Autour d'une conjecture de Serge Lang, *Invent. Math.* **94**, no. 3, 575–603 (1988).

[4] E, Hrushovski, The Manin-Mumford conjecture and the model theory of difference fields, *Annals of Pure and Applied Logic* **112**, 43–115 (2001).

[5] S. Lang, *Algebraic Number Theory*, Second Edition, Springer 1994.

[6] M. McQuillan, Division points on semi-abelian varieties, *Invent. Math.* **120**, no. 1, 143–159 (1995).

[7] J. Milne, Abelian Varieties, in *Arithmetic Geometry (Storrs, Conn., 1984)*, 103–150, Springer, New York 1986.

[8] R. Pink and D. Rössler, On psi-invariant subvarieties of semiabelian varieties and the Manin-Mumford conjecture, *J. Algebraic. Geom.* **13** (2004), no. 4, 771–798.

[9] J.-P. Serre, Quelques propriétés des groupes algébriques commutatifs, Appendix in *Astérisque* **69-70**, 191–202 (1979).

[10] T. Scanlon, Automatic uniformity, *Int. Math. Res. Not.* **62**, 3317–3326 (2004).

[11] P. Vojta, Integral points on subvarieties of semiabelian varieties, I, *Invent. Math.* **126**, no. 1, 133–181 (1996).

On the definitions of difference Galois groups

Zoé Chatzidakis[†]
CNRS - Université Paris 7

Charlotte Hardouin
Universität Heidelberg IWR INF 368

Michael F. Singer[‡]
University of North Carolina

Summary

We compare several definitions of the Galois group of a linear difference equation that have arisen in algebra, analysis and model theory and show, that these groups are isomorphic over suitable fields. In addition, we study properties of Picard-Vessiot extensions over fields with not necessarily algebraically closed subfields of constants.

1 Introduction

In the modern Galois theory of polynomials of degree n with coefficients in a field k[1], one associates to a polynomial $p(x)$ a splitting field K, that is a field K that is generated over k by the roots of $p(x)$. All such fields are k-isomorphic and this allows one to define the Galois group of $p(x)$ to be the group of k-automorphisms of such a K. If k is a differential field and $Y' = AY, A$ an $n \times n$ matrix with entries in k, one may be tempted to naively define a "splitting field" for this equation to be a differential field K containing k and generated (as a differential field) by the entries of a fundamental solution matrix Z of the differential equation[2]. Regrettably, such a field is not unique in general. For example,

† The author thanks the Isaac Newton Institute for Mathematical Sciences for its hospitality and financial support during spring 2005.

‡ The preparation of this paper was supported by NSF Grant CCR- 0096842 and by funds from the Isaac Newton Institute for Mathematical Sciences during a visit in May 2005.

---

1 All fields in this paper are assumed to be of characteristic zero.

2 that is, an invertible $n \times n$ matrix Z such that $Z' = AZ$. Note that the columns of Z form a basis of the solution space.

for the equation $y' = \frac{1}{2x}y$ over $k = \mathbb{C}(x), x' = 1$, the fields $k(x^{1/2})$ and $k(z), z$ transcendental over k and $z' = \frac{1}{2x}z$ are not k-isomorphic. If one insists that the constants $C_k = \{c \in k \mid c' = 0\}$ are algebraically closed and that K has no new constants, then Kolchin [16] showed that such a K exists (and is called the *Picard-Vessiot* associated with the equation) and is unique up to k-differential isomorphism. Kolchin [15] defined the Galois group of such a field to be the group of k-differential automorphisms of K and developed an appropriate Galois theory[3].

When one turns to difference fields k with automorphism σ and difference equations $\sigma Y = AY$, $A \in \mathrm{GL}_n(k)$, the situation becomes more complicated. One can consider difference fields K such that K is generated as a difference field by the entries of a fundamental solution matrix. If the field of constants $C_k = \{c \in k \mid \sigma(c) = c\}$ is algebraically closed and K has no new constants, then such a K is indeed unique and is again called a Picard-Vessiot extension ([23], Proposition 1.23 and Proposition 1.9). Unlike the differential case, there are equations for which such a field does not exist. In fact there are difference equations that do not have any nonzero solution in a difference field with algebraically closed constants. For example, let K be a difference field containing an element $z \neq 0$ such that $\sigma(z) = -z$. One then has that z^2 is a constant. If, in addition, the constants C_K of K are algebraically closed, then $z \in C_K$ so $\sigma(z) = z$, a contradiction. This example means that either one must consider "splitting fields" with subfields of constants that are not necessarily algebraically closed or consider "splitting rings" that are not necessarily domains. Both paths have been explored and the aim of this paper is to show that they lead, in essence, to the same Galois groups.

The field theoretic approach was developed by Franke[4] in [10] and succeeding papers. He showed that for Picard-Vessiot extension fields the Galois group is a linear algebraic group defined over the constants and that there is the usual correspondence between closed subgroups and intermediate difference fields. Franke notes that Picard-Vessiot extension fields do not always exist but does discuss situations when they do exist and results that can be used when adjoining solutions of a linear difference equation forces one to have new constants.

---

3 It is interesting to note that the Galois theory was developed before it was known if such K always exist. See the footnote on p.29 of [15].

4 Bialynicki-Birula [2] developed a general Galois theory for fields with operators but with restrictions that forced his Galois groups to be connected.

Another field theoretic approach is contained in the work of Chatzidakis and Hrushovski [4]. Starting from a difference field k, they form a certain large difference extension $\mathcal{U}$ having the properties (among others) that for any element in $\mathcal{U}$ but not in k, there is an automorphism of $\mathcal{U}$ that moves this element and that any set of difference equations (not necessarily linear) that have a solution in some extension of $\mathcal{U}$ already have a solution in $\mathcal{U}$. The subfield of constants $C_{\mathcal{U}}$ is not an algebraically closed field. Given a linear difference equation with coefficients in k, there exists a fundamental solution matrix with entries in $\mathcal{U}$. Adjoining the entries of these to $k(C_{\mathcal{U}})$ yields a difference field K. A natural candidate for a Galois group is the group of difference automorphisms of K over $k(C_{\mathcal{U}})$ and these do indeed correspond to points in a linear algebraic group. Equality of this automorphism group with the Galois group coming from Picard-Vessiot rings is shown in 4.15 under certain conditions (which are always verified when C_k is algebraically closed). Proofs are very algebraic in nature, and along the way produce some new algebraic results on Picard-Vessiot rings: we find numerical invariants of Picard-Vessiot rings of the equation $\sigma(X) = AX$, and show how to compute them (see 4.9 and 4.11). Furthermore, we show how to compute the number of primitive idempotents of a Picard-Vessiot ring when the field C_k is algebraically closed (4.13). This situation will be further discussed in Section 4.

The field theoretic approach also seems most natural in the analytic situation. For example, let $\mathcal{M}(\mathbb{C})$ be the field of functions $f(x)$ meromorphic on the complex plain endowed with the automorphism defined by the shift $\sigma(x) = x + 1$. Note that the constants $C_{\mathcal{M}(\mathbb{C})}$ are the periodic meromorphic functions. A theorem of Praagman [21] states that a difference equation with coefficients in $\mathcal{M}(\mathbb{C})$ will have a fundamental solution matrix with entries in $\mathcal{M}(\mathbb{C})$. If k is the smallest difference field containing the coefficients of the equation and $C_{\mathcal{M}(\mathbb{C})}$ and K is the smallest difference field containing k and the entries of fundamental solution matrix, then, in this context, the natural Galois group is the set of difference automorphisms of K over k. For example, the difference equation $\sigma(y) = -y$ has the solution $y = e^{\pi i x}$. This function is algebraic of degree 2 over the periodic functions $k = C_{\mathcal{M}(\mathbb{C})}$. Therefore, in this context the Galois group of $K = k(e^{\pi i x})$ over k is $\mathbb{Z}/2\mathbb{Z}$.

One can also consider the field $\mathcal{M}(\mathbb{C}^*)$ of meromorphic functions on the punctured plane $\mathbb{C}^* = \mathbb{C} \backslash \{0\}$ with q-automorphism $\sigma_q(x) = qx$, $|q| \neq 1$.

Difference equations in this context are q-difference equations and Praagman proved a global existence theorem in this context as well. The constants $C_{\mathcal{M}(\mathbb{C}^*)}$ naturally correspond to meromorphic functions on the elliptic curve $\mathbb{C}^*/q^{\mathbb{Z}}$ and one can proceed as in the case of the shift. One can also define local versions (at infinity in the case of the shift and at zero or infinity in the case of q-difference equations). In the local case and for certain restricted equations one does not necessarily need constants beyond those in $\mathbb{C}$ (see [9], [22], [23] as well as connections between the local and global cases. Another approach to q-difference equations is given by Sauloy in [26] and Ramis and Sauloy in [25] where a Galois group is produced using a combination of analytic and tannakian tools. The Galois groups discussed in these papers do not appear to act on rings or fields and, at present, it is not apparent how the techniques presented here can be used to compare these groups to other putative Galois groups.)

An approach to the Galois theory of difference equations with coefficients in difference fields based on rings that are not necessarily integral was presented in [23] (and generalized by André in [1] to include differential and difference equations with coefficients in fairly general rings as well). One defines a *Picard-Vessiot ring* associated with a difference equation $\sigma Y = AY$ with coefficients in a difference field k to be a simple difference ring (i.e., no σ-invariant ideals) R of the form $R = k[z_{i,j}, 1/\det(Z)]$ where $Z = (z_{i,j})$ is a fundamental solution matrix of $\sigma Y = AY$. Assuming that C_k is algebraically closed, it is shown in [23] that such a ring *always* exists and is unique up to k-difference isomorphism. A similar definition for differential equations yields a ring that is an integral domain and leads (by taking the field of quotients) to the usual theory of Picard-Vessiot extensions (see [24]). In the difference case, Picard-Vessiot rings need not be domains. For example, for the field $k = \mathbb{C}$ with the trivial automorphism, the Picard-Vessiot ring corresponding to $\sigma y = -y$ is $\mathbb{C}[Y]/(Y^2 - 1), \sigma(Y) = -Y$. Nonetheless, one defines the *difference Galois group of* $\sigma Y = AY$ to be the k-difference automorphisms of R and one can shows that this is a linear algebraic group defined over C_k. In the example above, the Galois group is easily seen to be $\mathbb{Z}/2\mathbb{Z}$. Furthermore, in general there is a Galois correspondence between certain subrings of the total quotient ring and closed subgroups of the Galois group.

The natural question arises: *How do these various groups relate to each other?* The example of $\sigma(y) = -y$ suggests that the groups may be the same. Our main result, Theorem 2.9, states that all these groups are isomorphic as algebraic groups over a suitable extension of the constants. This result has interesting ramifications for the analytic theory of difference equations. In [11], the second author gave criteria to insure that solutions, meromorphic in $\mathbb{C}^*$, of a first order q-difference equation over $\mathbb{C}(x)$ satisfy no algebraic differential relation over $C_{\mathcal{M}(\mathbb{C}^*)}(x)$, where $C_{\mathcal{M}(\mathbb{C}^*)}$ is the field of meromorphic functions on the elliptic curve $\mathbb{C}^*/q^{\mathbb{Z}}$. The proof of this result presented in [11] depended on knowing the dimension of Galois groups in the analytic (i.e., field-theoretic) setting. These groups could be calculated in the ring theoretic setting of [23] and the results of the present paper allow one to transfer this information to the analytic setting. Although we will not go into more detail concerning the results of [11], we will give an example of how one can deduce transcendence results in the analytic setting from their counterparts in the formal setting.

The rest of the paper is organized as follows. In Section 2, we show how results of [23] and [24] can be modified to prove the correspondence of various Galois groups. In Section 3 we prove this result again in the special case of q-difference equations over $\mathbb{C}(x)$ using tannakian tools in the spirit of Proposition 1.3.2 of [14]. In Section 4, we discuss the model-theoretic approach in more detail and, from this point of view, show the correspondence of the Galois groups. In addition, we consider some additional properties of Picard-Vessiot rings over fields with constant subfields that are not necessarily algebraically closed. The different approaches and proofs have points of contacts (in particular, Proposition 2.4) and we hope comparisons of these techniques are enlightening.

The authors would like to thank Daniel Bertrand for suggesting the approach of Section 3 and his many other useful comments concerning this paper.

2 A Ring-Theoretic Point of View

In this section we shall consider groups of difference automorphisms of rings and fields generated by solutions of linear difference equations and show that these groups are isomorphic, over the algebraic closure of the constants to the Galois groups defined in [24]. We begin by defining the rings and fields we will study.

Definition 2.1 Let K be a difference field with automorphism σ and let $A \in \mathrm{GL}_n(K)$.

a. We say that a difference ring extension R of K is a *weak Picard-Vessiot ring* for the equation $\sigma X = AX$ if

(i) $R = K[Z, \frac{1}{\det(Z)}]$ where $Z \in \mathrm{GL}_n(R)$ and $\sigma Z = AZ$ and
(ii) $C_R = C_K$.

b. We say that a difference field extension L of K is a *weak Picard-Vessiot field* for $\sigma X = AX$ if $C_L = C_K$ and L is the quotient field of a weak Picard-Vessiot ring of $\sigma X = AX$.

In [23], the authors define a *Picard-Vessiot ring* for the equation $\sigma Y = AY$ to be a difference ring R such that (i) holds and in addition R is simple as a difference ring, that is, there are no σ-invariant ideals except (0) and R. When C_K is algebraically closed, Picard-Vessiot rings exist, are unique up to K-difference isomorphisms and have the same constants as K ([23], Section 1.1). Therefore in this case, the Picard-Vessiot ring will be a weak Picard-Vessiot ring.

In general, even when the field of constants is algebraically closed, Example 1.25 of [23] shows that there will be weak Picard-Vessiot rings that are not Picard-Vessiot rings. Furthermore this example shows that the quotient field of a weak Picard-Vessiot integral domain R need not necessarily have the same constants as R so the requirement that $C_L = C_K$ is not superfluous.

The Galois theory of Picard-Vessiot rings is developed in [23] for Picard-Vessiot rings R over difference fields K with algebraically closed constants C_K. In particular, it is shown ([23], Theorem 1.13) that the groups of difference K-automorphisms of R over K corresponds to the set of C_K-points of a linear algebraic group defined over C_K. A similar result for differential equations is proven in ([24], Theorem 1.27). It has been observed by many authors beginning with Kolchin ([17], Ch. VI.3 and VI.6; others include [1], [7], [6], [14], [18] in a certain characteristic p setting for difference equations) that one does not need C_k to be algebraically closed to achieve this latter result. Recently, Dyckerhoff [8] showed how the proof of Theorem 1.27 of [24] can be adapted in the differential case to fields with constants that are not necessarily algebraically closed. We shall give a similar adaption in the difference case.

Proposition 2.2 *Let K be a difference field of characteristic zero and let $\sigma Y = AY, A \in \mathrm{GL}_n(K)$ be a difference equation over K. Let R be a weak Picard-Vessiot ring for this equation over K. The group of difference K-automorphisms of R can be identified with the C_K-points of a linear algebraic group G_R defined over C_K.*

Proof. We will define the group G_R by producing a representable functor from the category of commutative C_K-algebras to the category of groups (c.f., [27]).

First, we may write $R = K[Y_{i,j}, \frac{1}{\det(Y)}]/q$ as the quotient of a difference ring $K[Y_{i,j}, \frac{1}{\det(Y)}]$, where $Y = \{Y_{i,j}\}$ is an $n \times n$ matrix of indeterminates with $\sigma Y = AY$, by a σ-ideal q. Let $C = C_K$. For any C-algebra B, one defines the difference rings $K \otimes_C B$ and $R \otimes_C B$ with automorphism $\sigma(f \otimes b) = \sigma(f) \otimes b$ for $f \in K$ or R. In both cases, the ring of constants is B. We define the functor $\mathcal{G}_R$ as follows: the group $\mathcal{G}_R(B)$ is the group of $K \otimes_C B$-linear automorphisms of $R \otimes_C B$ that commute with σ. One can show that $\mathcal{G}_R(B)$ can be identified with the group of matrices $M \in \mathrm{GL}_n(B)$ such that the difference automorphism ϕ_M of $R \otimes_C B$, given by $(\phi_M Y_{i,j}) = (Y_{i,j})M$, has the property that $\phi_M(q) \subset (q)$ where (q) is the ideal of $K[Y_{i,j}, \frac{1}{\det(Y_{i,j})}] \otimes_C B$ generated by q.

We will now show that $\mathcal{G}_R$ is representable. Let $X_{s,t}$ be new indeterminates and let $M_0 = (X_{s,t})$. Let $q = (q_1, \ldots, q_r)$ and write $\sigma_{M_0}(q_i)$ mod $(q) \in R \otimes_C C[X_{s,t}, \frac{1}{\det(X_{s,t})}]$ as a finite sum

$$\sum_i C(M_0, i, j)e_i \text{ with all } C(M_0, i, j) \in C[X_{s,t}, \frac{1}{\det(X_{s,t})}] \,,$$

where $\{e_i\}_{i \in I}$ is a C-basis of R. Let I be the the ideal in $C[X_{s,t}, \frac{1}{\det(X_{s,t})}]$ generated by all the $C(M_0, i, j)$. We will show that

$$U := C[X_{s,t}, \frac{1}{\det(X_{s,t})}]/I$$

represents $\mathcal{G}_R$.

Let B be a C-algebra and $\phi \in \mathcal{G}_R(B)$ identified with ϕ_M for some

$M \in \mathrm{GL}_n(B)$. One defines the C-algebra homomorphism

$$\Phi : C[X_{s,t}, \frac{1}{\det(X_{s,t})}] \to B, \qquad (X_{s,t}) \mapsto M.$$

The condition on M implies that the kernel of Φ contains I. This then gives a unique C-algebra homomorphism

$$\Psi : U \to B, \qquad \Psi(M_0 \mod I) \mapsto M.$$

The Yoneda Lemma can now be used to show that $G_R = \mathrm{Spec}(U)$ is a linear algebraic group (see Appendix B, p. 382 of [24] to see how this is accomplished or Section 1.4 of [27]).   □

We will refer to G_R as the *Galois group* of R. When R is a Picard-Vessiot extension of K, we have the usual situation. We are going to compare the groups associated with a Picard-Vessiot extension and weak Picard-Vessiot field extensions for the same equation over different base fields. We will first show that extending a Picard-Vessiot ring by constants yields a Picard-Vessiot ring whose associated group is isomorphic to the original group over the new constants. In the differential case and when the new constants are algebraic over the original constants this appears in Dyckerhoff's work ([8], Proposition 1.18 and Theorem 1.26). Our proof is in the same spirit but without appealing to descent techniques. We will use Lemma 1.11 of [23], which we state here for the convenience of the reader:

Lemma 2.3 *Let R be a Picard-Vessiot ring over a field k with $C_R = C_k{}^5$ and A be a commutative algebra over C_k. The action of σ on A is supposed to be the identity. Let N be an ideal of $R \otimes_{C_k} A$ that is invariant under σ. Then N is generated by the ideal $N \cap A$ of A.*

Proposition 2.4 *Let $k \subset K$ be difference fields of characteristic zero and $K = k(C_K)$. Let R be a Picard-Vessiot ring over k with $C_R = C_k$ for the equation $\sigma X = AX, A \in \mathrm{GL}_n(k)$. If $R = k[Y, \frac{1}{\det(Y)}]/q$ where Y is an $n \times n$ matrix of indeterminates, $\sigma Y = AY$ and q is a maximal σ-ideal, then $S = K[Y, \frac{1}{\det(Y)}]/qK$ is a Picard-Vessiot extension of K for the same equation. Furthermore, $C_S = C_K$.*

5 The hypothesis $C_R = C_k$ is not explicitly stated in the statement of this result in [23] but is assumed in the proof.

Proof. First note that the ideal $qK \neq K[Y, \frac{1}{\det(Y)}]$. Secondly, Lemma 2.3 states that for R as above and A a commutative C_k algebra with identity, any σ-ideal N of $R \otimes_{C_k} A$ (where the action of σ on A is trivial) is generated by $N \cap A$. This implies that the difference ring $R \otimes_{C_k} C_K$ is simple. Therefore the map $\psi : R \otimes_{C_k} C_K \to S = K[Y, \frac{1}{\det(Y)}]/(q)K$ where $\psi(a \otimes b) = ab$ is injective. Let R' be the image of ψ. One sees that any element of S is of the form $\frac{a}{b}$ for some $a \in R', b \in k[C_k] \subset R'$. Therefore any ideal I in S is generated by $I \cap R'$ and so S is simple.

For any constant $c \in S$, the set $J = \{a \in R' \mid ac \in R'\} \subset R'$ is a nonzero σ-ideal so $c \in R'$. Since the constants of R' are C_K, this completes the proof. $\qquad\square$

Corollary 2.5 *Let R and S be as in Proposition 2.4. If G_R and G_S are the Galois groups associated with these rings as in Proposition 2.2, then G_R and G_S are isomorphic over C_K.*

Proof. We are considering G_R as the functor from C_k algebras A to groups defined by $G_R(A) := Aut(R \otimes_{C_k} A)$ where $Aut(..)$ is the group of difference $k \otimes A$-automorphisms. Let T_R be the finitely generated C_k-algebra representing G_R (i.e., the coordinate ring of the group). Similarly, let T_S be the C_K-algebra representing G_S. We define a new functor F from C_K-algebras to groups as $F(B) := Aut((R \otimes_{C_k} C_K) \otimes_{C_K} B)$. One checks that F is also a representable functor represented by $T_R \otimes_{C_k} C_K$. Using the embedding ψ of the previous proof, one sees that $F(B) = Aut(S \otimes_{C_K} B) = G_R(B)$ for any C_K-algebra B. The Yoneda Lemma implies that $T_R \otimes_{C_k} C_K \simeq T_S$. $\qquad\square$

In Proposition 2.7 we will compare Picard-Vessiot rings with weak Picard-Vessiot fields for the same difference equation. To do this we need the following lemma. A version of this in the differential case appears as Lemma 1.23 in [24].

Lemma 2.6 *Let L be a difference field. Let $Y = (Y_{i,j})$ be and $n \times n$ matrix of indeterminates and extend σ to $L[Y_{i,j}, \frac{1}{\det(Y)}]$ by setting $\sigma(Y_{i,j}) = Y_{i,j}$. The map $I \mapsto (I) = I \cdot L[Y_{i,j}, \frac{1}{\det(Y)}]$ from the set of ideals in $C_L[Y_{i,j}, \frac{1}{\det(Y)}]$ to the set of ideals of $L[Y_{i,j}, \frac{1}{\det(Y)}]$ is a bijection.*

Proof. One easily checks that $(I) \cap C_L[Y_{i,j}, \frac{1}{\det(Y)}] = I$. Now, let J be an ideal of $L[Y_{i,j}, \frac{1}{\det(Y)}]$ and let $I = J \cap C_L[Y_{i,j}, \frac{1}{\det(Y)}]$. Let $\{e_i\}$ be a basis of $C_L[Y_{i.j}, \frac{1}{\det(Y)}]$ over C_L. Given $f \in L[Y_{i.j}, \frac{1}{\det(Y)}]$, we may write f uniquely as $f = \sum f_i e_i$, $f_i \in L$. Let $\ell(f)$ be the number of i such that $f_i \neq 0$. We will show, by induction on $\ell(f)$, that for any $f \in J$, we have $f \in (I)$. If $\ell(f) = 0, 1$ this is trivial. Assume $\ell(f) > 1$. Since L is a field, we can assume that there exists an i_1 such that $f_{i_1} = 1$. Furthermore, we may assume that there is an $i_2 \neq i_1$ such that $f_{i_2} \in L\backslash C_L$. We have $\ell(f - \sigma(f)) < \ell(f)$ so $\sigma(f) - f \in (I)$. Similarly, $\sigma(f_{i_2}^{-1}f) - f_{i_2}^{-1}f \in (I)$. Therefore, $(\sigma(f_{i_2}^{-1}) - f_{i_2}^{-1})f = \sigma(f_{i_2}^{-1})(f - \sigma(f)) + (\sigma(f_{i_2}^{-1}f) - f_{i_2}^{-1}f) \in (I)$. This implies that $f \in (I)$. $\qquad\square$

The following is a version of Proposition 1.22 of [24] modified for difference fields taking into account the possibility that the constants are not algebraically closed.

Proposition 2.7 *Let K be a difference field with constants C and let $A \in \mathrm{GL}_n(K)$. Let $S = K[U, \frac{1}{\det(U)}]$, $U \in \mathrm{GL}_n(S)$, $\sigma(U) = AU$ be a Picard-Vessiot extension of K with $C_S = C_k$ and let $L = K(V)$, $V \in GL_n(L)$, $\sigma(V) = AV$ be a weak Picard-Vessiot field extension of K. Then there exists a K-difference embedding $\rho : S \to L \otimes_C \overline{C}$ where $\overline{C}$ is the algebraic closure of C and σ acts on $L \otimes_C \overline{C}$ as $\sigma(v \otimes c) = \sigma(v) \otimes c$.*

Proof. Let $X = (X_{i,j})$ be an $n \times n$ matrix of indeterminates over L and let $S_0 := K[X_{i,j}, \frac{1}{\det(X)}] \subset L[X_{i,j}, \frac{1}{\det(X)}]$. We define a difference ring structure on $L[X_{i,j}, \frac{1}{\det(X)}]$ by setting $\sigma(X) = AX$ and this gives a difference ring structure on S_0. Abusing notation slightly, we may write $S = S_0/p$ where p is a maximal σ-ideal of S_0. Define elements $Y_{i,j} \in L[X_{i,j}, \frac{1}{\det(X)}]$ via the formula $(Y_{i,j}) = V^{-1}(X_{i,j})$. Note that $\sigma Y_{i,j} = Y_{i,j}$ for all i, j and that $L[X_{i,j}, \frac{1}{\det(X)}] = L[Y_{i,j}, \frac{1}{\det(Y)}]$. Define $S_1 := C[Y_{i,j}, \frac{1}{\det(Y)}]$. The ideal $p \subset S_0 \subset L[Y_{i,j}, \frac{1}{\det(Y)}]$ generates an ideal (p) in $L[Y_{i,j}, \frac{1}{\det(Y)}]$. We define $\tilde{p} = (p) \cap S_1$. Let m be a maximal ideal in S_1 such that $\tilde{p} \subset m$. We then have a homomorphism

$$S_1 \to S_1/m \to \overline{C}.$$

We can extend this to a homomorphism

$$\psi : L[Y_{i,j}, \frac{1}{\det(Y)}] = L \otimes_C S_1 \to L \otimes_C \overline{C}.$$

Restricting ψ to S_0, we have a difference homomorphism

$$\psi : S_0 \to L \otimes_C \overline{C}$$

whose kernel contains p. Since p is a maximal σ-ideal we have that this kernel is p. Therefore ψ yields an embedding

$$\rho : S = S_0/p \to L \otimes_C \overline{C}.$$

$\square$

Corollary 2.8 *Let $K, C, \overline{C}, S, L, \rho$ be as above and let $T = K[V, \frac{1}{\det(V)}]$. Then ρ maps $S \otimes_C \overline{C}$ isomorphically onto $T \otimes_C \overline{C}$. Therefore the Galois group G_S is isomorphic to G_T over $\overline{C}$.*

Proof. In Proposition 2.7, we have that $\rho(U) = V(c_{i,j})$ for some $(c_{i,j}) \in \mathrm{GL}_n(\overline{C})$. Therefore ρ is an isomorphism. The isomorphism of G_S and G_T over $\overline{C}$ now follows in the same manner as the conclusion of Corollary 2.5. $\square$

We can now prove the following result.

Theorem 2.9 *Let*

1. *k be a difference field with algebraically closed field of constants C,*
2. *$\sigma Y = AY$ be a difference equation with $A \in \mathrm{GL}_n(k)$ and let R be the Picard-Vessiot ring for this equation over k,*
3. *K a difference field extension of k such that $K = k(C_K)$*
4. *L a weak Picard-Vessiot field for the equation $\sigma(Y) = AY$ over K.*

Then

a. *If we write $L = K(V)$ where $V \in \mathrm{GL}_n(L)$ and $\sigma V = AV$ then $R \otimes_C \overline{C}_K \simeq K[V, \frac{1}{\det(V)}] \otimes_{C_K} \overline{C}_K$ where $\overline{C}_K$ is the algebraic closure of C_K. Therefore $K[V, \frac{1}{\det(V)}]$ is also a Picard-Vessiot extension of K.*

b. *The Galois groups of R and $K[V, \frac{1}{\det(V)}]$ are isomorphic over $\overline{C}_K$.*

Proof. Let $Y = (Y_{i,j})$ be an $n \times n$ matrix of indeterminates and write $R = k[Y_{i,j}, \frac{1}{\det(Y)}]/(p)$, where (p) is a maximal σ-ideal. Assumptions 1. and 2. imply that $C_R = C_k$ ([23],Lemma 1.8) so Propostion 2.4 implies that $S = K[Y_{i,j}, \frac{1}{\det(Y)}]/(p)K$ is a Picard-Vessiot ring with constants C_K. Corollary 2.5 implies that its Galois group G_R is isomorphic over C to G_S. Corolary 2.8 finishes the proof. $\square$

3 A Tannakian Point of View

In this section we shall give another proof of Theorem 2.9 for q-difference equations in the analytic situation. Let $\mathcal{M}(\mathbb{C}^*)$ be the field of functions $f(x)$ meromorphic on $\mathbb{C}^* = \mathbb{C}\backslash\{0\}$ with the automorphism $\sigma(f(x)) = f(qx)$ where $q \in \mathbb{C}^*$ is a fixed complex number with $|q| \neq 1$. As noted before, the constants $C_{\mathcal{M}(\mathbb{C}^*)}$ in this situation correspond to meromorphic functions $\mathcal{M}(E)$ on the elliptic curve $E = \mathbb{C}^*/q^{\mathbb{Z}}$. We shall show how the theory of tannakian categories also yields a proof of Theorem 2.9 when $k = \mathbb{C}(x)$ and $K = k(C_{\mathcal{M}(\mathbb{C}^*)})$.

We shall assume that the reader is familiar with some basic facts concerning difference modules ([23], Ch. 1.4) and tannakian categories ([7],[6]; see [24], Appendix B or [3] for an overview). We will denote by $\mathcal{D}_k = k[\sigma, \sigma^{-1}]$ (resp. $\mathcal{D}_K = K[\sigma, \sigma^{-1}]$) the rings of difference operators over k (resp. K). Following ([23], Ch. 1.4), we will denote by $Diff(k, \sigma)$ (resp. by $Diff(K, \sigma)$) the category of difference-modules over k (resp. K). The ring of endomorphisms of the unit object is equal to $\mathbb{C}$ (resp. $C_K = C_{\mathcal{M}(\mathbb{C}^*)} = \mathcal{M}(E)$) the field of constants of k (resp. K).

Let M be a $\mathcal{D}_k$-module of finite type over k. We will denote by $M_K = M \otimes_k K$ the $\mathcal{D}_K$-module constructed by extending the field k to K. We will let $\{\{M\}\}$ (resp. $\{\{M_K\}\}$) denote the full abelian tensor subcategory of $Diff(k, \sigma)$ (resp. $Diff(K, \sigma)$) generated by M (resp. M_K) and its dual M^* (resp. $M_K{}^*$).

Theorem 1.32 of [23] gives a fiber functor ω_M over $\mathbb{C}$ for $\{\{M\}\}$. In [21], Praagman gave an existence theorem (see Section 1) for q-difference equations which can be used to construct a fiber functor ω_{M_K} for $\{\{M_K\}\}$ over C_K (described in detail in Proposition 3.9 below). In particular, $\{\{M\}\}$ and $\{\{M_K\}\}$ are neutral tannakian categories over $\mathbb{C}$ and C_K respectively. The main task of this section is to compare the Galois groups associated to the fiber functors ω_M and ω_{M_K}. We will prove the following theorem:

Theorem 3.1 *Let $M \in Diff(k, \sigma)$ be a $\mathcal{D}_k$-module of finite type over k.*
Then
$$Aut^{\otimes}(\omega_M) \otimes_{\mathbb{C}} \overline{C_K} \simeq Aut^{\otimes}(\omega_{M_K}) \otimes_{C_K} \overline{C_K}.$$

The proof is divided in two parts. In the first part, we will construct a fiber functor $\tilde{\omega}_M$ from $\{\{M_K\}\}$ to $Vect_{C_K}$, which *extends* ω_M and we will compare its Galois group to that associated to ω_M. In the second part, we will compare the Galois group associated to ω_{M_K} and the Galois group associated to $\tilde{\omega}_M$, and finally relate these groups to the Galois groups considered in Theorem 2.9.b).

3.1 The action of $\mathrm{Aut}(C_K/\mathbb{C})$ on $\{\{M_K\}\}$

A module $M_K = M \otimes_k K$ is constructed from the module M essentially by extending the scalars from $\mathbb{C}$ to C_K. In order to compare the subcategories $\{\{M\}\}$ and $\{\{M_K\}\}$ they generate, it seems natural therefore to consider an action of the automorphism group $\mathrm{Aut}(C_K/\mathbb{C})$ on M_K as well as on $\{\{M_K\}\}$. Before we define this action we state some preliminary facts.

Lemma 3.2 *We have:*

1. *The fixed field $C_K^{\mathrm{Aut}(C_K/\mathbb{C})}$ is $\mathbb{C}$.*
2. *$K \simeq C_K(X)$ where $C_K(X)$ denotes the field of rational functions with coefficients in C_K. This isomorphism maps $\mathbb{C}(X)$ isomorphically onto k.*

Proof. 1. For all $c \in \mathbb{C}^*$, the restriction to C_K of the map σ_c which associates to $f(x) \in C_K$ the function $\sigma_c(f)(x) = f(cx)$ defines an element of $\mathrm{Aut}(C_K/\mathbb{C})$. Let $\phi \in C_K^{\mathrm{Aut}(C_K/\mathbb{C})}$, the fixed field of $\mathrm{Aut}(C_K/\mathbb{C})$. Because $\sigma_c(\phi) = \phi$ for any $c \in \mathbb{C}^*$, ϕ must be constant.

2. For any $f(X) \in C_K[X]$, put $\phi(f) = f(z)$, viewed as a meromorphic function of the variable $z \in \mathbb{C}^*$. Then, ϕ is a morphism from $C_E[X]$ to K_E. We claim that ϕ is injective. Indeed, let us consider a dependence relation:

$$(1) \qquad \sum c_i(z)k_i(z) = 0, \forall z \in \mathbb{C}$$

where $c_i \in C_E$ and $k_i \in K$. Using Lemma II of ([5], p. 271) or the Lemma of ([9], p. 5) the relation (1) implies that

$$(2) \qquad \sum c_i(z)k_i(X) = 0, \forall z \in \mathbb{C}.$$

So ϕ extends to the function field $C_K(X)$, whose image is the full K. Notice that, by definition of ϕ, $\mathbb{C}(X)$ maps isomorphically on k. □

Since $\mathrm{Aut}(C_K/\mathbb{C})$ acts on $C_K(X)$ (via its action on coefficients), we can consider its action on K.

Lemma 3.3   *1. The action of* $\mathrm{Aut}(C_K/\mathbb{C})$ *on* K *extends the natural action of* $\mathrm{Aut}(C_K/\mathbb{C})$ *on* C_K. *Moreover the action of* $\mathrm{Aut}(C_K/\mathbb{C})$ *on* K *is trivial on* k.

2. $K^{\mathrm{Aut}(C_K/\mathbb{C})} = k$.

3. The action of $\mathrm{Aut}(C_K/\mathbb{C})$ *on* K *commutes with the action of* σ_q.

Proof. 1. This comes from the definition of the action of $\mathrm{Aut}(C_K/\mathbb{C})$ on K. Because $\mathrm{Aut}(C_K/\mathbb{C})$ acts trivially on $\mathbb{C}(X)$, its action on k is also trivial.

2. Because of Lemma 3.2, $C_K^{\mathrm{Aut}(C_K/\mathbb{C})} = \mathbb{C}$. Thus, by construction $K^{\mathrm{Aut}(C_K/\mathbb{C})} = k$.

3. Let i be a natural integer and $f(X) = cX^i$ where $c \in C_K$. Then

$$\tau(\sigma_q(f)) = \tau(cq^i X^i) = \tau(c)q^i X^i = \sigma_q(\tau(f))$$

with $\tau \in \mathrm{Aut}(C_K/\mathbb{C})$. Thus, the action of $\mathrm{Aut}(C_K/\mathbb{C})$ commutes with σ_q on $C_K[X]$. It therefore commutes on $C_K(X) = K$.   □

Before we finally define the action of $\mathrm{Aut}(C_K/\mathbb{C})$ on $\{\{M_K\}\}$, we need one more definition.

Definition 3.4 Let F be a field of caracteristic zero and V be a F-vector space of finite dimension over F. We denote by $Constr_F(V)$ any *construction of linear algebra* applied to V inside $Vect_F$, that is to say any vector space over F obtained by tensor products over F, direct sums, symetric and antisymetric products on V and its dual $V^* := Hom_{F-lin}(V, F)$.

Lemma 3.5 *Let* V *be a vector space of finite dimension over* k *(respectively over* $\mathbb{C}$*). Then,* $Constr_k(V) \otimes_k K = Constr_K(V \otimes K)$ *(respectively* $Constr_\mathbb{C}(V) \otimes C_K = Constr_{C_K}(V \otimes C_K)$*). In other words, the constructions of linear algebra commute with the scalar extension.*

Proof. Consider for instance $Constr_k(V) = Hom_{k-lin}(V, k)$. Because V is of finite dimension over k, we have

$$Hom_{k-lin}(V, k) \otimes_k K = Hom_{K-lin}(V \otimes_k K, K).$$

□

To define the action of $\mathrm{Aut}(C_K/\mathbb{C})$ on $\{\{M_K\}\}$ we note that for any object N of $\{\{M_K\}\}$, there exists, by definition, a construction $M' = Constr_k(M)$ such that $N \subset M' \otimes_k K$. Let now $M' = Constr_k(M)$ be a construction of linear algebra applied to M. The Galois group $\mathrm{Aut}(C_K/\mathbb{C})$ acts on $M'_K = M' \otimes_k K$ via the semi-linear action $(\tau \to id \otimes \tau)$. It therefore permutes the objects of $\{\{M_K\}\}$. It remains to prove that this permutation is well defined and is independent of the choice of the construction in which these objects lie. If there exist M_1 and M_2 two objects of $Constr_k(M)$ such that $N \subset M_1 \otimes_k K$ and $N \subset M_2 \otimes_k K$. Then, by a diagonal embedding $N \subset (M_1 \oplus M_2) \otimes_k K$. The action of $\mathrm{Aut}(C_K/\mathbb{C})$ on $(M_1 \oplus M_2) \otimes_k K$ is the direct sum of the action of $\mathrm{Aut}(C_K/\mathbb{C})$ on $M_1 \otimes_k K$ with the action of $\mathrm{Aut}(C_K/\mathbb{C})$ on $M_2 \otimes_k K$. This shows that the restriction of the action of $\mathrm{Aut}(C_K/\mathbb{C})$ on $M_1 \otimes_k K$ to N is the same as the restriction of the action of $\mathrm{Aut}(C_K/\mathbb{C})$ on $M_2 \otimes_k K$ to N. Thus, the permutation is independent of the choice of the construction in which these objects lie.

3.2 Another fiber functor $\tilde{\omega}_M$ for $\{\{M_K\}\}$

We now extend ω_M to a fiber functor $\tilde{\omega}_M$ on the category $\{\{M_K\}\}$. For this purpose, we appeal to Proposition 2.4 to conclude that *if R be a Picard-Vessiot ring for M over k and $\sigma X = AX, A \in \mathrm{GL}_n(k)$ be an equation of M over k. If $R = K[Y, \frac{1}{det(Y)}]/I$ where Y is an $n \times n$ matrix of indeterminates, $\sigma Y = AY$ and I is a maximal σ-ideal, then $R_K = R \otimes_k K$ is a weak Picard-Vessiot ring for M_K over K.*

We then have the following proposition-definition:

Proposition 3.6 *For any object N of $\{\{M_K\}\}$ let*

$$\tilde{\omega}_M(N) = \mathrm{Ker}(\sigma - Id, R_K \otimes_K N).$$

Then $\tilde{\omega}_M : \{\{M_K\}\} \to Vect_{C_K}$ is a faithful exact, C_K-linear tensor functor. Moreover, $\tilde{\omega}_M(N \otimes K) = \omega_M(N) \otimes C_K$ for every $N \in \{\{M\}\}$.

Proof. Because of the existence of a fundamental matrix with coefficients in R_K, $\tilde{\omega}_M(M_K)$ satisfies $R_K \otimes_{K_K} M_K = R_K \otimes_{C_K} \tilde{\omega}_M(M_K)$. Let $\sigma X = AX, A \in \mathrm{GL}_n(k)$ be an equation of M over k and $R = k[Y, \frac{1}{det(Y)}]/I$ be its corresponding Picard-Vessiot ring over k. Let M' be a construction of linear algebra applied to M over k. Then R_K contains a fundamental matrix of $M' \otimes K$. This comes from the fact that an equation of M' is obtained from the same construction of linear algebra applied to A. Moreover, if $N \in \{\{M_K\}\}$, then R contains also a fundamental matrix for N. Indeed, there exists M', a construction of linear algebra applied to M over k, such that $N \subset M' \otimes K$. Now, R_K contains the entries of a fundamental solution matrix of N and this matrix is invertible because its determinant divides the determinant of a fundamental matrix of solutions of $M' \otimes K$. Thus, $R_K \otimes_K N = R_K \otimes_{C_K} \tilde{\omega}_M(N)$. We deduce from this fact, that $\tilde{\omega}_M$ is a faithful, exact, C_K-linear tensor functor.

For every $N \in \{\{M\}\}$, we have a natural inclusion of C_K-vector spaces of solutions $\omega_M(N) \otimes C_K \subset \tilde{\omega}_M(N \otimes K)$. Since their dimensions over C_K are both equal to the dimension of N over k, they must coincide. $\square$

3.3 Comparison of the Galois groups

Let $M' = Constr_K(M)$ be a construction of linear algebra applied to M. The group $\mathrm{Aut}(C_K/\mathbb{C})$ acts on $\tilde{\omega}_M(M'_K) = \omega_M(M') \otimes_{\mathbb{C}} C_K$ via the semi-linear action $(\tau \to id \otimes \tau)$. It therefore permutes the objects of the tannakian category generated by $\omega_M(M) \otimes_{\mathbb{C}} C_K$ inside $Vect_{C_K}$.

Lemma 3.7 *Let N be an object of $\{\{M_K\}\}$ and τ be an element of* $\mathrm{Aut}(C_K/\mathbb{C})$. *Then, for the actions of* $\mathrm{Aut}(C_K/\mathbb{C})$ *defined as above and in Section 3.1, we have:*

$$\tau(\tilde{\omega}_M(N)) = \tilde{\omega}_M(\tau(N))$$

(equality inside $\omega_M(M') \otimes_{\mathbb{C}} C_K$ for any $M' = Constr_k(M)$ such that $N \subset M' \otimes K$.)

Proof. Let $M' = Constr_K M$ be such that $N \subset M' \otimes_k K$ and consider the action of $\mathrm{Aut}(C_K/\mathbb{C})$ on $R \otimes_k (M' \otimes_k K)$ defined by $id \otimes id \otimes \tau$.

This allows us to consider the action of $\mathrm{Aut}(C_K/\mathbb{C})$ on $R_K \otimes_K N = R \otimes_k N$. By definition, we have

$$\tau(R_K \otimes_K N) = R \otimes_k (\tau(N)) = R_K \otimes_K \tau(N)$$

for all $\tau \in \mathrm{Aut}(C_K/\mathbb{C})$. Moreover inside $R \otimes_k (M' \otimes K)$, the action of $\mathrm{Aut}(C_K/\mathbb{C})$ commutes with the action of σ_q (see Lemma 3.3). Therefore

$$\tau(\mathrm{Ker}(\sigma_q - Id, R_K \otimes_K N)) = \mathrm{Ker}(\sigma_q - Id, R_K \otimes_K \tau(N)).$$

□

The next proposition is Corollary 2.5, but we shall now give a tannakian proof of it, following the proof of ([14], Lemma 1.3.2).

Proposition 3.8 $Aut^{\otimes}(\omega_M) \otimes C_K = Aut^{\otimes}(\tilde{\omega}_M)$.

Proof. By definition, $Aut^{\otimes}(\tilde{\omega}_M) = Stab(\tilde{\omega}_M(W), \ W \in \{\{M_K\}\})$ is the stabilizer inside $\mathcal{Gl}(\tilde{\omega}_M(M_K)) = \mathcal{Gl}(\omega_M(M)) \otimes C_K$ of the fibers of all the sub-equations W of M_K. Similarly, $Aut^{\otimes}(\omega_M) = Stab(\omega_M(W)), \ W \in \{\{M\}\}$), so that the following inclusion holds:

$$Aut^{\otimes}(\tilde{\omega}_M) \subset Aut^{\otimes}(\omega_M) \otimes C_K.$$

The semi-linear action of $\mathrm{Aut}(C_K/\mathbb{C})$ permutes the sub-$\mathcal{D}_K$-modules W of $\{\{M_K\}\}$ and the fixed field of C_K of Γ_E is $\mathbb{C}$ (see Lemma 3.2.1). Therefore $Aut^{\otimes}(\tilde{\omega}_M)$ *is defined* over $\mathbb{C}$, i.e., it is of the form $G \otimes C_K$ for a unique subgroup $G \subset Aut^{\otimes}(\omega_M)$. By Chevalley's theorem, G is defined as the stabilizer of one $\mathbb{C}$-subspace V of $\omega_M(M')$ for some construction $M' = Constr_k(M)$.

We must show that V is stable under $Aut^{\otimes}(\omega_M)$, i.e., we must show that V is of the form $\omega_M(N)$ for $N \in \{\{M\}\}$. Because $G \otimes C_K = Aut^{\otimes}(\tilde{\omega}_M)$ leaves $V \otimes C_K$ stable, we know that there exists $N \in \{\{M_K\}\}$ with $\tilde{\omega}_M(N) = V \otimes C_K$. For any $\tau \in \mathrm{Aut}(C_K/\mathbb{C})$,

$$\tilde{\omega}_M(N) = V \otimes C_K = \tau(V \otimes C_K) = \tau(\tilde{\omega}_M(N)) = \tilde{\omega}_M(\tau(N)),$$

in view of Lemma 3.7. We therefore deduce from Proposition 3.6 that $\tau(N) = N$ for any $\tau \in \mathrm{Aut}(C_K/\mathbb{C})$. Consequently, N is *defined over* K (see Lemma 3.3.3)), i.e., it is of the form $N \otimes K$, where $N \in \{\{M\}\}$. □

We need to define one more functor before we finish the proof of Theorem 3.1.

Proposition 3.9 *Let*

$$(3) \qquad\qquad \sigma_q Y = AY$$

be an equation of M with $A \in \mathrm{GL}_n(K)$. There exists a fundamental matrix of solutions U of (3) with coefficients in the field $\mathcal{M}(\mathbb{C}^*)$ of functions meromorphic on $\mathbb{C}^*$. Moreover, if V is another fundamental matrix of solutions of (3), there exists $P \in \mathrm{GL}_n(C_K)$ such that $U = PV$.

Let L be the subfield of $\mathcal{M}(\mathbb{C}^*)$ generated over K by the entries of U. For any object N of $\{\{M_K\}\}$ let

$$\omega_{M_K}(N) = \mathrm{Ker}(\sigma - Id, L \otimes N).$$

Then $\omega_{M_K} : \{\{M_K\}\} \to Vect_{C_K}$ is a faithful exact, C_K-linear tensor functor.

Proof. For the existence of U see [21] Theorem 3. Since the field of constants of L is C_K, L is a *weak Picard Vessiot* field for M_K, and the proof that ω_{M_K} is a fiber functor on $\{\{M_K\}\}$ is the same as that of Proposition 3.6. $\qquad\square$

We now turn to the

Proof of Theorem. By Propositions 3.6 on the one hand and 3.9 on the other hand, there exists two fiber functors $\tilde{\omega}_M$ and ω_{M_K} on $\{\{M_K\}\}$ which is a neutral tannakian category over C_E. A fundamental theorem of Deligne ([7], Theorem 3.2) asserts that for any field C of caracteristic zero, two fiber functors of a neutral tannakian category over C become isomorphic over the algebraic closure of C. Taking $C = C_K$ and combining this with Proposition 3.8, we therefore have

$$Aut^{\otimes}(\omega_M) \otimes \overline{C_K} \simeq Aut^{\otimes}(\omega_{M_K}) \otimes \overline{C_K}.$$

$\square$To show the connection between Theorem 3.1 and Theorem 2.9 we must show that the group of difference k (rep. K)-automorphisms of R (resp. F) can be identified with the $\mathbb{C}$ (resp. C_K)-points of $Aut^{\otimes}(\omega_M)$ (resp. $Aut^{\otimes}(\omega_{M_K})$). In the first case, this has been shown in Theorem 1.32.3 of [23]; the second case can be established in a similar manner. This enables us to deduce, in our special case, Theorem 2.9 from Theorem 3.1.

We conclude with an example to show that these considerations can be used to show the algebraic independence of certain classical functions.

Example 3.10 Consider the q-difference equation

$$(4) \qquad\qquad \sigma_q(y) = y + 1.$$

In ([23], Section 12.1) the authors denote by l the formal solution of 4, i.e. the formal q-logarithm. It is easily seen that the Galois group, in the sense of [23], of (4) is equal to $(\mathbb{C}, +)$ and therefore that the dimension of the Galois group $G_{R/\mathbb{C}}$ is equal to 1. We deduce from Theorem 2.9 that the dimension of the Galois group G_{S/C_K} is also equal to 1. In particular, the field generated over K by the meromorphic solutions of (4) has transcendence degree 1 over K.

The classical Weiestrass ζ function associated to the elliptic curve $E = \mathbb{C}^*/q^{\mathbb{Z}}$ satisfies the equation (4). Therefore, if $\wp$ is the Weierstrass function of E, we obtain that $\zeta(z)$ is transcendental over the field $\mathbb{C}(z, \wp(z))$.

4 A Model-Theoretic Point of View

4.1 Preliminary model-theoretic definitions and results

Definition 4.1 Let K be a difference field with automorphism σ.

1. K is *generic* iff
   (∗)   every **finite** system of difference equations with coefficients in K and which has a solution in a difference field containing K, already has a solution in K.
2. A *finite σ-stable extension* M of K is a finite separably algebraic extension of K such that $\sigma(M) = M$.
3. The *core of L over K*, denoted by $\mathrm{Core}(L/K)$, is the union of all finite σ-stable extensions of K which are contained in L.

One of the difficulties with difference fields, is that there are usually several non-isomorphic ways of extending the automorphism to the algebraic closure of the field. An important result of Babbitt (see [5]) says that once we know the behaviour of σ on $\mathrm{Core}(\overline{K}/K)$, then we know how σ behaves on the algebraic closure $\overline{K}$ of K.

Fix an infinite cardinal κ which is larger than all the cardinals of structures considered (e.g., in our case, we may take $\kappa = |\mathbb{C}|^+ = (2^{\aleph_0})^+$). In what follows we will work in a generic difference field $\mathcal{U}$, which we will assume *sufficiently saturated*, i.e., which has the following properties:

(i) (∗) *above holds for every system of difference equations* **of size** $< \kappa$ *(in infinitely many variables).*

(ii) *(1.5 in [4]) If f is an isomorphism between two algebraically closed difference subfields of $\mathcal{U}$ which are of cardinality $< \kappa$, then f extends to an automorphism of $\mathcal{U}$.*

(iii) *Let $K \subset L$ be difference fields of cardinality $< \kappa$, and assume that $K \subset \mathcal{U}$. If every finite σ-stable extension of K which is contained in L K-embeds in $\mathcal{U}$, then there is a K-embedding of L in $\mathcal{U}$.*

Note that the hypotheses of (iii) are always verified if K is an algebraically closed subfield of $\mathcal{U}$. If K is a difference field containing the algebraic closure $\overline{\mathbb{Q}}$ of $\mathbb{Q}$, then K will embed into $\mathcal{U}$, if and only if the difference subfield $\overline{\mathbb{Q}}$ of K and the difference subfield $\overline{\mathbb{Q}}$ of $\mathcal{U}$ are isomorphic. This might not always be the case. However, every difference field embeds into some sufficiently saturated generic difference field.

Let us also recall the following result (1.12 in [4]): Let n be a positive integer, and consider the field $\mathcal{U}$ with the automorphism σ^n. Then $(\mathcal{U}, \sigma^n)$ is a generic difference field, and satisfies the saturation properties required of $(\mathcal{U}, \sigma)$.

Notation. We use the following notation. Let R be a difference ring. Then, as in the previous sections, C_R denotes the field of "constants" of R, i.e., $C_R = \{a \in R \mid \sigma(a) = a\}$. We let $D_R = \{a \in R \mid \sigma^m(a) = a \text{ for some } m \neq 0\}$. Then D_R is a difference subring of R, and if R is a field, D_R is the relative algebraic closure of C_R in R. We let D'_R denote the difference ring with same underlying ring as D_R and on which σ acts trivially. Thus $C_{\mathcal{U}}$ is a pseudo-finite field (see 1.2 in [4]), and $D_{\mathcal{U}}$ is its algebraic closure (with the action of σ), $D'_{\mathcal{U}}$ the algebraic closure of $C_{\mathcal{U}}$ on which σ acts trivially.

Later we will work with powers of σ, and will write $Fix(\sigma^n)(R)$ for $\{a \in R \mid \sigma^n(a) = a\}$ so that no confusion arises. If $R = \mathcal{U}$, we will simply write $Fix(\sigma^n)$. Here are some additional properties of $\mathcal{U}$ that we will use.

Let $K \subset M$ be difference subfields of $\mathcal{U}$, with M algebraically closed, and let a be a tuple of $\mathcal{U}$. By 1.7 in [4]:

(iv) *If the orbit of a under $\mathrm{Aut}(\mathcal{U}/K)$ is finite, then $a \in \overline{K}$ (the algebraic closure of K).*

We already know that every element of $\mathrm{Aut}(M/KC_M)$ extends to an automorphism of $\mathcal{U}$. More is true: using 1.4, 1.11 and Lemma 1 in the appendix of [4]:

(v) *every element of* $\mathrm{Aut}(M/KC_M)$ *can be extended to an element of* $\mathrm{Aut}(\mathcal{U}/KC_{\mathcal{U}})$.

Recall that a definable subset S of $\mathcal{U}^n$ is *stably embedded* if whenever $R \subset \mathcal{U}^{nm}$ is definable with parameters from $\mathcal{U}$, then $R \cap S^m$ is definable using parameters from S. An important result ([4] 1.11)) shows that $C_{\mathcal{U}}$ is stably embedded. Let $d \geq 1$. Then, adding parameters from $Fix(\sigma^d)$, there is a definable isomorphism between $Fix(\sigma^d)$ and $C_{\mathcal{U}}^d$. Hence,

(vi) *for every* $d > 0$, $Fix(\sigma^d)$ *is stably embedded, and*

(vii) *if* θ *defines an automorphism of* $D_{\mathcal{U}}$ *which is the identity on* D_M, *then* θ *extends to an automorphism of* $\mathcal{U}$ *which is the identity on* M.

We also need the following lemma. The proof is rather model-theoretic and we refer to the Appendix of [4] for the definitions and results. Recall that if K is a difference subfield of $\mathcal{U}$, then its definable closure, $\mathrm{dcl}(K)$, is the subfield of $\mathcal{U}$ fixed by $\mathrm{Aut}(\mathcal{U}/K)$. It is an algebraic extension of K, and is the subfield of the algebraic closure $\overline{K}$ of K which is fixed by the subgroup $\{\tau \in \mathcal{G}al(\overline{K}/K) \mid \sigma^{-1}\tau\sigma = \tau\}$.

Lemma 4.2 *Let* K *be a difference field, and* M *be a finite* σ-*stable extension of* $KC_{\mathcal{U}}$. *Then* $M \subset \overline{K}D_{\mathcal{U}}$, *i.e., there is some finite* σ-*stable extension* M_0 *of* K *such that* $M \subset M_0 D_{\mathcal{U}}$.

Proof. Fix an integer $d \geq 1$. Then, in the difference field $(\mathcal{U}, \sigma^d)$, $Fix(\sigma^d)$ is stably embedded, $\mathrm{dcl}(\overline{K}) = \overline{K}$ and $\mathrm{dcl}(Fix(\sigma^d)) = Fix(\sigma^d)$. Denoting types in $(\mathcal{U}, \sigma^d)$ by tp_d, this implies

$$(\sharp) \qquad tp_d(\overline{K}/\overline{K} \cap Fix(\sigma^d)) \vdash tp_d(\overline{K}/Fix(\sigma^d)).$$

Assume by way of contradiction that $KC_{\mathcal{U}}$ has a finite σ-stable extension M which is not contained in $\overline{K}D_{\mathcal{U}}$. We may assume that M is Galois over $\overline{K}C_{\mathcal{U}}$ (see Thm 7.16.V in [5]), with Galois group G. Choose d large enough so that σ^d commutes with all elements of G, and $M = M_0 D_{\mathcal{U}}$, where M_0 is Galois over $\overline{K}Fix(\sigma^d)$. Then there are several non-isomorphic ways of extending σ^d to M. As $tp_d(\overline{K}/Fix(\sigma^d))$ describes in particular the $\overline{K}Fix(\sigma^d)$-isomorphism type of the σ^d-difference field M, this contradicts $(\sharp)$ (see Lemmas 2.6 and 2.9 in [4]). $\square$

4.2 The Galois group

From now on, we assume that all fields are of characteristic 0. Most of the statements below can be easily adapted to the positive characteristic case. Let K be a difference subfield of $\mathcal{U}$, $A \in \mathrm{GL}_n(K)$, and consider the set $\mathcal{S} = \mathcal{S}(\mathcal{U})$ of solutions of the equation

$$\sigma(X) = AX, \ \det(X) \neq 0.$$

Consider $H = \mathrm{Aut}(K(\mathcal{S})/KC_\mathcal{U})$. We will call H the *Galois group of* $\sigma(X) = AX$ *over* $KC_\mathcal{U}$[6].

Then H is the set of $C_\mathcal{U}$-points of some algebraic group $\mathbb{H}$ defined over $KC_\mathcal{U}$. To see this, we consider the ring $R = K[Y, \det(Y)^{-1}]$ (where $Y = (Y_{i,j})$ is an $n \times n$ matrix of indeterminates), extend σ to R by setting $\sigma(Y) = AY$, and let L be the field of fractions of R. Then L is a regular extension of K, and there is a K-embedding φ of L in $\mathcal{U}$, which sends C_L to a subfield of $C_\mathcal{U}$, and D_L to a subfield of $D_\mathcal{U}$. We let $T = \varphi(Y)$. Then every element $g \in H$ is completely determined by the matrix $M_g = T^{-1}g(T) \in \mathrm{GL}_n(C_\mathcal{U})$, since if $B \in \mathcal{S}$, then $B^{-1}T \in \mathrm{GL}_n(C_\mathcal{U})$. Moreover, since $KC_{\varphi(L)}(T)$ and $KC_\mathcal{U}$ are linearly disjoint over $KC_{\varphi(L)}$, the algebraic locus W of T over $KC_\mathcal{U}$ (an algebraic subset of GL_n) is defined over $KC_{\varphi(L)}$, and H is the set of elements of $\mathrm{GL}_n(C_\mathcal{U})$ which leave W invariant. It is therefore the set of $C_\mathcal{U}$-points of an algebraic group $\mathbb{H}$, defined over $KC_{\varphi(L)}$. We let $\mathbb{H}'$ denote the Zariski closure of $\mathbb{H}(C_\mathcal{U})$. Then $\mathbb{H}'$ is defined over $C_\mathcal{U}$, and it is also clearly defined over $\overline{K\varphi(C_L)}$, so that it is defined over $C_\mathcal{U} \cap \overline{K\varphi(C_L)} = C_\mathcal{U} \cap \overline{\varphi(C_L)}$.

Proposition 4.3 *Let $\mathbb{H}^0$ denote the connected component of $\mathbb{H}$, and let M_0 be the relative algebraic closure of $K\varphi(C_L)$ in $\varphi(L)$, M its Galois closure over $K\varphi(C_L)$.*

1. $\dim(\mathbb{H}) = \mathrm{tr.deg}(L/KC_L)$.
2. M_0 *is a finite σ-stable extension of $K\varphi(C_L)$ and $[\mathbb{H} : \mathbb{H}^0]$ divides $[M : K\varphi(C_L)]$*
3. $[\mathbb{H}' : \mathbb{H}^0] = [\mathbb{H}(C_\mathcal{U}) : \mathbb{H}^0(C_\mathcal{U})]$ *equals the number of left cosets of $\mathcal{G}al(M/M_0)$ in $\mathcal{G}al(M/K\varphi(C_L))$ which are invariant under the action of σ by conjugation.*
4. *If the algebraic closure of C_K is contained in $C_\mathcal{U}$, then the element $\sigma \in \mathcal{G}al(D_L/C_L)$ lifts to an element of $\mathrm{Aut}(KC_\mathcal{U}(T)/KC_\mathcal{U})$.*

---

6 Warning: This is not the usual Galois group defined by model theorists, please see the discussion in subsection 4.4.

Proof. 1. Choose another K-embedding φ' of L into $\mathcal{U}$ which extends φ on the relative algebraic closure of KC_L in L, and is such that $\varphi'(L)$ and $\varphi(L)$ are linearly disjoint over M_0. Then $B = \varphi'(Y)^{-1}T \in \mathbb{H}(C_{\mathcal{U}})$, and $\text{tr.deg}(\varphi(KC_L)(B))/\varphi(KC_L)) = \text{tr.deg}(L/KC_L)$. Thus $\dim(\mathbb{H}) = \text{tr.deg}(L/KC_L)$.

2. As $M_0 \subset K\varphi(L)$, we obtain that $[M_0 : K\varphi(C_L)]$ is finite and $\sigma(M_0) = M_0$. Furthermore, $\sigma(M) = M$ (see Thm 7.16.V in [5]). The algebraic group $\mathbb{H}$ is defined as the set of matrices of GL_n which leaves the algebraic set W (the algebraic locus of T over $K\varphi(C_L)$) invariant.

Hence $\mathbb{H}^0$ is the subgroup of $\mathbb{H}$ which leaves all absolutely irreducible components of W invariant. Its index in $\mathbb{H}$ therefore must divide $[M : K\varphi(C_L)]$.

3. The first equality follows from the fact that $\mathbb{H}^0(C_{\mathcal{U}})$ and $\mathbb{H}'(C_{\mathcal{U}})$ are Zariski dense in $\mathbb{H}^0$ and $\mathbb{H}'$ respectively. Some of the (absolutely irreducible) components of W intersect $\mathcal{S}$ in the empty set. Indeed, let W_0 be the component of W containing T, let W_1 be another component of W and $\tau \in \mathcal{G}al(M/K\varphi(C_L))$ such that $W_1 = W_0^\tau$. Then W_1 is defined over $\tau(M_0)$. If τ defines a (difference-field) isomorphism between M_0 and $\tau(M_0)$, then τ extends to an isomorphism between $K\varphi(L)$ and a regular extension of $K\varphi(C_L)\tau(M_0)$, and therefore $W_1 \cap \mathcal{S} \neq \emptyset$. Conversely, if $B \in W_1 \cap \mathcal{S}$, then $B^{-1}T \in \mathbb{H}(C_{\mathcal{U}})$, so that B is a generic of W_1. The difference fields $K\varphi(C_L)(B)$ and $K\varphi(L)$ are therefore isomorphic (over $K\varphi(C_L)$), and $\tau(M_0) \subset K\varphi(C_L)(B)$. Hence the difference subfields M_0 and $\tau(M_0)$ of M are $K\varphi(C_L)$-isomorphic. One verifies that M_0 and $\tau(M_0)$ are isomorphic over $K\varphi(C_L)$ if and only if $\sigma^{-1}\tau^{-1}\sigma\tau \in \mathcal{G}al(M/M_0)$, if and only if the coset $\tau\mathcal{G}al(M/M_0)$ is invariant under the action of σ by conjugation.

4. We know that the algebraic closure $\overline{K}$ of K and $D_{\mathcal{U}}$ are linearly disjoint over $C_{\overline{K}} = \overline{C_K}$. Let $a \in \varphi(D_L)$ generates $\varphi(D_L)$ over $\varphi(C_L)$. By 4.1(vi), $tp(a/\overline{K}C_{\mathcal{U}}) = tp(\sigma(a)/\overline{K}C_{\mathcal{U}})$, and therefore there is θ in $\text{Aut}(\mathcal{U}/\overline{K}C_{\mathcal{U}})$ such that $\theta(a) = \sigma(a)$. Thus $T^{-1}\theta(T) \in H$. $\qquad\square$

Remarks 4.4 1. Even when the algebraic closure of C_K is contained in $C_{\mathcal{U}}$, we still cannot in general conclude that $\mathbb{H}' = \mathbb{H}$.

2. The isomorphism type of the algebraic group $\mathbb{H}$ only depends on the isomorphism type of the difference field K (and on the matrix A). The isomorphism type of the algebraic group $\mathbb{H}'$ does however depend on the embedding of K in $\mathcal{U}$, that is, on the isomorphism

type of the difference field $\mathrm{Core}(\overline{K}/K)$. Indeed, while we know
the isomorphism type of the difference field M_0 over $K\varphi(C_L)$, we
do not know the isomorphism type of the difference field M over
$K\varphi(C_L)$, and in view of 4.3.3, if $\mathcal{G}al(M/K\varphi(C_L))$ is not abelian,
it may happen that non-isomorphic extensions of σ to M yield
different Galois groups.

3. Assume that σ acts trivially on $\mathcal{G}al(\mathrm{Core}(\overline{K}/K)/K)$, and that
$\mathcal{G}al(\mathrm{Core}(\overline{K}/K)/K)$ is abelian. Then

$$\mathbb{H} = \mathbb{H}' \quad \text{and} \quad [\mathbb{H} : \mathbb{H}^0] = [M_0 : K\varphi(C_L)].$$

Indeed, by Lemma 4.2, M_0 is Galois over $K\varphi(C_L)$ with abelian
Galois group G and σ acts trivially on G. The result follows by
4.3.3. Thus we obtain equality of $\mathbb{H}$ and $\mathbb{H}'$ in two important
classical cases:

    a. $K = \mathbb{C}(t)$, $C_K = \mathbb{C}$ and $\sigma(t) = t + 1$.

    b. $K = \mathbb{C}(t)$, $C_K = \mathbb{C}$ and $\sigma(t) = qt$ for some $0 \neq q \in \mathbb{C}$, q
    not a root of unity.

4. If $B \in \mathcal{S}$, then the above construction can be repeated, using
B instead of T. We then obtain an algebraic group $\mathbb{H}_1$, with
$\mathbb{H}_1(C_\mathcal{U}) \simeq \mathrm{Aut}(KC_\mathcal{U}(\mathcal{S})/KC_\mathcal{U})$. Since $KC_\mathcal{U}(B) = KC_\mathcal{U}(T)$, the
algebraic groups $\mathbb{H}_1$ and $\mathbb{H}$ are isomorphic (via $B^{-1}T$).

5. In the next subsection, we will show that the algebraic group $\mathbb{H}$
and the algebraic group $G_{R'}$ introduced in section 2 are isomorphic when $C_{R'} = C_K = D_K$.

4.3 More on Picard-Vessiot rings

Throughout the rest of this section, we fix a difference ring K, some
$A \in \mathrm{GL}_n(K)$, $R = K[Y, \det(Y)^{-1}]$ as above, with $\sigma(Y) = AY$, and
$R' = R/q$ a Picard-Vessiot ring for $\sigma(X) = AX$ over K. We denote the
image of Y in R' by y. We keep the notation introduced in the previous
subsections.

If q is not a prime ideal, then there exists ℓ and a prime σ^ℓ-ideal p of
R which is a maximal σ^ℓ-ideal of R, such that $q = \bigcap_{i=0}^{\ell-1} \sigma^i(p)$, and
$R' \simeq \bigoplus_{i=0}^{\ell-1} R_i$, where $R_i = R/\sigma^i(p)$ (see Corollary 1.16 of [23]. One verifies that the second proof does not use the fact that C_K is algebraically
closed). Thus the σ^ℓ-difference ring R_0 is a Picard-Vessiot ring for the
difference equation $\sigma^\ell(X) = \sigma^{\ell-1}(A) \cdots \sigma(A)AX$ over K. We denote

$\sigma^{\ell-1}(A) \cdots \sigma(A)A$ by A_ℓ.

We will identify R' with $\oplus_{i=0}^{\ell-1} R_i$, and denote by e_i the primitive idempotent of R' such that $e_i R' = R_i$. Then $e_i = \sigma^i(e_0)$. We will denote by R^* the ring of quotients of R', i.e., $R^* = \oplus_{i=0}^{\ell-1} R_i^*$, where R_i^* is the field of fractions of R_i. The difference ring R^* is also called the *total Picard-Vessiot ring of* $\sigma(X) = AX$ *over* K. There are two numerical invariants associated to R': the number $\ell = \ell(R')$, and the number $m(R')$ which is the product of $\ell(R')$ with $[D_{R_0^*} : D_K C_{R_0^*}]$. We call $m(R')$ the *m-invariant of* R'. We will be considering other Picard-Vessiot rings for $\sigma(X) = AX$, and will use this notation for them as well.

Recall that the *Krull dimension* of a ring S is the maximal integer n (if it exists) such that there is a (strict) chain of prime ideals of S of length n. We denote it by $\mathrm{Kr.dim}(S)$. If S is a domain, and is finitely generated over some subfield k, then $\mathrm{Kr.dim}(S)$ equals the transcendence degree over k of its field of fractions. Observe that if S is a domain of finite Krull dimension, and $0 \neq I$ is an ideal of S, then $\mathrm{Kr.dim}(S) > \mathrm{Kr.dim}(S/I)$. Also, if $S = \oplus_i S_i$, then $\mathrm{Kr.dim}(S) = \sup\{\mathrm{Kr.dim}(S_i)\}$.

Lemma 4.5   1. $C_{R'}$ *is a finite algebraic extension of* C_K, *and is linearly disjoint from* K *over* C_K *(inside* R'*).*

     2. *If* $C_{R'} \otimes_{C_K} D_K$ *is a domain, then* R' *is a Picard-Vessiot ring for* $\sigma(X) = AX$ *over* $KC_{R'}$.

Proof. 1. We know by Lemma 1.7 of [23] that $C_{R'}$ is a field. Assume by way of contradiction that $C_{R'}$ and K are not linearly disjoint over C_K, and choose n minimal such that there are $a_1, \ldots, a_n \in C_{R'}$ which are C_K-linearly independent, but not K-linearly independent. Let $0 \neq c_1, \ldots, c_n \in K$ be such that $\sum_{i=1}^n a_i c_i = 0$. Multiplying by c_1^{-1}, we may assume $c_1 = 1$. Then $\sigma(\sum_{i=1}^n a_i c_i) = \sum_{i=1}^n a_i \sigma(c_i) = 0$, and therefore $\sum_{i=2}^n a_i(\sigma(c_i) - c_i) = 0$. By minimality of n, all $(\sigma(c_i) - c_i)$ are 0, i.e., all $c_i \in C_K$, which gives us a contradiction.

Observe that $e_0 C_{R'} \subset Fix(\sigma^\ell)(R_0)$, and we may therefore replace R' by the domain R_0. Since R_0 is a finitely generated K-algebra, we know that its Krull dimension equals the transcendence degree over K of its field of fractions. Thus R_0 cannot contain a subfield which is transcendental over K, i.e., the elements of $Fix(\sigma^\ell)(R_0)$ are algebraic over K. his furthermore implies that $Fix(\sigma^\ell)(R_0)$ is an algebraic extension of $Fix(\sigma^\ell)(K)$. Since the latter field is an algebraic extension of C_K, we have the conclusion.

2. Our hypothesis implies that $K[C_{R'}]$ is a field. Hence R' is a simple difference ring containing $KC_{R'}$, and is therefore a Picard-Vessiot ring for $\sigma(X) = AX$ over $KC_{R'}$. □

Lemma 4.6   *1.* $C_{R'} = C_{R^*}$.
  2. $Fix(\sigma^\ell)(e_0 R^*) = e_0 C_{R'}$.
  3. $D_{R^*} = \oplus_{i=0}^{\ell-1} D_{e_i R^*}$.

Proof.   1. If $c \in C_{R^*}$, then c can be represented by some ℓ-tuple $(\frac{a_0}{b_0}, \ldots, \frac{a_{\ell-1}}{b_{\ell-1}})$, where $a_i, b_i \in R_i$, and $b_i \neq 0$. Thus the ideal $I = \{d \in R' \mid dc \in R'\}$ is a σ-ideal of R' and contains the element $b = (b_0, \ldots, b_{\ell-1}) \neq 0$. Since R' is simple, $1 \in I$, i.e., $c \in R'$.

2. Assume $a \in e_0 R^*$ satisfies $\sigma^\ell(a) = a$. Then $a = e_0 a$, $\sum_{i=0}^{\ell-1} \sigma^i(e_0 a)$ is fixed by σ, and therefore belongs to $C_{R'}$. Hence $a \in e_0 C_{R'}$.

3. If $a \in R^*$ satisfies $\sigma^m(a) = a$ for some m, then $\sigma^{m\ell}(e_i a) = e_i a$.   □

Remark 4.7 Observe that ℓ and the isomorphism type of the K-σ^ℓ-difference algebra R_0 completely determine the isomorphism type of the difference algebra R'. Indeed, for each $i = 1, \ldots, \ell - 1$, one chooses a copy R_i of the domain R_0, together with an isomorphism $f_i : R_0 \to R_i$ which extends σ^i on K. This f_i then induces an automorphism σ^ℓ of R_i. One then defines σ on $\oplus_{i=0}^{\ell-1} R_i$ by setting $\sigma(a_0, \ldots, a_{\ell-1}) = (f_1(a_0), f_2 f_1^{-1}(a_1), \ldots, \sigma^\ell f_{\ell-1}^{-1}(a_{\ell-1}))$.

Proposition 4.8 *Let $K \subset K_1$ be difference fields of characteristic* 0 *where $K_1 = K(C_{K_1})$, and assume that $C_K = D_K$. Then $R' \otimes_K K_1 = \oplus_{i=1}^d R_i'$, where each R_i' is a Picard-Vessiot ring for $\sigma(X) = AX$ over K_1, and $d \leq [C_{R'} : C_K]$. Moreover, each R_i' has the same Krull-dimension and m-invariant as R'.*

Proof. Our assumption implies that $K \otimes_{C_K} C_{K_1}$ is a domain. Let C be the relative algebraic closure of C_K in C_{K_1}. Then $K(C) = K[C]$, and $R' \otimes_K K(C) \simeq R' \otimes_{C_K} C$.

Let $a \in C_{R'}$ be such that $C_{R'} = C_K(a)$ and let $f(X) \in C_K[X]$ be its minimal polynomial over C_K. Let $g_1(X), \ldots, g_d(X)$ be the irreducible factors of $f(X)$ over C. Then $f(X) = \prod_{i=1}^d g_i(X)$, and $C_R' \otimes_{C_K} C \simeq \oplus_{i=1}^d C_i$, where C_i is generated over C by a root of $g_i(X) = 0$. Indeed, identifying C with $1 \otimes C$, every prime ideal of $C_{R'} \otimes_{C_K} C$ must contain some $g_i(a \otimes 1)$; on the other hand, each $g_i(a \otimes 1)$ generates a maximal ideal of $C_{R'} \otimes_{C_K} C$. Thus

$$R' \otimes_{C_K} C \simeq R' \otimes_{C_{R'}} (C_{R'} \otimes_{C_K} C) \simeq \oplus_{i=1}^d R' \otimes_{C_{R'}} C_i.$$

By Lemmas 2.3 and 4.5, each $R' \otimes_{C_{R'}} C_i = R'_i$ is a simple difference ring, with field of constants C_i. Hence R'_i is a Picard-Vessiot ring for $\sigma(X) = AX$ over KC (and also over KC_i). Note that $d \leq \deg(f) = [C_{R'} : C_K]$, and that $\mathrm{Kr.dim}(R'_i) = \mathrm{Kr.dim}(R')$ (because KC is algebraic over K, and R'_i is finitely generated over K).

By Proposition 2.4, $R'_i \otimes_{KC_i} K_1 C_i$ is a Picard-Vessiot ring. Because C_i and K_1 are linearly disjoint over C, and C_i is algebraic over C, $KC_i \otimes_{KC} K_1 \simeq K_1 C_i$, and therefore

$$R'_i \otimes_{KC} K_1 \simeq R'_i \otimes_{KC_i} K_1 C_i.$$

This shows that $R' \otimes_K K_1$ is the direct sum of Picard-Vessiot rings over K_1.

Identifying $C_{R'}$ with $e_j C_{R'} = C_{R_j}$, we obtain

$$R'_i = (\oplus_{j=0}^{\ell-1} R_j) \otimes_{C_{R'}} C_i \simeq \oplus_{j=0}^{\ell-1} R_j \otimes_{C_{R'}} C_i.$$

Each R_j being a Picard-Vessiot ring for $\sigma^\ell(X) = A_\ell X$, we know by Proposition 2.4 that $R_j \otimes_{C_{R'}} C_i$ is also a Picard-Vessiot ring for $\sigma^\ell(X) = A_\ell X$. Thus $R_0 \otimes_{C_{R'}} C_i = \sum_{j=0}^{s-1} S_j$, where each S_j is a simple $\sigma^{\ell s}$-difference ring, and a domain. Because all rings R_j are isomorphic over $C_{R'}$, and all S_j are isomorphic over $C_{R'}$, $m(R'_i)$ is the product of ℓs with $m(S_0) = [D_{S_0^*} : C_{S_0^*}]$, where S_0^* is the field of fractions of S_0. To show that $m(R'_i) = m(R')$, it therefore suffices to show that $sm(S_0) = m(R_0)$. By Lemma 4.5.2,

$$Fix(\sigma^{\ell s})(S_0^*) = Fix(\sigma^\ell)(R_0^* \otimes_{C_{R'}} C_i) = Fix(\sigma)(R' \otimes_{C_{R'}} C_i) = C_i.$$

We know that $D_{R_0^*}$ is a (cyclic) Galois extension of $C_{R'} = Fix(\sigma^\ell)(R_0^*)$, and is therefore linearly disjoint from C_i over $D_{R_0^*} \cap C_i = C'_i$. Write $C'_i = C_{R'}(\alpha)$, and let $a, b \in R_0$, $b \neq 0$, be such that (inside R_0^*), $C_{R'}(a/b) = C'_i$. The minimal prime ideals of $R_0 \otimes_{C_{R'}} C_i$ are the ideals $Q_0, \ldots, Q_{r-1}$, where $r = [C'_i : C_{R'}]$ and Q_k is generated by $\sigma^{k\ell}(a) \otimes 1 - \sigma^{k\ell}(b) \otimes \alpha$. This shows that $r = s$, since s is also the number of minimal prime ideals of $R_0 \otimes_{C_{R'}} C_i$.

Let e be a primitive idempotent of $R_0 \otimes_{C_{R'}} C_i$ such that $S_0 = e(R_0 \otimes_{C_{R'}} C_i)$. Then $eC_i D_{R_0^*}$ is a subfield of S_0^*, contained in $D_{S_0^*}$, and its degree over $eC_i = Fix(\sigma^{\ell s})(S_0^*)$ is the quotient of $[D_{R_0^*} : C_{R'}]$ by $[C_i' : C_{R'}]$, i.e., equals $m(R_0)/s$. To finish the proof, it therefore suffices to show that $D_{S_0^*} = eC_i D_{R_0}^*$.

Assume that $c \in R_0^* \otimes_{C_{R'}} C_i$ satisfies $\sigma^m(c) = c$ for some $m \neq 0$. Write $c = \sum_k a_k \otimes c_k$, where the a_k are in R_0^*, and the c_k are in C_i and are linearly independent over $C_{R'}$. Then $\sigma^m(c) = c = \sum_k \sigma^m(a_k)c_k$, which implies $\sigma^m(a_k) = a_k$ for all k, and all a_k's are in $D_{R_0^*}$. As every element of $D_{S_0^*}$ is of the form ec for such a c (Lemma 4.6.3), this shows that $D_{S_0^*} = eC_i D_{R_0}^*$. This finishes the proof that $m(R_i') = m(R')$.

Consider now $R' \otimes_{KC} K_1$. It is the direct sum of ℓs $\sigma^{\ell s}$-difference rings, each one being isomorphic to $S_0 \otimes_{KC} K_1$. Because K_1 is a regular extension of KC, $S_0 \otimes_{KC} K_1$ is a domain, of Krull dimension equal to $\mathrm{Kr.dim}(S_0) = \mathrm{Kr.dim}(R')$. Inside its field of fractions (a $\sigma^{\ell s}$-difference field) K_1 and S_0^* are linearly disjoint over KC, which implies that $C_{K_1}C_i$ is the field of constants of $S_0 \otimes_{KC} K_1$, $C_{K_1}D_{S_0^*}$ is the field of elements fixed by some power of σ, and $[C_{K_1}D_{S_0^*} : C_{K_1}C_i] = [D_{S_0}^* : C_i] = m(S_0)$. This shows that $m(R_i' \otimes_{KC} K_1) = m(R')$ and finishes the proof. $\square$

Proposition 4.9 *Assume that $C_K = D_K$. Then all Picard-Vessiot rings for $\sigma(X) = AX$ over K have the same Krull dimension and the same m-invariant.*

Proof. Let C be the algebraic closure of C_K, and let R'' be a Picard-Vessiot ring for $\sigma(X) = AX$ over K. By Proposition 4.8, $R' \otimes_K KC$ is the direct sum of finitely many Picard-Vessiot rings for $\sigma(X) = AX$ over KC, and each of these rings has the same Krull dimension and m-invariant as R'. The same statement holds for R''. On the other hand, by Proposition 1.9 of [23], all Picard-Vessiot rings over KC are isomorphic. $\square$

Corollary 4.10 *Assume $D_K = C_K$. Let $R'' = K[V, \det(V)^{-1}]$, where $\sigma(V) = AV$, and assume that $\mathrm{Kr.dim}(R'') = \mathrm{Kr.dim}(R')$ and that R'' has no nilpotent elements. Then R'' is a finite direct sum of Picard-Vessiot rings for $\sigma(X) = AX$.*

Proof. Because R'' has no nilpotent elements and is Noetherian, (0) is the intersection of the finitely many prime minimal ideals of R''. Let $\mathcal{P}$ be the set of minimal prime ideals of R''. Then the intersection of any proper subset of $\mathcal{P}$ is not (0), i.e., no element of $\mathcal{P}$ contains the intersection of the other elements of $\mathcal{P}$. Also, if $P \in \mathcal{P}$, then $\sigma(P) \in \mathcal{P}$, and there exists $m > 0$ such that $\sigma^m(P) = P$. Then $I_P = \bigcap_{i=0}^{m-1} \sigma^i(P)$ is a σ-ideal, which is proper if the orbit of P under σ is not all of $\mathcal{P}$. Observe that for each $P \in \mathcal{P}$, $\mathrm{Kr.dim}(R''/P) \leq \mathrm{Kr.dim}(R''/I_P) \leq \mathrm{Kr.dim}(R'') = \mathrm{Kr.dim}(R')$, and that for some P we have equality.

If I is a maximal σ-ideal of R'', then $\mathrm{Kr.dim}(R''/I) = \mathrm{Kr.dim}(R') = \mathrm{Kr.dim}(R'')$ by Proposition 4.8, and this implies that I is contained in some $P \in \mathcal{P}$. Hence $I = I_P$ and R''/I_P is a Picard-Vessiot ring. If $I = (0)$, then we are finished. Otherwise, $\mathcal{P}$ contains some element P_1 not in the orbit of P under σ. Observe that I_{P_1} is contained in some maximal σ-ideal of R'', and is therefore maximal, by the same reasoning. Since the intersection of any proper subset of $\mathcal{P}$ is non-trivial, $I_P + I_{P_1}$ is a σ-ideal of R'' which contains properly I_P, and therefore equals 1. If $P_1, \ldots, P_r$ are representatives from the σ-orbits in $\mathcal{P}$, the Chinese Remainder Theorem then yields $R'' \simeq \oplus_{i=1}^r R''/I_{P_i}$. $\qquad\square$

Proposition 4.11 *Assume* $C_K = D_K$. *Then* $KC_L[R]$ *is a Picard-Vessiot ring for* $\sigma(X) = AX$ *over* KC_L,

$$\mathrm{Kr.dim}(R') = \mathrm{tr.deg}(L/KC_L), \quad and \quad [D_L : C_L] = m(R').$$

Proof. Let us first assume that R' is a domain. There is some generic difference field $\mathcal{U}$ containing R' and its field of fractions R^*, and which is sufficiently saturated. Because L is a regular extension of K, there is some K-embedding φ of L into $\mathcal{U}$, and we will denote by T the image of Y in $\mathcal{U}$, and by y the image of Y in R'. Then $\varphi(C_L) \subset C_\mathcal{U}$, and there is some $B \in \mathrm{GL}_n(C_\mathcal{U})$ such that $T = yB$. Hence

$$KC_\mathcal{U}[T, \det(T)^{-1}] = KC_\mathcal{U}[y, \det(y)^{-1}].$$

By Proposition 4.8, $R' \otimes_K KC_\mathcal{U}$ is a direct sum of Picard-Vessiot rings of $\sigma(X) = AX$ over $KC_\mathcal{U}$, and clearly one of those is the domain $KC_\mathcal{U}[y, \det(y)^{-1}]$. Thus

$$\mathrm{Kr.dim}(R') = \mathrm{tr.deg}(R^*/K) = \mathrm{tr.deg}(L/KC_L),$$
$$D_{R^*} C_\mathcal{U} = \varphi(D_L)C_\mathcal{U}, \text{ and } m(R') = [D_L : C_L].$$

This implies also that $K\varphi(C_L)[T, \det(T)^{-1}]$ is a simple difference ring, and therefore a Picard-Vessiot ring for $\sigma(X) = AX$ over $K\varphi(C_L)$. Hence $KC_L[R]$ is a Picard-Vessiot extension for $\sigma(X) = AX$ over KC_L.

In the general case, we replace R' by R_0, σ by σ^ℓ, find some generic sufficiently saturated σ^ℓ-difference field $\mathcal{U}$ containing R_0, and a K-embedding φ of the σ^ℓ-difference domain L into $\mathcal{U}$, and conclude as above that $KFix(\sigma^\ell)[R_0] = KFix(\sigma^\ell)[\varphi(R)]$, that the Krull dimension of R' equals tr.deg(L/KC_L), and that $m(R_0) = [Fix(\sigma^\ell)(\varphi(D_L)) : Fix(\sigma^\ell)]$.

Because K and D_L are linearly disjoint over C_K, $[KD_L : KC_L] = [D_L : C_L]$, whence $D_{KC_L} = KC_L$, and by Corollary 4.10, the difference domain $KC_L[R]$ is a simple difference ring, i.e., a Picard-Vessiot ring for $\sigma(X) = AX$ over KC_L. By Proposition 4.8 $m(R') = [D_L : C_L]$.

We have $m(R') = \ell m(R_0)$, and $m(R_0)$ is the quotient of $[D_L : C_L]$ by the greatest common divisor of $[D_L : C_L]$ and ℓ. $\square$

Corollary 4.12 *Assume that $C_K = D_K$. Let $R'' = K[V, \det(V)^{-1}]$ be a difference domain, where $\sigma(V) = AV$, with field of fractions L_1, and assume that C_{L_1} is a finite algebraic extension of C_K. Then R'' is a Picard-Vessiot ring for $\sigma(X) = AX$ over K.*

Proof. Let $\mathcal{U}$ be a sufficiently saturated generic difference field containing R'', and let φ be a K-embedding of L into $\mathcal{U}$. Then $KC_{\mathcal{U}}[\varphi(R)] = KC_{\mathcal{U}}[R'']$. Hence Kr.dim$(R'') =$ Kr.dim(R') and R'' is a Picard-Vessiot ring by Corollary 4.10. $\square$

Corollary 4.13 *Assume that C_K is algebraically closed. Then $\ell(R') = [D_L : C_L]$.*

Proof. Immediate from Proposition 4.11 and the fact that $D_{R^*} = C_{R'} = C_K$. $\square$

Corollary 4.14 *The difference ring $KC_L[R]$ is a Picard-Vessiot ring for $\sigma(X) = AX$ over KC_L. All Picard-Vessiot rings for $\sigma(X) = AX$ over K have the same Krull dimension, which equals tr.deg(L/KC_L).*

Proof. Let $m = [D_K : C_K]$. Note that replacing σ by some power of σ does not change the fields D_K or D_L, and that $Fix(\sigma^m)(K) = D_K$. Therefore we can apply the previous results to the equation $\sigma^m(X) = A_m X$ over K. By Corollary 4.12 and because $KC_L[R]$ is a domain, $KC_L[R]$ is a Picard-Vessiot ring for $\sigma^m(X) = A_m X$ over KC_L, and therefore a simple σ^m-difference ring, whence a simple σ-difference ring, and finally a Picard-Vessiot ring for $\sigma(X) = AX$ over K.

Let $R' = R/q$ be a Picard-Vessiot ring for $\sigma(X) = AX$ over K. Assume first that R' is a domain, and let $\mathcal{U}$ be a generic difference field containing it. Because L is a regular extension of K, there is a K-embedding φ of L into $\mathcal{U}$, and from $KC_{\mathcal{U}}[\varphi(R)] = KC_{\mathcal{U}}[R']$ and Lemma 4.5.1, we obtain the result. If R' is not a domain, then we reason in the same fashion, replacing R' by R_0 and σ by σ^ℓ, to obtain the result. $\qquad\square$

Proposition 4.15 *Assume that $C_{R'} = C_K = D_K$ and $K \subset \mathcal{U}$. Then $G_{R'}$ and $\mathbb{H}$ are isomorphic.*

Proof. By Proposition 2.4, we may replace R' by $R' \otimes_K KD'_{\mathcal{U}}$, and consider the ring $K\varphi(C_L)[T, \det(T)^{-1}] \otimes_{K\varphi(C_L)} KD'_{\mathcal{U}}$, which is a Picard-Vessiot ring by Proposition 4.11 and Corollary 4.10. We identify $1 \otimes KD'_{\mathcal{U}}$ with $KD'_{\mathcal{U}}$. These two rings are isomorphic over $KD'_{\mathcal{U}}$ by Proposition 1.9 of [23], and it therefore suffices to show that

$$\mathrm{Aut}(\varphi(L) \otimes_{K\varphi(C_L)} KD'_{\mathcal{U}} / KD'_{\mathcal{U}}) = \mathbb{H}(D'_{\mathcal{U}}).$$

Inside $\varphi(L) \otimes_{K\varphi(C_L)} KD'_{\mathcal{U}}$, $\varphi(L) \otimes 1$ and $KD'_{\mathcal{U}}$ are linearly disjoint over $K\varphi(C_L)$. Hence, the algebraic loci of $(T, \det(T)^{-1})$ over $K\varphi(C_L)$ and over $KD'_{\mathcal{U}}$ coincide. As $\mathbb{H}$ was described as the subgroup of GL_n which leaves this algebraic set invariant, we get the result. $\qquad\square$

4.4 Concluding remarks

4.16 Model-theoretic Galois groups: definition and a bit of history. Model-theoretic Galois groups first appeared in a paper by Zilber [28] in the context of $\aleph_1$-categorical theories, and under the name of *binding groups*. Grosso modo, the general situation is as follows: in a saturated model M we have definable sets D and C such that, for some finite tuple b in M, $D \subset \mathrm{dcl}(C, b)$ (one then says that D is C-internal). The group $\mathrm{Aut}(M/C)$ induces a group of (elementary) permutations of D, and it is this group which one calls the *Galois group of D over C*. In Zilber's context, this group and its action on D are definable in M. One issue is therefore to find the correct assumptions so that these Galois groups and their action are definable, or at least, an intersection of definable groups. Hrushovski shows in his PhD thesis ([12]) that this is the case when the ambient theory is stable.

Poizat, in [20], recognized the importance of elimination of imaginaries in establishing the Galois correspondence for these Galois groups. He also noticed that if M is a differentially closed field of characteristic 0 and D is the set of solutions of some linear differential equation over some differential subfield K of M, and C is the field of constants of M, then the model-theoretic Galois group coincides with the differential Galois group introduced by Kolchin [15]. This connection was further explored by Pillay in a series of papers, see [19]. Note that because the theory of differentially closed fields of characteristic 0 eliminates quantifiers, this Galois group does coincide with the group of KC-automorphisms of the differential field $KC(D)$.

Since then, many authors studied or used Galois groups, under various assumptions on the ambient theory, and in various contexts, either purely model-theoretic (e.g., simple theories) or more algebraic (e.g. fields with Hasse derivations). In the context of generic difference fields, (model-theoretic) Galois groups were investigated in (5.11) of [4] (a slight modification in the proof then gives the Galois group described in section 4.1 of this paper). In positive characteristic p, the results generalize easily to twisted difference equations of the form $\sigma(X) = AX^{p^m}$, the field $Fix(\sigma)$ being then replaced by $Fix(\tau)$, where $\tau : x \mapsto \sigma(x)^{p^{-m}}$.

Recent work of Kamensky ([13]) isolates the common ingredients underlying all the definability results on Galois groups, and in particular *very much weakens the assumptions* on the ambient theory (it is not even assumed to be complete). With the correct definition of C-internality of the definable set D, he is able to show that a certain group of permutations of D is definable in M. These are just permutations, do not a priori preserve any relations of the language other than equality. From this group, he is then able to show that subgroups which preserve a (fixed) finite set of relations are also definable, and that the complexity of the defining formula does not increase, or not too much. For details, see section 3 of [13].

This approach of course applies to the set D of solutions of a linear system of difference equations (over a difference field K), and Kamensky also obtains the result that $\mathrm{Aut}(KFix(\sigma)(D)/KFix(\sigma))$ is definable (see section 5 in [13]).

4.17 A question arises in view of the proof of the general case of Proposition 4.11. When R' is not a domain, we found an embedding of the σ^ℓ-difference ring R_0 into a generic σ^ℓ-difference field $\mathcal{U}$. It may however happen that K is not relatively algebraically closed in R_0^*, even when $D_{R_0} = C_K$. Thus one can wonder: can one always find a generic difference field $\mathcal{U}$ containing K, and such that there is a K-embedding of the σ^ℓ-difference ring R_0 into $(\mathcal{U}, \sigma^\ell)$? Or are there Picard-Vessiot rings for which this is impossible?

4.18 Issues of definability. It is fairly clear that the algebraic group $\mathbb{H}$ is defined over $\varphi(KC_L)$. On the other hand, using the saturation of $\mathcal{U}$ and the fact that L is a regular extension of K, we may choose another K-embedding φ_1 of L in $\mathcal{U}$, and will obtain an algebraic group $\mathbb{H}_1$, which will be isomorphic to $\mathbb{H}$ (via some matrix $C \in \mathrm{GL}_n(C_\mathcal{U})$). It follows that $\mathbb{H}$ is K-isomorphic to an algebraic group $\mathbb{H}_0$ defined over the intersections of all possible $\varphi(KC_L)$, i.e., over K.

Observe that the isomorphism between $\mathbb{H}$ and $\mathbb{H}_1$ yields an isomorphism between $\mathbb{H}(C_\mathcal{U})$ and $\mathbb{H}_1(C_\mathcal{U})$, so that we will also have an isomorphism between $\mathbb{H}_0(C_\mathcal{U})$ and $\mathbb{H}(C_\mathcal{U})$, i.e., $\mathbb{H}'$ is K-isomorphic to an algebraic subgroup of $\mathbb{H}_0$ which is defined over $\overline{C_K} \cap C_\mathcal{U}$. Thus when C_K is algebraically closed, it will be defined over C_K.

The Galois duality works as well for subgroups of $\mathbb{H}(C_\mathcal{U})$ defined by equations (i.e., corresponding to algebraic subgroups of $\mathbb{H}'$, whose irreducible components are defined over $C_\mathcal{U}$). It works less well for arbitrary definable subgroups of $\mathbb{H}(C_\mathcal{U})$. In order for it to work, we need to replace $K(\mathcal{S})$ by its definable closure $\mathrm{dcl}(K\mathcal{S})$, i.e., the subfield of $\mathcal{U}$ which is fixed by all elements of $\mathrm{Aut}_{el}(\mathcal{U}/K\mathcal{S})$. Because the theory of $\mathcal{U}$ eliminates imaginaries (1.10 in [4]), any orbit of an element of $\mathcal{S}$ under the action of a definable subgroup of $\mathbb{H}(C_\mathcal{U})$ has a "code" inside $\mathrm{dcl}(K\mathcal{S})$.

4.19 Problems with the algebraic closure. Assume that $\mathcal{U}$ is a generic difference field containing K, and sufficiently saturated. Then if K is not relatively algebraically closed in the field of fractions of R_0, we may not be able to find a K-embedding of R_0 into the σ^ℓ-difference field $\mathcal{U}$. Thus in particular, a priori not all Picard-Vessiot domains K-embed into $\mathcal{U}$. This problem of course does not arise if we assume that K is algebraically closed, or, more precisely, if we assume that
All extensions of the automorphism σ to the algebraic closure of K define K-isomorphic difference fields.

This is the case if K has no finite (proper) σ-stable extension, for instance when $K = \mathbb{C}(t)$, with $\sigma(t) = t + 1$ and σ the identity on $\mathbb{C}$. However, in another classical case, this problem does arise: let $q \in \mathbb{C}$ be non-zero and not a root of unity, and let $K = \mathbb{C}(t)$, where σ is the identity on $\mathbb{C}$ and $\sigma(t) = qt$. Then K has non-trivial finite σ-stable extensions, and they are obtained by adding n-th roots of t.

Let us assume that, inside $\mathcal{U}$, we have $\sigma(\sqrt{t}) = \sqrt{q}\sqrt{t}$. Let us consider the system

$$\sigma(Y) = -\sqrt{q}Y, \quad Y \neq 0$$

over K. Then the Picard-Vessiot ring is $R' = K(y)$, where $y^2 = t$ and $\sigma(y) = -\sqrt{q}y$. Clearly R' does not embed in $\mathcal{U}$. If instead we had considered this system over $K(\sqrt{t})$, then the new Picard-Vessiot ring R'' is not a domain anymore, because it will contain a non-zero solution of $\sigma(X)+X = 0$ (namely, $y/\sqrt{t}$). In both cases however the Galois group is $\mathbb{Z}/2\mathbb{Z}$. And because R' embeds in R'', it also embeds in $K(T)\otimes_{\varphi(C_L)} D'_{\mathcal{U}}$.

This suggests that, when $C_K = D_K$, if one takes $\mathcal{M}$ to be the subfield of $\mathcal{U}$ generated over $KC_{\mathcal{U}}$ by all tuples of $\mathcal{U}$ satisfying some linear difference equation over K, then $\mathcal{M}\otimes_{C_{\mathcal{U}}} D'_{\mathcal{U}}$ is a universal (full) Picard-Vessiot ring of $KD'_{\mathcal{U}}$. This ring is not so difficult to describe in terms of $\mathcal{M}$. Observe that $\mathcal{M}$ contains $D_{\mathcal{U}}$. Thus $\mathcal{M}\otimes_{C_{\mathcal{U}}} D'_{\mathcal{U}}$ is isomorphic to $\mathcal{M}\otimes_{D_{\mathcal{U}}} (D_{\mathcal{U}}\otimes_{C_{\mathcal{U}}} D'_{\mathcal{U}})$. It is a regular ring, with prime spectrum the Cantor space $\mathcal{C}$ (i.e., the prime spectrum of $D_{\mathcal{U}} \otimes_{C_{\mathcal{U}}} D'_{\mathcal{U}}$), and σ acting on $\mathcal{C}$. As a ring, it is isomorphic to the ring of locally constant functions from $\mathcal{C}$ to $\mathcal{M}$.

It would be interesting to relate this ring to the universal Picard-Vessiot rings defined in [23].

4.20 Saturation hypotheses. The saturation hypothesis on $\mathcal{U}$ is not really needed to define the model-theoretic Galois group, since we only need $\mathcal{U}$ to contain a copy of L to define it. We also used it in the proof of Proposition 4.11, when we needed a K-embedding of L into $\mathcal{U}$. Thus, to define the model-theoretic Galois group, we only need $\mathcal{U}$ to be a generic difference field containing K. Its field of constants will however usually be larger than C_K. Indeed, the field $C_{\mathcal{U}}$ is always a pseudo-finite field (that is, a perfect, pseudo-algebraically closed field, with Galois group isomorphic to $\hat{\mathbb{Z}}$). However, one can show that if F is a pseudo-finite field of characteristic 0, then there is a generic difference field $\mathcal{U}$ containing F and such that $C_{\mathcal{U}} = F$. Thus, the field of constants of $\mathcal{U}$ does not need to be much larger than C_K. In the general case, a general non-sense construction allows one to find a pseudo-finite field F containing C_K and of transcendence degree at most 1 over C_K.

4.21 A partial description of the maximal σ^ℓ-ideal p of R. We keep the notation of the previous subsections, and will first assume that the Picard-Vessiot ring $R' = R/q$ is a domain contained in $\mathcal{U}$.

We will describe some of the elements of q. Write $C_L = C_K(\alpha_1, \ldots, \alpha_m)$, and $\alpha_i = f_i(Y)/g_i(Y)$, where $f_i(Y), g_i(Y) \in K[Y]$ are relatively prime. Then $\sigma(f_i)(AY)$ and $\sigma(g_i)(AY)$ are also relatively prime. Looking at the divisors defined by these polynomials, we obtain that there is some $k_i \in K$ such that $\sigma(f_i)(AY) = k_i f_i(Y)$ and $\sigma(g_i)(AY) = k_i g_i(Y)$. Then $(q, f_i(Y))$ and $(q, g_i(Y))$ are σ-ideals. By the maximality of q, this implies that either $f_i(Y)$ and $g_i(Y)$ are both in q, or else, say if $f_i(Y) \notin q$, that there is some $c_i \in C_{R'}$ such that $g_i(y) = c_i f_i(y)$, because $f_i(y)$ is invertible in R'. If $P_i(Z)$ is the minimal monic polynomial of c_i over C_K and is of degree r, then $g_i(Y)^r P_i(g_i(Y)/f_i(Y)) \in q$. In case $C_{R'} = C_K$ (this is the case for instance if C_K is algebraically closed), then $c_i \in C_K$, and $g_i(Y) - c_i f_i(Y)$ will belong to q. (Note also that if $k_i = k_j$, then also for some $d_j \in C_K$ we will have $f_j(Y) - d_j f_i(Y) \in q$, and $g_j(Y) - c_j d_j f_i(Y) \in q$). The σ-ideal I generated by all these polynomials in R could all of q. In any case one shows easily that q is a minimal prime ideal containing it (because $KC_L[Y, \det(Y)^{-1}]$ and R/I have the same Krull dimension, which is also the Krull dimension of R').

A better result is obtained by Kamensky in [13] Proposition 33: if $C_{R'} = C_K$, and instead of looking at a generating set of C_L over C_K one applies the same procedure to all elements of C_L, one obtains a generating set of the ideal q.

In case R' is not a domain, we reason in the same fashion to get a partial description of the σ^ℓ-ideal p.

References

[1] Y. André, Différentielles non commutatives et théorie de Galois différentielle ou aux différences, *Ann. Sci. École Norm. Sup. (4)*, 34 (2001), no. 5, 685–739.

[2] A. Bialynicki-Birula, On Galois theory of fields with operators, *Amer. J. Math.*, 84 (1962), 89–109.

[3] L. Breen, Tannakian categories, In U. Jannsen and et al, editors, *Motives*, volume 55 of *Proceedings of Symposia in Pure Mathematics*, American Mathematical Society, 1994, 337–376.

[4] Z. Chatzidakis and E. Hrushovski, Model theory of difference fields, *Trans. Amer. Math. Soc.*, 351(1999) no. 8, 2997–3071, .

[5] R. Cohn, *Difference Algebra*, Tracts In Mathematics, Number 17, Interscience Press, New York, 1965.

[6] P. Deligne, Catégories tannakiennes, In P. Cartier et al., *The*

Grothendieck Festschrift, Vol. 2, pages 111–195, Progress in Mathematics, Vol. 87, Birkhäuser, Boston, MA, 1990.

[7] P. Deligne and J. Milne, Tannakian categories, In P. Deligne et al., *Hodge cycles, motives and Shimura varieties*, pages 101–228, Lecture Notes in Mathematics, Vol. 900, Springer-Verlag, Berlin-New York, 1982.

[8] T. Dyckerhoff, Picard-Vessiot extensions over number fields, Diplomarbeit, Fakultät für Mathematik und Informatik der Universität Heidelberg, 2005.

[9] P. I. Etingof, Galois groups and connection matrices of q-difference equations, *Electron. Res. Announc. Amer. Math. Soc.*, 1(1995) no.1, 1–9 (electronic).

[10] C. H. Franke, Picard-Vessiot theory of linear homogeneous difference equations, *Transactions of the AMS*, 108 (1963), 491–515.

[11] C. Hardouin, *Structure galoisienne des extensions itérées de modules différentiels*, Ph.D. thesis, Paris 6, 2005, available at http://www.insti tut.math.jussieu.fr/theses/2005/hardouin/.

[12] E. Hrushovski, *Contributions to stable model theory*, Ph.D. thesis, Berkeley, 1986.

[13] M. Kamensky, Definable groups of partial automorphisms, Preprint available at http://arxiv.org/abs/math.LO/0607718, 2006.

[14] N. Katz, On the calculation of some differential Galois groups, *Inventiones Mathematicae*, 87 (1987), 13–61.

[15] E. R. Kolchin, Algebraic matrix groups and the Picard-Vessiot theory of homogeneous linear ordinary differential equations, *Annals of Mathematics*, 49 (1948), 1–42.

[16] E. R. Kolchin, Existence theorems connected with the Picard-Vessiot theory of homogeneous linear ordinary differential equations, *Bull. Amer. Math. Soc.*, 54 (1948), 927–932.

[17] E. R. Kolchin, *Differential algebra and algebraic groups*, Academic Press, New York, 1976.

[18] M. A. Papanikolas, Tannakian duality for Anderson-Drinfeld motives and algebraic independence of Carlitz logarithms, Preprint, Texas A&M University, 2005, available at http://arxiv.org/abs/math.NT/0506078.

[19] A. Pillay, Differential Galois theory I, *Illinois J. Math.* 42 (1998), 678–699.

[20] B. Poizat, Une théorie de Galois imaginaire, *J. of Symb. Logic*, 48 (1983), 1151–1170.

[21] C. Praagman, Fundamental solutions for meromorphic linear difference equations in the complex plane, and related problems, *J. Reine Angew. Math.*, 369 (1986), 101–109.

[22] M. van der Put and M. Reversat, Galois theory of q-difference equations, Preprint 2005, available at http://arxiv.org/abs/math.QA/0507098.

[23] M. van der Put and M. F. Singer, *Galois theory of difference equations*, volume 1666 of *Lecture Notes in Mathematics*, Springer-Verlag, Heidelberg, 1997.

[24] M. van der Put and M. F. Singer, *Galois theory of linear differential equations*, volume 328 of *Grundlehren der mathematischen Wissenshaften*. Springer, Heidelberg, 2003.

[25] J.-P. Ramis and J. Sauloy, The q-analogue of the wild fundamental group (I), in *Algebraic, analytic and geometric aspects of complex differential equations and their deformations. Painlevé hierarchies*, 167–193, RIMS

Kôkyûroku Bessatsu, B2, Res. Inst. Math. Sci. (RIMS), Kyoto, 2007. Preprint available at http://arxiv.org/abs/math.QA/0611521, 2006.

[26] J. Sauloy, Galois theory of Fuchsian q-difference equations, *Ann. Sci. École Norm. Sup. (4)*, 36 (2003) no.6, 925–968.

[27] W. C. Waterhouse, *Introduction to affine group schemes*, volume 66 of *Graduate Texts in Mathematics*, Springer-Verlag, New York, 1979.

[28] B. I. Zi'lber, Totally categorical theories; structural properties and the non-finite axiomatizability. in L. Pacholski et al. editors, *Model theory of algebra and arithmetic (Proc. Conf., Karpacz, 1979)*, pages 381–410, Lecture Notes in Mathematics volume 834, Springer Verlag, Berlin Heidelberg 1980.

Differentially valued fields are not differentially closed

Thomas Scanlon[†]

Summary

In answer to a question of M. Aschenbrenner and L. van den Dries, we show that no differentially closed field possesses a differential valuation.

1 Introduction

In connection with their work on H-fields [1], M. Aschenbrenner and L. van den Dries asked whether a differentially closed field can admit a nontrivial (Rosenlicht) differential valuation.

If K is a field and v is a Krull valuation on K and L/K is an extensions field, then there is at least one extension of v to a valuation on L. It is known that the analogous statement for differential specializations on differential fields is false. Indeed, anomalous properties of specializations of differential rings were observed already by Ritt [11] and examples of nonextendible specializations are known (see Exercise 6(c) of Section 6 of Chapter IV of [7] and [4, 5, 9] for a fuller account).

In this short note, we answer their question negatively by exhibiting a class of equations which cannot be solved in any differentially valued field even though they have solutions in differentially closed fields. In a forthcoming work of Aschenbrenner, van den Dries and van der Hoeven [2], the main results of this note are explained via direct computations.

I thank M. Aschenbrenner and L. van den Dries for bringing this question to my attention and discussing the matter with me and the Isaac Newton Institute for providing a mathematically rich setting for those discussions.

† Partially supported by NSF CAREER grant DMS-0450010

2 Logarithmic derivatives and differential valuations

In this section we recall Rosenlicht's notion of a differential valuation and show how the elliptic logarithmic derivative construction can be used to answer Aschenbrenner and van den Dries' question.

In what follows, if v is a valuation on a field K, then we write $\mathcal{O} := \{x \in K \mid v(x) \geq 0\}$ for the v-integers and if ∂ is a derivation on K, then we write $\mathcal{C} := \{x \in K \mid \partial(x) = 0\}$ for the differential constants.

Definition 2.1 A *differential valuation (in the sense of Rosenlicht)* v on a differential field (K, ∂) is a valuation for which the differential constants form a field of representatives in the sense that $\mathcal{C}^\times \subseteq \mathcal{O}^\times$ and for any $x \in K$ there is some $y \in \mathcal{C}$ with $v(x - y) > 0$ and an abstract version of L'Hôpital's Rule holds in the sense that if $v(x) > 0$ and $v(y) > 0$, then $v(y'x/x') > 0$.

Remark 2.2 It should be noted that Rosenlicht's notion of a differential valuation does not agree with Blum's [4].

Rosenlicht proved that to a differential valuation there is an associated *asymptotic couple*: the value group of the valuation, Γ, given together with a function $\psi : \Gamma \smallsetminus \{0\} \to \Gamma$ defined by $\psi(v(a)) = v(\partial(a)/a)$ for $a \in K^\times$ with $v(a) \neq 0$ [12]. For us, the most important property of this asymptotic couple is that if $\alpha, \beta \in \Gamma \smallsetminus \{0\}$, then $\psi(\alpha) < \psi(\beta) + |\beta|$. From this property it follows that if a differential field admits a differential valuation, then the derivation may be scaled so that the resulting derivation preserves the ring of integers.

Lemma 2.3 *If (K, ∂) is a differential field and v is a nontrivial differential valuation on K, then there is some $b \in K^\times$ so that if $\widetilde{\partial} = b\partial$, then v is a differential valuation on $(K, \widetilde{\partial})$ for which $\widetilde{\partial}(\mathcal{O}) \subseteq \mathcal{O}$.*

Proof Let $a \in K^\times$ be any element with $v(a) \neq 0$ and set $b := a/\partial(a)$. If $x \in \mathcal{O}$, then we may write $x = c + y$ where $\partial(c) = 0$ and $v(y) > 0$. Using Rosenlicht's inequality, we have $\psi(v(a)) = -v(b) < \psi(v(y)) + v(y) = v(\partial(y)) = v(\partial(c + y)) = -v(b) + v(\widetilde{\partial}(x))$. In particular, $v(\widetilde{\partial}(x)) > 0$. $\qquad \square$

With the next lemma we note that scaling a derivation does not change the property of the differential field being differentially closed.

Lemma 2.4 *If (K, ∂) is a differentially closed field, $b \in K^\times$ is nonzero, and $\widetilde{\partial}$, then $(K, \widetilde{\partial})$ is also differentially closed.*

Proof Using the Blum axioms for differentially closed fields [3], we must show that if $P(X_0, \ldots, X_n) \in K[X_0, \ldots, X_n]$ is an irreducible polynomial over K in $n+1$ variables and $G(X_0, \ldots, X_{n-1}) \in K[X_0, \ldots, X_{n-1}]$ is a nonzero polynomial in fewer variables, then there is some $a \in K$ with $P(a, \widetilde{\partial}(a), \ldots, \widetilde{\partial}^n(a)) = 0$ and $G(a, \ldots, \widetilde{\partial}^{n-1}(a)) \neq 0$.

In the ring $K\langle \partial \rangle$ of linear differential operators in ∂ over K, for each positive integer m we may write $(b\partial)^m = b^m \partial^m + \sum_{i=1}^{m-1} d_i^{(m)} \partial^i$ for some $d_i^{(m)} \in K$. Indeed, the base case of $m = 1$ is trivial, and

$$
\begin{aligned}
(b\partial)^{m+1} &= b\partial(b^m \partial^m + \sum_{i=1}^{m-1} d_i^{(m)} \partial^i) \\
&= b(b^m \partial^{m+1} + mb^{m-1}\partial(b)\partial^m + \sum_{i=1}^{m-1}(\partial(d_i^{(m)})\partial^i + d_i^{(m)}\partial^{i+1})) \\
&= b^{m+1}\partial^{m+1} + \sum_{i=1}^{m} d_i^{(m+1)} \partial^j
\end{aligned}
$$

where $d_m^{(m+1)} = mb^m\partial(b) + bd_{m-1}^{(m)}$ and $d_i^{(m+1)} = \partial(d_i^{(m)}) + d_{j-1}^{(m)}$ for $1 \leq j < m$.

The map $\rho : K[X_0, \ldots, X_n] \to K[X_0, \ldots, X_n]$ given by $X_0 \mapsto X_0$ and $X_i \mapsto b^i X_i + \sum_{j=1}^{i-1} d_j^{(i)} X_i$ is an automorphism for which for any $F \in K[X_0, \ldots, X_n]$ and $c \in K$ we have $\rho(F)(c, \partial(c), \ldots, \partial^n(c)) = F(c, \widetilde{\partial}(c), \ldots, \widetilde{\partial}^n(c))$. As P is irreducible and ρ is an automorphism, $\rho(P)$ is irreducible. Visibly, $\rho(K[X_0, \ldots, X_{n-1}]) \subseteq K[X_0, \ldots, X_{n-1}]$. So, $\rho(G) \in K[X_0, \ldots, X_{n-1}]$. As (K, ∂) is differentially closed there is some $d \in K$ with $0 = \rho(P)(d, \partial(d), \ldots, \partial^n(d)) = P(d, \widetilde{\partial}(d), \ldots, \widetilde{\partial}^n(d))$ and $0 \neq \rho(G)(d, \ldots, \partial^{n-1}(d)) = G(d, \ldots, \widetilde{\partial}^{n-1}(d))$. That is, $(K, \widetilde{\partial})$ is differentially closed. $\qquad\square$

Theorem 2.5 *Suppose that (K, ∂) is a differentially closed field and v is a valuation on K for which the derivation preserves the ring of integers in the sense that $\partial(\mathcal{O}) \subseteq \mathcal{O}$. Then v is trivial.*

Proof Let E be any elliptic curve over $\mathcal{C} \cap \mathcal{O}$. If v is trivial on $\mathbb{Q}$, we can take E to be any elliptic curve over $\mathbb{Q}$. Otherwise, v restricts to a p-adic valuation on $\mathbb{Q}$ and we can take E to be a model of an elliptic curve over $\mathbb{Z}$ having good reduction at p.

Consider the elliptic logarithmic derivative $\partial \log_E : E(K) \to \mathbb{G}_a(K)$. The reader should consult section 22 of chapter 5 of [7] for a thorough development of the theory of logarithmic differentiation. For the sake of completeness we recall the construction of $\partial \log_E$.

There is a group homomorphism $\nabla : E(K) \to TE(K)$ from the K-rational points of E to the K-rational points of the tangent bundle of E defined in coordinates by $(x_1, \ldots, x_n) \mapsto (x_1, \ldots, x_n; \partial(x_1), \ldots, \partial(x_n))$. As we will need to keep track of integrality conditions, it will help to see ∇ more conceptually. Using the Weil restriction of scalars construction, one can identify $TE(K)$ with $E(K[\epsilon]/(\epsilon^2))$, or more generally, $TE(R)$ with $E(R[\epsilon]/(\epsilon^2))$ for any commutative $\mathcal{C} \cap \mathcal{O}$-algebra R. If we have a derivation $\delta : R \to R$ on the algebra R, then there is a ring homomorphism $\exp(\delta) : R \to R[\epsilon]/(\epsilon^2)$ given by $x \mapsto x + \delta(x)\epsilon$. The map $\nabla : E(R) \to TE(R)$ corresponds to the map on points $E(R) \to E(R[\epsilon]/(\epsilon^2))$ induced by $\exp(\delta)$. In particular, ∇ takes R-rational points to R-rational points.

The tangent bundle TE splits as $s : TE \to E \times T_0 E$ where $T_0 E$ is the tangent space to E at the origin via the map $(P, w) \mapsto (P, d\tau_{-P} w)$ where $\tau_{-P} : E \to E$ is the translation map $x \mapsto x - P$ on E. If $\pi : E \times T_0 E \to \mathbb{G}_a$ is the projection onto the second coördinate followed by an isomorphism between $T_0 E$ and the additive group $\mathbb{G}_a$, then the elliptic logarithmic derivative is $\partial \log_E = \pi \circ s \circ \nabla$.

As K is differentially closed, the map $\partial \log_E : E(K) \to \mathbb{G}_a(K) = K$ is surjective. Indeed, one can see this in several ways. For instance, we can work with the Lascar rank (see [8]). The kernel of $\partial \log_E$ is $E(\mathcal{C})$ and as such has Lascar rank 1 whilst the Lascar rank of $E(K)$ is ω. Hence, the Lascar rank of the image of $\partial \log_E$ is also ω which is the same as that of the connected group $\mathbb{G}_a(K)$. Hence, $\partial \log_E$ is surjective. Alternatively, one could simply apply the geometric axioms of [10]. Let $P \in \mathbb{G}_a(K)$ be any point. Relative to the above trivialization of TE, we define a section $s_P : E \to TE$ by $Q \mapsto (Q, P)$. By the geometric axiom, there is a point $Q \in E(K)$ with $s_P(Q) = s \circ \nabla(Q)$. That is, $P = \partial \log_E(Q)$.

Since each of the maps forming $\partial \log_E$ takes $\mathcal{O}$-rational points to $\mathcal{O}$-rational points, the image of $\partial \log_E$ on $E(\mathcal{O})$ is contained in $\mathbb{G}_a(\mathcal{O}) = \mathcal{O}$. As E is proper, $E(\mathcal{O}) = E(K)$ (or, really, the image of $E(\mathcal{O})$ in $E(K)$ under the map induced by $\mathcal{O} \hookrightarrow K$ is all of $E(K)$). Hence, $\mathcal{O} = K$. That is, v is a trivial valuation. $\qquad \square$

Combining Lemmata 2.3 and 2.4 with Theorem 2.5 we conclude with a negative answer to Aschenbrenner and van den Dries' question.

Corollary 2.6 *No differentially closed field admits a nontrivial differential valuation.*

Remark 2.7 As with Buium's construction of examples of Ritt's anomaly of the differential dimension of an intersection [6], our construction is based on the observation that projective algebraic varieties may admit nonconstant differential regular functions. Indeed, as the reader can readily verify, our argument shows that if (K, ∂) is a differential field admitting a nontrivial valuation whose ring of integers is preserved by ∂, and X is a projective scheme over $\mathcal{O}$ whose Albanese map is injective, then by composing the Albanese map with a component of a Manin homomorphism one produces a nonconstant differential regular function $f : X(K) \to \mathbb{A}^1(K)$ whose image is contained in $\mathcal{O}$.

References

[1] M. Aschenbrenner and L. van den Dries, Liouville closed H-fields, *J. Pure Appl. Algebra* **197** (2005), no. 1-3, 83–139.

[2] M. Aschenbrenner, L. van den Dries, and J. van der Hoeven, Linear differential equations over H-fields, in preparation.

[3] L. Blum, *Generalized Algebraic Structures: A Model Theoretic Approach*, PhD dissertation, Massachussetts Institute of Technology, Cambridge, MA (1968).

[4] P. Blum, Complete models of differential fields, *Trans. Amer. Math. Soc.* **137** (1969) 309–325.

[5] P. Blum, Extending differential specializations, *Proc. Amer. Math. Soc.* **24** (1970) 471–474.

[6] A. Buium, Geometry of differential polynomial functions II: Algebraic curves, *Amer. J. Math.* **116** (1994), no. 4, 785–818.

[7] E. R. Kolchin, *Differential Algebra and Algebraic Groups*, Pure and Applied Mathematics **54**, Academic Press, New York-London, 1973.

[8] D. Marker, Model theory of differential fields, in *Model Theory of Fields* (D. Marker, M. Messmer and A. Pillay, eds.) Lecture Notes in Logic **5**, Springer-Verlag, Berlin, 1996.

[9] S. D. Morrison, Extensions of differential places, *Amer. J. Math.* **100** (1978), no. 2, 245–261.

[10] D. Pierce and A. Pillay, A note on the axioms for differentially closed fields of characteristic zero, *J. Algebra* **204** (1998), no. 1, 108–115.

[11] J. F. Ritt, On a type of algebraic differential manifold. *Trans. Amer. Math. Soc.* **48** (1940) 542–552.

[12] M. Rosenlicht, Differential valuations, *Pacific J. Math.* **86**, no. 1 (1980), 301 – 319.

Complex analytic geometry in a nonstandard setting

Ya'acov Peterzil
University of Haifa

Sergei Starchenko[†]
University of Notre Dame

Summary

Given an arbitrary o-minimal expansion of a real closed field **R**, we develop the basic theory of definable manifolds and definable analytic sets, with respect to the algebraic closure of **R**, along the lines of classical complex analytic geometry. Because of the o-minimality assumption, we obtain strong theorems on removal of singularities and strong finiteness results in both the classical and the nonstandard settings.

We also use a theorem of Bianconi to characterize all complex analytic sets definable in $\mathbb{R}_{exp}$.

1 Introduction

Let **R** be a real closed field and **K** its algebraic closure, identified with **R**^{2} (after fixing a square-root of -1). In [13] we investigated the notion of a **K**-holomorphic function from (subsets of) **K** into **K** which are definable in o-minimal expansions of **R**. Examples of such functions are abundant, especially in the case when **R** is the field of real numbers and **K** is the complex field. In [14] we extended this investigation to functions of several variables and began examining the notions of a **K**-manifold and **K**-analytic set, modeled after the classical notions. Here we return to this last question, while modifying slightly the definitions of a **K**-manifold and a **K**-analytic set from [14].

The theorems in this paper are of different kinds. First, we give a

† The first author thanks the Logic group at University of Illinois Urbana-Champaign for its warm hospitality during 2003-2004. The second author was partially supported by the NSF. Both thank the Newton Mathematical Institute for its hospitality during Spring 2005. Both authors thank the anonymous referee for a thorough reading of the paper.

117

rigorous treatment of the theory of complex analytic geometry in this nonstandard setting, along the lines of the classical theory. From this point of view the paper can be read as a basic textbook in complex analytic geometry, written from the point of view of a model theorist (note that since every germ of a holomorphic function is definable in the o-minimal structure $\mathbb{R}_{an}$, the results proved here cover parts of classical complex geometry as well). As is often the case, the loss of local compactness of the underlying fields **R** and **K**, which might be nonarchimedean, is compensated by o-minimality.

However, we do more than just recover analogues of the classical theory. In some cases o-minimality yields stronger theorems than the classical ones, even when the underlying field is that of the complex numbers. Indeed, in [16] we showed how, working over the real and complex fields, o-minimality implies strong closure theorems for locally definable complex analytic sets. The same theorems hold in the nonstandard settings as well.

We also prove here several finiteness results which were not treated in [16]. For example, it follows from our results that any definable locally analytic subset A of a definable complex manifold can be covered by finitely many definable open sets, on each of which A is the zero set of finitely many definable holomorphic functions (see Theorem 4.14). Similarly, we formulate and prove a finite version of the classical Coherence Theorem (see Theorem 11.1). Here again, one replaces compactness assumptions on the underlying manifolds with definability in an o-minimal structure.

The main topological tool for most of the theorems is a general result (see Theorem 2.14), interesting on its own right, which allows us to move from an arbitrary $2d$-dimensional definable set in $\mathbf{K}^n$ to a set whose projection on the first d **K**-coordinates is "definably proper" over its image.

In the appendix to the paper we use a theorem of Bianconi [3] to characterize all definable locally analytic subsets of $\mathbb{C}^n$ in the structure $\mathbb{R}_{exp} = \langle \mathbb{R}, <, +, \cdot, e^x \rangle$. We also observe there that, given a holomorphic function f of n variables, definable in some o-minimal expansion of the real field, its real an imaginary parts can be extended to holomorphic functions (of $2n$ variables) which are definable in the same structure.

Although the theorems in [16] were formulated in the context of the real and complex fields, most of the proofs there were written with the nonstandard setting in mind and hence carry over almost verbatim to our setting. We therefore refer at times to [16] for proofs of theorems in our paper. Also, we let ourselves refer at times to proofs from Whitney's book [19], when we found that there was no advantage in copying them into this paper. This book, as well as Chirka's book [6] were of great help to us when we came to learn the basics of complex analytic geometry. For a reference on o-minimal structures we suggest van den Dries' book [7].

The structure of the paper is as follows: In Section 2 we consider the analogous notion in our setting to local compactness and proper maps and prove the result about a certain finite covering of definable locally closed sets. In Section 3 we discuss **K**-manifolds and submanifolds. In section 4 we define **K**-analytic subsets of **K**-manifolds and establish their basic properties. In Section 5 we prove a strong version of Chow's Theorem. In Section 6 we show that the set of singular points of a **K**-analytic set is **K**-analytic itself. In Section 7 we prove a strong version of the Remmert Proper Mapping Theorem. In Section 8 we discuss the relationship to model theory and show, that just like Zil'ber's result in the classical case, every definably compact **K**-manifold, equipped with all **K**-analytic subsets of its cartesian products, is a structure of finite Morley Rank. In Section 9 we discuss **K**-meromorphic maps. In Section 10 we formulate (and refer to a proof of) the analogue of the Campana-Fujiki Theorem in our nonstandard setting. In Section 11 we formulate and prove our finite version to the Coherence Theorem. Finally, in the Appendix we prove the result about definable complex analytic sets in the structure $\mathbb{R}_{exp}$.

Throughout the paper we work in a fixed o-minimal expansion of a real closed field **R**. *We use the term "definable" sets to mean definable in this fixed o-minimal structure, possibly with parameters.*

2 Topological preliminaries

2.1 "Real" and "complex" dimensions

As in complex analysis, we will sometimes prefer to view definable subsets of $\mathbf{K}^n$ as subsets of $\mathbf{R}^{2n}$. As such, every definable set $A \subseteq \mathbf{K}^n$ has its o-minimal dimension, which we denote by $\dim_{\mathbf{R}} A$. **K**-analytic sets and **K**-manifolds will also be associated a dimension with respect to **K**, which we will denote by $\dim_{\mathbf{K}} A$. We say "L is a d-dimensional **K**-linear subspace of $\mathbf{K}^n$" when the dimension of L, as a **K**-vector space, equals

d (i.e., $\dim_{\mathbf{K}} L = d$). When both make sense, it is immediate to see that $\dim_{\mathbf{R}} A = 2 \dim_{\mathbf{K}} A$.

2.2 Locally closed sets and definably proper maps

Definition 2.1 Recall that *a definable C^0 $\mathbf{R}$–manifold of dimension n, with respect to $\mathbf{R}$* is a set X, covered by finitely many nonempty sets $U_1, \ldots, U_k$, and for each $i = 1, \ldots, k$ there is a set-theoretic bijection $\phi_i : U_i \to V_i$, where V_i is definable and open in $\mathbf{R}^n$ and such that each $\phi_i(U_j \cap U_j)$ is definable an open and the transition maps are definable and continuous. Moreover, the topology induced on X by this covering is Hausdorff.

We call such a manifold *a definable C^p $\mathbf{R}$-manifold* if in addition the transition maps are C^p with respect to the field $\mathbf{R}$.

Although there is no a priori assumption that X is a definable set it follows (see discussion in Section 4, [1]) that X, with its manifold topology, can be realized as a definable subset of $\mathbf{R}^k$ for some k, with the subspace topology.

Let X be a definable subset of $\mathbf{R}^n$. We recall that X is called *definably compact* if for every definable continuous $\gamma : (0,1) \to X$ the limit of $\gamma(t)$, as t tends to 0 in $\mathbf{R}$, exists in X. This is equivalent to X being closed and bounded in $\mathbf{R}^n$.

Definition 2.2 We say that a definable set $X \subseteq \mathbf{R}^n$ is *locally definably compact* if every $x \in X$ has a definable neighborhood $V \subseteq X$ (i.e, V contains an X-open set around x) which is definably compact.

$X \subseteq \mathbf{R}^n$ is *locally closed* if there is a (definable) open set $U \subseteq \mathbf{R}^n$ containing X such that X is relatively closed in U.

Let U be an open subset of $\mathbf{R}^n$. For $X \subseteq U$, the *frontier of X in U* is defined as $Fr_U(X) = Cl_U(X) \setminus X$, where $Cl_U(X)$ is the closure of X in U. If $U = \mathbf{R}^n$ then we write $Fr(X)$ instead of $Fr_{\mathbf{R}^n}$.

The following is easy to verify:

Lemma 2.3 *Let X be a definable subset of $\mathbf{R}^n$. Then the following are equivalent:*
(i) X is locally definably compact.
(ii) X is locally closed in $\mathbf{R}^n$.
(iii) $Fr(X)$ is a closed subset of $\mathbf{R}^n$.

Now, assume that $X \subseteq \mathbf{R}^n$ is locally closed in $\mathbf{R}^n$ and definably homeomorphic to a set $Y \subseteq \mathbf{R}^m$. It follows from the lemma that Y is also locally closed in $\mathbf{R}^n$ (since the notion of "locally definably compact" is invariant under definable homeomorphism).

Definition 2.4 Let f be a definable continuous map from a definable $X \subseteq \mathbf{R}^n$ into $Y \subseteq \mathbf{R}^k$.

For $b \in Y$, we say that f *is definably proper over* b if for every definable curve $\gamma : (0,1) \to X$ such that $\lim_{t \to 0} f(\gamma(t)) = b$, $\gamma(t)$ tends to some limit in X as t tends to 0 (in [7] this is called "γ is completable"). If $f : X \to Y$ is definably proper over every $b \in Y$ then we say that f *is definably proper over* Y or just f *is definably proper*.

For $A \subseteq X$, we say that $f|A$ is *definably proper over its image* if $f|A : A \to f(A)$ is definably proper over $f(A)$.

We say that f is *bounded over* $b \in Y$ if there is a neighborhood $W \subseteq Y$ of b such that $f^{-1}(W)$ is a bounded subset of $\mathbf{R}^n$.

In [7], an equivalent definition for definable properness is given and it is shown (Section 6, Lemma 4.5) that a definable and continuous $f : X \to Y$ is definably proper (over Y) in the sense of definition 2.4 if and only if the preimage of every closed and bounded set in $\mathbf{R}^k$ is closed and bounded in $\mathbf{R}^n$.

The following lemma, which is easy to verify, implies that the set of all $y \in Y$ such that f is definably proper over y is itself definable.

Lemma 2.5 *For $f : X \to Y$ a definable continuous map, $X \subseteq \mathbf{R}^n$, and $y \in Y$, the following are equivalent:*
(i) f is definably proper over y.
(ii) f is bounded over y and the intersection of the closure of the graph of f in $\mathbf{R}^n \times Y$ with $Fr(X) \times \{y\}$ is empty.

Lemma 2.6 *Let $X \subseteq \mathbf{R}^n$ be a definable, locally closed set, $f : X \to \mathbf{R}^k$ a definable continuous map. Then,*
(i) The set of all $y \in \mathbf{R}^k$ such that f is definably proper over y is open in $\mathbf{R}^k$.
(ii) If f is definably proper over $f(X)$ then $f(X)$ is a locally closed set.

Proof. (i) is a corollary of Lemma 2.5.

(ii) Let $W = \{y \in \mathbf{R}^k : f \text{ is definably proper over } y\}$. By (i), W is a definable open set and by our assumption $f(X) \subseteq W$. It is easy to see that $f(X)$ is relatively closed in W. $\qquad\square$

2.3 Linear and affine subspaces

We assume here that our structure is ω_1-saturated, but any statement which does not mention generic points holds in every elementarily equivalent structure.

Definition 2.7 Let $H \subseteq \mathbf{K}^n$ be a d-dimensional **K**-subspace of $\mathbf{K}^n$. We say that H *is generic over a set* $C \subseteq \mathbf{R}$ if the following holds: Let $\{H_s : s \in S\}$ be some C-definable parametrization of all d-dimensional **K**-linear subspaces of $\mathbf{K}^n$. Then $H = H_s$ for some s generic in S over C.

The following is easy to verify using the dimension formula.

Fact 2.8 *Let H be a d-dimensional **K**-subspace of $\mathbf{K}^n$, $C \subseteq \mathbf{R}$. Then the following are equivalent:*

(1) *H is generic over C.*
(2) *H has a generic basis $\{v_1, \ldots, v_d\}$ over C. Namely, it is a **K**-linear basis for H where $\dim_{\mathbf{R}}(v_i/Cv_1, \ldots, v_{i-1}) = 2n$ for every $i = 1, \ldots, d$.*
(3) *H has a generic-orthogonal basis $\{v_1, \ldots, v_d\}$ with respect to the standard dot product on $\mathbf{K}^n$ induced by $\mathbf{R}$. Namely, for every $i = 1, \ldots, d$, v_i is generic over $Cv_1 \ldots, v_{i-1}$ in the orthogonal complement of $sp_{\mathbf{K}}\{v_1, \ldots, v_{i-1}\}$ in $\mathbf{K}^n$.*

Lemma 2.9 *For $C \subseteq \mathbf{K}$, let $H \subseteq \mathbf{K}^n$ be a d-dimensional **K**-subspace of $\mathbf{K}^n$ which is generic over C.*
(i) Let B be a generic basis for H over C. Then for every $0 \neq v \in H$,

$$\dim_{\mathbf{R}}(v/C) - (2n - 2d) \geqslant \dim_{\mathbf{R}}(v/BC).$$

(ii) For every $0 \neq v \in H$, $\dim_{\mathbf{R}}(v/C) \geqslant 2n - 2d$.

Proof. (i) Let $B = \{v_1, \ldots, v_d\} \subseteq \mathbf{K}^n$ be a generic basis for H over C (see 2.8).

Now, given a nonzero $v \in H$, we may assume, after reordering B, that $B_1 = \{v_1, \ldots, v_{d-1}, v\}$ is a basis for H. In particular, H is defined over B_1, hence, $\dim_{\mathbf{R}}(v_d/B_1C) \leqslant \dim_{\mathbf{R}} H \leqslant 2d$. By the dimension formula,

$$\dim_{\mathbf{R}}(v_d/B_1C) + \dim_{\mathbf{R}}(v/C, v_1, \ldots v_{d-1})$$
$$= \dim_{\mathbf{R}}(v/BC) + \dim_{\mathbf{R}}(v_d/C, v_1, \ldots, v_{d-1}).$$

Since $\dim_{\mathbf{R}}(v_d/C, v_1, \ldots, v_{d-1}) = 2n$, we have

$$\dim_{\mathbf{R}}(v/a) \geqslant \dim_{\mathbf{R}}(v/Cv_1, \ldots, v_{d-1}) \geqslant 2n - 2d + \dim_{\mathbf{R}}(v/BC).$$

(ii) This is immediate from (i). □

Corollary 2.10 *Given* $a \in \mathbf{K}^n$, $H \subseteq \mathbf{K}^n$ *a d-dimensional* $\mathbf{K}$-*subspace which is generic over* a, *and* $A \subseteq \mathbf{K}^n$ *an a-definable set, we have:*
(i) If $\dim_{\mathbf{R}}(A) \geqslant codim_{\mathbb{R}}(H)(= 2n - 2d)$ *then* $\dim_{\mathbf{R}}(A \cap a + H) \leqslant \dim_{\mathbf{R}} A - (2n - 2d)$.
(ii) If $\dim_{\mathbf{R}}(A) < codim_{\mathbb{R}}(H)$ *then* $A \cap (a + H) \subseteq \{a\}$.

Proof. Notice that if $A \cap (a + H) \subseteq \{a\}$ then both (i) and (ii) hold (recall that the $\mathbf{R}$-dimension of the empty set is $-\infty$), so we assume that $A \cap a + H$ contains at least one element different than a.

Fix B a generic basis of H over a, and let $w \neq a$ be a generic element in $A \cap a + H$ over aB. The translated set $A - a$ is still definable over a, hence $(A - a) \cap H$ is defined over aB. By Lemma 2.9(i),

$$\dim_{\mathbf{R}}(w - a/aB) \leqslant \dim_{\mathbf{R}}(w - a/a) - (2n - 2d).$$

Because $\dim_{\mathbf{R}}((w - a)/aB) = \dim_{\mathbf{R}}(w/aB)$ and $\dim_{\mathbf{R}}((w - a)/a) = \dim_{\mathbf{R}}(w/a)$, we have

$$\dim_{\mathbf{R}}(w/aB) \leqslant \dim_{\mathbf{R}}(w/a) - (2n - 2d) \leqslant \dim_{\mathbf{R}}(A) - (2n - 2d).$$

This implies both (i) and (ii). □

2.4 Generic projections

For H a d-dimensional $\mathbf{K}$-subspace of $\mathbf{K}^n$, we say that an orthogonal projection $\pi : \mathbf{K}^n \to H$ is *generic over* C if H is generic over C.

Definition 2.11 A definable subset A of $\mathbf{R}^k$ is called *an m-dimensional* C^p $\mathbf{R}$-*submanifold of* $\mathbf{R}^k$ if for every $x \in A$ there is a definable open neighborhood U of x and definable C^p (with respect to to $\mathbf{R}$) map $f : U \to \mathbf{R}^{n-m}$ such that $A \cap U = f^{-1}(0)$ and the rank of the $\mathbf{R}$-differential df_x of f at x equals $n - m$.

(We chose here a local definition of a submanifold. However, as in the semialgebraic case, every definable $\mathbf{R}$-submanifold can be realized as a definable $\mathbf{R}$-manifold with finitely many charts, see Proposition 9.3.10 in [2]. We will not make use of this fact in the text).

Lemma 2.12 *For $a \in \mathbf{K}^n$, let $H \subseteq \mathbf{K}^n$ be an s-dimensional $\mathbf{K}$-subspace of $\mathbf{K}^n$ which is generic over a and let $A \subseteq \mathbf{K}^n$ be an a-definable set such that $\dim_{\mathbf{R}} A = r = 2n - 2s$.*

If $x_0 \in (a + H) \cap A$, $x_0 \neq a$, then A is a C^1 $\mathbf{R}$-submanifold of $\mathbf{K}^n$ at x_0, of dimension r, and $a + H$ intersects A transversally at x_0.

Proof. By 2.10, the set $a + H \cap A$ is finite and every point $\neq a$ in this intersection is generic in A over a, thus A is a C^1 $\mathbf{R}$-submanifold of $\mathbf{K}^n$ of $\mathbf{R}$-dimension r, at every point of intersection. By restricting ourselves to a neighborhood $V \subseteq \mathbf{K}^n$ of x_0, we may assume that A is a C^1 $\mathbf{R}$-submanifold of $\mathbf{K}^n$.

The family of all $\mathbf{K}$-subspaces of $\mathbf{K}^n$ of $\mathbf{K}$-dimension s has $\mathbf{R}$-dimension $2s(2n - 2s)$. Using, say, the Grassmanian construction, there is a $\emptyset$-definable C^1 $\mathbf{R}$-manifold G whose dimension is $2s(2n - 2s)$ which parametrizes all these subspaces.

Let $\{H_g : g \in G\}$ be the family of all these subspaces. Then $H = H_{g_0}$ where g_0 is generic in G over a. Fix $H' \subseteq H_{g_0}$ a $\mathbf{K}$-subspace of H_{g_0} of $\mathbf{K}$-dimension $s - 1$ which is generic over $a g_0$ among all such subspaces.

Claim If B is a generic basis for H' then $\dim_{\mathbf{R}}(g_0/aB) = 2n - 2s$.

This follows from the fact that H_{g_0}/H' is a one-dimensional subspace of $\mathbf{K}^n/H'$ which is generic over aB, and because the set of all 1-dimensional $\mathbf{K}$-subspaces of $\mathbf{K}^n/H'$ has dimension $2n - 2s$.

Let $G_1 = \{g \in G : H' \subseteq H_g\}$. It is B-definable, and by working in the quotient $\mathbf{K}^n/H'$ we can endow G_1 with a structure of a definable C^1 $\mathbf{R}$-manifold of dimension $2n - 2s$. We now have: for all $x \in \mathbf{K}^n \setminus H'$ there is a unique $g = h(x) \in G_1$ such that $x \in H_g$. Moreover, the $\mathbf{R}$-manifold structure on G_1 can be chosen so that $h : \mathbf{K}^n \setminus H' \to G_1$ is a C^1 map with respect to $\mathbf{R}$.

By 2.10, $a + H' \cap A$ contains at most the point a, hence there is an open neighborhood $U \subseteq \mathbf{K}^n$ of x_0 such that $U \cap a + H' = \emptyset$. We define $F : U \to G_1$ by $F(x) = h(x - a)$. Namely, $g = F(x)$ if $x \in a + H_g$. We now proceed just like in the proof of Lemma 3.6 in [13]:

Notice that each $H_g \cap U$ is a level set of F and that F is $\mathbf{R}$-differentiable on U. It follows that $ker(dF_x) = H_{F(x)}$ (where dF_x denotes the $\mathbf{R}$-differential of F at x). Since $A \cap U$ is a C^1 $\mathbf{R}$-submanifold of $\mathbf{K}^n$ of dimension $r = 2n - 2s$, there is an a-definable open $V \subseteq \mathbf{R}^r$ and an a-definable $\sigma : V \to \mathbf{R}^{2n}$ such that $x_0 \in \sigma(V) \subseteq A \cap U$ and σ is an immersion. It follows that for every $y \in V$, $d\sigma_y$ gives an isomorphism between $\mathbf{R}^r$ and $T_{\sigma(y)}A$, and by the chain rule, $T_{\sigma(y)}A \cap H_{F(\sigma(y))}$ is nonzero if and only if $ker(d(F \circ \sigma)_y)$ is nonzero.

Assume now that $a + H_{g_0}$ intersects A at x_0 non-transversally. Then $T_{x_0}(A) \cap H_{g_0}$ is nonzero and since g_0 is generic in G_1 over aB, there is some neighborhood $W \subseteq G_1$ of g_0 such that for every $g \in W$, $a + H_g$ intersects A non-transversally. If we let $V_1 = \sigma^{-1}(F^{-1}(W))$ then for every $y \in V_1$, $ker(d(F \circ \sigma)_y)$ is nonzero. This implies that the rank of $d(F \circ \sigma)_y$ is less than r and hence $\dim_{\mathbf{R}}(F \circ \sigma(V_1)) < r$. Since $\dim V_1 = r$ it follows that for some $g \in W$, there are infinitely many $y \in V_1$ such that $F \circ \sigma(y) = g$, or said differently, $H_g \cap A$ is infinite. But, since $H_{g_0} \cap A$ is finite and g_0 generic in G_1 we could have chosen $W \subseteq G_1$ such that for every $g \in W$, $H_g \cap A$ is finite. Contradiction. $\qquad\square$

For $A \subseteq \mathbf{K}^n$ definable, we write $Fr(A)$ for the frontier of A in $\mathbf{K}^n$, identified with $\mathbf{R}^{2n}$. By o-minimality, $\dim_{\mathbf{R}} Fr(A) \leqslant \dim_{\mathbf{R}}(A) - 1$.

Lemma 2.13 *Let $A \subseteq \mathbf{K}^n$ be a $\emptyset$-definable locally closed set of $\mathbf{R}$-dimension $2d < 2n$ and let L be a generic over $\emptyset$ d-dimensional $\mathbf{K}$-subspace of $\mathbf{K}^n$. Let $\pi : \mathbf{K}^n \to L$ be the orthogonal projection. Then:*
(i) For all $y \in L$, $\pi|A$ is bounded over y.
(ii) If $a \in \mathbf{K}^n \setminus Fr(A)$ and π is generic over a then $\pi|A$ is definably proper over $\pi(a)$.
(iii) If A is closed in $\mathbf{K}^n$ then $\pi|A$ is definably proper over all of L.

Proof. (i) This can be shown either by working in the projective space $\mathbb{P}^n(\mathbf{K})$, or directly as follows:

Let S^{2n-1} be the unit sphere in $\mathbf{R}^{2n}$, and let

$$A^* = \left\{ x \in S^{2n-1} : \forall t > 0 \, \exists x' \in A \left(\| x' \| > t \& \left\| x - \frac{x'}{\| x' \|} \right\| < 1/t \right) \right\}.$$

Notice that A^* can also be obtained as follows:

First intersect $Cl(\{(\frac{1}{\|x'\|}, \frac{x'}{\|x'\|}) : x' \in A\})$ with the set $\{0\} \times S^{2n-1}$ and then project onto the second coordinate. It follows from o-minimality that $\dim_{\mathbf{R}}(A^*) \leqslant 2d - 1$.

Let $H \subseteq \mathbf{K}^n$ be the orthogonal complement of L. It is not hard to see that if, for some $y \in L$, $\pi|A$ were not bounded over y then the intersection of H with A^* is nonempty. However, by Lemma 2.10 (ii), $H \cap A^* = \emptyset$, contradiction.

For (ii) we define,

$$I(\pi) = \{y \in L : \pi|A \text{ is not definably proper over } y\}.$$

By Lemma 2.6, $I(\pi)$ is a definable closed set.

Since A is locally closed, we have (see Lemma 2.5), for all $y \in L$, $y \in I(\pi)$ if and only if either $(y + H) \cap Fr(A) \neq \emptyset$ or $\pi|A$ is not bounded over y.

Lemma 2.10 (ii) implies that $\pi(a) + H \cap FrA = \emptyset$, and by (i), $\pi|A$ is bounded over $\pi(a)$ thus $\pi|A$ is definably proper over $\pi(a)$.

(iii) is an immediate corollary of (i) and (ii). □

2.5 The main covering theorem

In classical complex analytic geometry, given a complex analytic set A and $p \in A$, one moves to a local coordinates system, chosen properly, in order to ensure that the projection map from A onto (some of) the coordinates is a proper map in this neighborhood. The following theorem is probably the most significant advantage of the o-minimal setting in comparison with the general classical setting. It allows us to work "globally" rather than "locally".

Theorem 2.14 *Let A be a definable, locally closed subset of $\mathbf{K}^n$, such that $\dim_{\mathbf{R}} A = 2d < 2n$.*

Then there are definable open sets $U_1, \ldots, U_k \subseteq \mathbf{K}^n$ and d-dimensional $\mathbf{K}$-linear subspaces $L_1, \ldots, L_k$ such that

(1) *$\bigcup_{i=1}^{k} U_i = \mathbf{K}^n \setminus Fr_{\mathbf{K}^n}(A)$ (in particular A is contained in the union of the U_i's).*

(2) *For each $i = 1, \ldots, k$, if $\pi_i : \mathbf{K}^n \to L_i$ is the orthogonal projection, then $\pi_i|U_i \cap A$ is definably proper over $\pi_i(U_i)$.*

(3) *If furthermore A is a C^1 $\mathbf{R}$-submanifold of $\mathbf{K}^n$ then the U_i's and L_i's can be chosen to satisfy also: For every $i = 1 \ldots, k$ the map $\pi_i|A \cap U_i$ is a local diffeomorphism and there is $m = m_i$ such that every $y \in \pi_i(U_i)$ has exactly m preimages under π_i in $A \cap U_i$.*

Proof. Note that since the statement of the theorem is first-order we may prove it in an ω_1-saturated elementary extension. We assume that A is $\emptyset$-definable.

Let $\pi_1, \ldots, \pi_{2n+1}$ be an independent sequence of generic orthogonal projections onto d-dimensional subspaces $L_1, \ldots, L_{2n+1}$, respectively (i.e, each π_k is generic over $\pi_1, \cdots, \pi_{k-1}$). We first claim that for every $a \in \mathbf{K}^n$ there is $i = 1, \ldots, 2n + 1$ such that π_i is generic over a.

Indeed, if not then each π_i is not generic over a, which implies, by the dimension formula,

$$\dim_{\mathbf{R}}(a/\emptyset) > \dim_{\mathbf{R}}(a/\pi_1) > \dim_{\mathbf{R}}(a/\pi_1, \pi_2) > \cdots$$
$$> \dim_{\mathbf{R}}(a/\pi_1, \cdots, \pi_{2n+1}).$$

This is clearly impossible since $\dim_{\mathbf{R}}(a/\emptyset) \leqslant 2n$ (by $\dim(a/\pi_1, \cdots, \pi_t)$ we mean $\dim(a/g_1, \ldots, g_t)$ where g_i is a parameter for L_i in some 0-definable parametrization of all d-dimensional $\mathbf{K}$-subspaces of $\mathbf{K}^n$).

For every $i = 1, \ldots, 2n + 1$, let U_i be the definable open set $U_i = \pi_i^{-1}(L_i \setminus I(\pi_i))$ (see the proof of 2.13 for the definition of $I(\pi)$). By Lemma 2.6 (ii), U_i is an open set.

We claim that $\mathbf{K}^n \setminus Fr(A) = \bigcup_i U_i$.

For one inclusion, notice that if $x \in Fr(A)$ then no π_i is definably proper over $\pi(x)$ and hence x is not in any of the U_i's. For the opposite inclusion, consider $x \in \mathbf{K}^n \setminus Fr(A)$. By the above observations, there is an $i = 1, \ldots, n$ such that π_i is generic over x. But then, by Lemma 2.13 (ii), $\pi_i|A$ is definably proper over $\pi_i(x)$ and hence $x \in U_i$.

By definition of the U_i's, each $\pi_i|(A \cap U_i)$ is definably proper over $\pi_i(U_i)$, thus proving (2).

For (3), assume that A is a C^1 $\mathbf{R}$-submanifold of $\mathbf{K}^n$ and choose the π_i's as above. For every $i = 1, \ldots, 2n + 1$, let

$$B_i' = I(\pi_i) \quad = \quad \{y \in L_i : \pi_i|A \text{ not definably proper over } y \},$$
$$B_i'' \quad = \quad \{y \in L_i : \exists x \in A \; \pi_i(x) = y$$
$$\& \; \pi_i|A \text{ not a local diffeomorphism at } x\},$$

and let $B_i = B_i' \cup B_i''$.

Claim 1 B_i is a closed subset of L_i.

Indeed, take $y \in L_i \setminus B_i$. We want to show that there is a neighborhood of y that is disjoint from B_i. Since $y \notin I(\pi_i)$, $\pi_i|A$ is definably proper over y. But then either $y \in \pi_i(A)$ or there is a neighborhood V of y which is disjoint from $\pi_i(A)$, and in particular then $V \cap B_i = \emptyset$. We assume then that $y \in \pi_i(A)$. Since $y \notin B_i$, for all $x \in \pi_i^{-1}(y) \cap A$, $\pi_i|A$ is a local diffeomorphism near x. In particular, $\pi^{-1}(y)$ is finite, and by the properness of $\pi_i|A$ over y (and hence over a neighborhood of y), there is an open neighborhood $V \subseteq L_i$ of y such that the restriction of π_i to $A \cap \pi_i^{-1}(V)$ is a finite-to-one surjection onto V, which is also a local diffeomorphism. We have then $V \cap B_i = \emptyset$, completing the proof that B_i is a closed subset of L_i.

For every $i = 1, \ldots, 2n + 1$, let $U_i = \pi_i^{-1}(L_i \setminus B_i)$.

Claim 2 $\bigcup_i U_i = \mathbf{K}^n \setminus Fr(A)$.

As in (2), we need to see that if π_i is generic over $a \in \mathbf{K}^n \setminus Fr(A)$ then $a \in U_i$, or equivalently, $\pi_i(a) \notin B_i$.

By Lemma 2.13, π_i is definably proper over $\pi_i(a)$ and hence $\pi_i(a) \notin I(\pi_i)$. It is left to verify that for all $x \in A \cap \pi_i^{-1}\pi_i(a)$, $\pi_i|A$ is a local diffeomorphism at x.

It follows from Lemma 2.12 that for every $x_0 \neq a$ such that $\pi_i(x_0) = \pi_i(a)$, $\pi_i|A$ is a local diffeomorphism at x_0. This is sufficient to ensure that $a \in U_i$ in case that a itself is not in A. If $a \in A$, then, since A is a C^1 $\mathbf{R}$-submanifold at a, and π_i is generic over a it is easy to see that π_i is a local diffeomorphism at a as well and hence $a \in U_i$, thus proving the claim.

After possibly partitioning each U_i to its definably connected components, we may assume that each U_i is definably connected. Notice that $\pi_i|(A \cap U_i)$ is still a local diffeomorphism at every point of these components and that its restriction to each component is still definably proper over $\pi_i(U_i)$.

It is left to show the following:

Claim 3 Assume that A is a relatively closed subset of U, a definable open subset of $\mathbf{K}^n$, such that the projection $\pi : U \to \mathbf{K}^d$ satisfies:
(i) $\pi(U)$ is definably connected, and
(ii) $\pi|A : A \to \pi(U)$ is a local diffeomorphism which is moreover definably proper over $\pi(U)$.

Then there is an m such every element in $\pi(U)$ has exactly m preimages under π in $A \cap U$.

Since π is a local homeomorphism, it is everywhere a finite-to-one map. By o-minimality, there is an r such that it is at most r-to-1. For $i = 1, \ldots, r$, let W_i be the set of y in $\pi(U)$ such that $\pi^{-1}(y) \cap A$ contains exactly i elements. The sets $W_1, \ldots, W_r$ form a partition of $\pi(U)$. Definable properness implies that each W_i is an open set and therefore closed as well. By the definable connectedness of $\pi(U)$, there is exactly one such W_i. $\qquad\square$

Remark We presume that a similar theorem to 2.14 can be proved with respect to projection onto **R**-subspaces of $\mathbf{R}^n$. However, since generic **K**-subspaces of $\mathbf{K}^n$ are not generic as **R**-subspaces of $\mathbf{R}^{2n}$, such a theorem will not directly imply Theorem 2.14.

3 K-manifolds and submanifolds

The notions of a **K**-manifold and its **K**-analytic subsets were first treated in [14]. However, here we simplify the definitions. Instead of requiring uniform definability of the data which constitutes the **K**-manifold or the **K**-analytic subset, we require all data to be finite. Lemma 3.3 and Corollary 4.14 justify the changes.

We first recall the basic notions of **K**-holomorphicity from [13] and [14].

For $U \subseteq \mathbf{K}$ a definable open set and $z_0 \in U$, we say that a definable function $f : U \to \mathbf{K}$ is **K**-*differentiable at* z_0 if $\lim_{h\to 0}(f(z_0 + h) - f(z_0))/h$ exists in **K**. If f is **K**-differentiable at every point on U then we say that f is **K**-*holomorphic in* U. If U is a definable open subset of $\mathbf{K}^n$ and $f : U \to \mathbf{K}$ is a definable function then f is called **K**-*holomorphic in* U if it is continuous and in addition **K**-holomorphic in each of the variables separately. A **K**-*holomorphic map* $f : U \to \mathbf{K}^m$ is a definable map, each of whose coordinate maps is a **K**-holomorphic function.

Definition 3.1 *A d-dimensional* **K**-*manifold is a set* X, *together with a finite cover* $X = \bigcup_{i=1}^{r} U_i$, *and for each* $i = 1, \ldots, r$ *a set-theoretic bijection* $\phi_i : U_i \to V_i$, *with* V_i *a definable subset open of* $\mathbf{K}^d$, *such that for each* i, j, *the set* $\phi_i(U_i \cap U_j)$ *is a definable open subset of* V_i *and the transition maps* $\phi_j \phi_i^{-1}$ *are* K-holomorphic (in particular definable). Moreover, the naturally induced topology is Hausdorff.

The notion of *a* **K**-*holomorphic map between* **K**-*manifolds* is defined using the transition maps and charts just like in classical differential topology.

Notice that every **K**-manifold is in particular a definable C^1 **R**-manifold. Therefore, as was discussed earlier, we may assume that the underlying set X, with its manifold topology, is a definable subset of $\mathbf{R}^k$ for some k. Also, when **K** is the field of complex numbers then "**K**-holomorphic" just means "holomorphic and definable in some o-minimal expansion of the real field".

Definition 3.2 Let N be a **K**-manifold. A definable $M \subseteq N$ is called *a d-dimensional* **K**-*submanifold of* N if for every $a \in M$ there is a definable open set $U \subseteq N$ containing a and a **K**-holomorphic $f : U \to \mathbf{K}^{n-d}$ such that $M \cap U = f^{-1}(0)$ and $Rank_\mathbf{K}(Df_a) = n - d$ (where Df_x is the **K**-differential of f at x.

It is not clear a priori that a **K**-submanifold can itself be covered by a finite atlas, but just like definable **R**-submanifolds, it turns out to be true:

Lemma 3.3 *Let N be a* **K**-*manifold, $M \subseteq N$ a* **K**-*submanifold. Then M is a* **K**-*manifold as well. More precisely, there is a* **K**-*manifold M_1 and a* **K**-*holomorphic embedding $f : M_1 \to N$ such that $f(M_1) = M$.*

Proof. Since N is covered by finitely many charts we may assume that $N = U$ a definable open subset of $\mathbf{K}^n$, and that $M \subseteq U$ is a **K**-submanifold of dimension d. Moreover, after considering all the possible projections on d of the **K**-coordinates, we may assume that the projection π onto the first d coordinates is a local homeomorphism. We now use Proposition 9.3.9 from [2] and conclude that there exists an finite open definable cover $M = \bigcup_{i=1}^{r} U_i$ such that for every $i = 1, \ldots, r$, the projection map $\pi|U_i$ is a homeomorphism onto $\pi(U_i)$ (the result from [2] is stated for semialgebraic sets but the proof just uses the triangulation and trivialization theorems, both true in our o-minimal setting, see [7]). We therefore may assume that the projection of M onto the first d coordinates is a homeomorphism.

Hence M is the graph of a continuous function ϕ from $\pi(M)$ into $\mathbf{K}^{n-d}$. By the implicit function theorem, ϕ is **K**-holomorphic at generic points and since it is everywhere continuous, it follows from Theorem 2.14 in [14] that ϕ is **K**-holomorphic. This is sufficient in order to obtain the necessary charts $\qquad\qquad\square$

Lemma 3.4 *Let $\mathcal{M}$ be a definably connected d-dimensional* **K**-*manifold and $f : M \to \mathbf{K}$ a* **K**-*holomorphic function. Then the zero set of f is either the whole of M or a subset of M whose* **R**-*dimension is $2d - 2$.*

Proof. We may assume that M is an open subset of $\mathbf{K}^d$. Fix an arbitrary $a = (a_1, \ldots, a_d)$ in Z the zero set of f. If Z is not the whole of M then using a change of coordinates we may assume that f is regular in the last variable, and hence we may apply the Weierstrass Preparation Theorem (see Theorem 2.20 in [14] and the definitions preceding it). It follows that in a neighborhood of a the set Z is the zero set of a definable **K**-holomorphic function $h(z_1, \ldots, z_{d-1}, y)$ which is a monic polynomial in $(y - a_d)$ and coefficients which are **K**-holomorphic functions of $z_1, \ldots, z_{d-1}$. By Theorem 2.3 (5) in [14], there is a neighborhood U_1 of $(a_1, \ldots, a_{d-1})$ such that for every $(z_1, \ldots, z_{d-1})$ in U_1 the function h has a fixed finite number of zeroes, counted with multiplicity, near a_d. It follows that $\dim_{\mathbf{R}}(Z) = 2d - 2$ in a neighborhood of a. $\qquad\square$

Lemma 3.5 *Let M, N be definably connected **K**-manifolds of dimension d and assume that $f : M \to N$ is a **K**-holomorphic function, such that $\dim_{\mathbf{R}} f(M) = 2d$.*

Let

$$M_f = \{x \in M : f \text{ is a local } \mathbf{K}\text{-biholomorphism at } x\}.$$

Then

$$\dim_{\mathbf{R}}(M \setminus M_f) \leqslant 2d - 2.$$

In particular, M_f is definably connected.

Proof. By working locally, we may assume that M is an open, definably connected subset of $\mathbf{K}^d$, which we call U, and that $N = \mathbf{K}^d$. We write U_f for M_f. Notice that $U \setminus U_f$ is contained in the zero set of the function $|Df| : U \to \mathbf{K}$ (where $|Df_x|$ is the determinant of the corresponding **K**-linear function at x). Since this last function is **K**-holomorphic in U its zero set, by 3.4, is either the whole of U or of **R**-dimension $2d - 2$. If $|Df_x|$ vanished on the whole of U then the dimension of $f(M)$ would be smaller than $2d$. It thus follows from our assumption that the latter must hold, proving the lemma. $\qquad\square$

Lemma 3.6 *Let M be a definably connected d-dimensional **K**-submanifold of $\mathbf{K}^n$, $\pi : \mathbf{K}^n \to \mathbf{K}^d$ the projection on the first d coordinates, and assume that $\dim_{\mathbf{R}} \pi(M) = 2d$. Then*

(1) *Outside a definable subset of M of **R**-dimension $2d - 2$, π is a local homeomorphism.*

(2) *There are definable open, pairwise disjoint $V_1, \ldots, V_r \subseteq \pi(M)$, such that $\dim_{\mathbf{R}}(\pi(M) \setminus \bigcup_i V_i) \leqslant 2d - 1$ and for each i, there are definable **K**-holomorphic functions $\phi_{i,1}, \ldots, \phi_{i,i} : V_i \to \mathbf{K}^{n-d}$, taking distinct values at every $x \in V_i$, such that for all $x \in V_i$ and $y \in \mathbf{K}^{n-d}$, $(x, y) \in M$ iff $y = \phi_{i,j}(x)$ for some $j = 1, \ldots, i$.*

(3) *If moreover $\pi|M$ is definably proper over $\pi(M)$ then there is a definable closed $S \subseteq \mathbf{K}^d$ of dimension $2d-2$ and $m \in \mathbb{N}$ such that for every $x \in \pi(M) \setminus S$, $\pi^{-1}(x) \cap M$ contains exactly m distinct elements, at each of which π_M is a local $\mathbf{K}$-biholomorphism.*

Proof. The first part of the lemma follows easily from Lemma 3.5.

For (2), note that the existence of such sets and definable *continuous* functions follows from o-minimality. The functions are $\mathbf{K}$-holomorphic at generic points by the implicit function theorem, and by continuity they are $\mathbf{K}$-holomorphic on their whole domain (see Theorem 2.14 in [14]).

(3) Assume now that $\pi|M$ is definably proper over $\pi(M)$, and let M_π be the set of all points in M where π is a local $\mathbf{K}$-biholomorphism. Let $S = Cl_{\mathbf{K}^d}(\pi(M \setminus M_\pi))$ and $V = \pi(M) \setminus S$. By Lemma 3.5 and o-minimality, $\dim_{\mathbf{R}} S \leqslant 2d-2$, and therefore V is a definably connected open set. Moreover, every element in V has a finite number of preimages in M, at each of which π is a local homeomorphism.

Thus, if we let $A = \pi^{-1}(V) \cap M$ then A satisfies the assumptions of the last claim from the proof of Theorem 2.14 (3). It follows that there is an m such that every point in V has exactly m preimages in M.   □

The following technical lemma will play an important role in later results.

Lemma 3.7 *Let M be a definably connected $\mathbf{K}$-submanifold of $\mathbf{K}^n$, and let $f : M \to \mathbf{K}$ be a $\mathbf{K}$-holomorphic function. Assume that Z is the set of all $z_0 \in Cl_{\mathbf{K}}(M)$ such that the limit of $f(z)$ exists and equals 0 as z approaches z_0 in M. If $\dim_{\mathbf{R}} Z \geqslant \dim_{\mathbf{R}} M - 1$ then f vanishes on all of M.*

Proof. This is the exact analogue of Theorem 3.1 from [16]. The proof there (as well as lemma 2.10 which it uses) goes through word-for-word in the current setting.   □

4 K-analytic sets

Definition 4.1 If A is a definable subset of a $\mathbf{K}$-manifold M, then A is a *locally $\mathbf{K}$-analytic subset of M* if for every $a \in A$ there is a definable open $V \subseteq M$ containing a and a $\mathbf{K}$-holomorphic map $f : V \to \mathbf{K}^m$, for some m, such that $A \cap V = f^{-1}(0)$. A is called *a $\mathbf{K}$-analytic subset of M* if in addition to the above A is closed in M.

A definable closed $A \subseteq M$ is *finitely* **K**-*analytic subset of* M if it can be covered by finitely many definable open subsets of M, on each of which A is the zero set of some **K**-holomorphic map.

A **K**-analytic subset A of M is called *reducible* if it can be written as $A = A_1 \cup A_2$, where A_1 and A_2 are **K**-analytic in M and none of the A_i's is contained in the other. When A is not reducible it is called *irreducible*.

Remark Because a **K**-holomorphic map remains **K**-holomorphic in every elementary extension, the notions of a **K**-manifold and a finitely **K**-analytic subset are invariant under elementary extensions. It is not so clear a priori that a **K**-analytic subset of a **K**-manifold remains so in every elementary extension (this was the reason that we originally defined a **K**-analytic subset in [14], using uniformly definable data). However, as we eventually show, every **K**-analytic set is necessarily finitely **K**-analytic thus making this notion first-order as well. Moreover, we will give an explicit first-order characterization of **K**-analytic sets (see 4.14 below).

Note that, by definition, every (locally) **K**-analytic set is definable in our ambient o-minimal structure.

Examples

1. Any **K**-algebraic subset of $\mathbf{K}^n$ is clearly a **K**-analytic subset of $\mathbf{K}^n$. More generally, the intersection of any such algebraic set with a semialgebraic open set $U \subseteq \mathbf{K}^n$ is a **K**-analytic subset of U. The same is true for projective algebraic subsets of $\mathbb{P}^n(\mathbf{K})$ (which is itself a **K**-manifold).

2. Consider $\mathbb{R}_{an}$, the expansion of the real field by all real analytic functions on the closed unit cubes. Every compact complex manifold M can be realized as a **K**-manifold in this structure (with $\mathbf{K} = \mathbb{C}$) and every analytic subset of M is a definable **K**-analytic subset (see discussion in Section 2.2, p.8, of [14]). Moreover, as we pointed in [14], elementary extensions of $\mathbb{R}_{an}$ give rise to new and interesting **K**-analytic subsets of such manifolds (see 3.3 there).

3. Consider the structure $\mathbb{R}_{exp} = \langle \mathbb{R}, <, +, \cdot, e^x \rangle$, which is the expansion of the real field by the real exponential function. This is an o-minimal structure ([20]) but, as is shown in [3], every germ of a holomorphic map which is definable in $\mathbb{R}_{exp}$ is semialgebraic. Using this fact we prove in the appendix (see Theorem 12.6):

Theorem 4.2 *Let* $G \subseteq \mathbb{C}^n$ *be an open set and let* X *be a* $\mathbb{C}$-*analytic* subset of G such that both X and G are definable in $\mathbb{R}_{exp}$. Then there is a complex algebraic set $A \subseteq \mathbb{C}^n$ such that X is the union of several irreducible components of $A \cap G$.

Remark Note that it is not true, under the above assumptions, that there is an algebraic $A \subseteq \mathbf{K}^n$ such that $X = A \cap G$. Take for instance X to be the algebraic irreducible set $\{(x, y) \in \mathbb{C}^2 : y^2 = x^2(x + 1)\}$. Locally, in a small neighborhood U of $(0, 0)$, this analytic set has two irreducible component, each given as the vanishing set of one of branches of $y \pm x\sqrt{x + 1}$. Both components are semialgebraic but there is no algebraic subset of $\mathbb{C}^2$ whose intersection with U gives only one of these components.

Before we continue developing the theory of **K**-analytic sets we prove a corollary to Lemma 3.7:

Lemma 4.3 *Let M be a definably connected $\mathbf{K}$-submanifold of some definable open $U \subseteq \mathbf{K}^n$ and let A be a $\mathbf{K}$-analytic subset of U. If $\dim_{\mathbf{R}}(A \cap Cl(M)) \geqslant \dim_{\mathbf{R}} M - 1$ then M is contained in A.*

Proof. Pick first a generic z_0 in $A \cap Cl(M)$ and consider a **K**-holomorphic function f, in some neighborhood W of z_0, which vanishes on A. We may choose W so that every definably connected component of $M \cap W$ has z_0 in its closure. The function f, considered as a function on each component of $W \cap M$, satisfies the assumptions of Lemma 3.7 and therefore vanishes on $W \cap M$. The same is true for all the functions which define A near z_0 and hence M is contained in A, in some neighborhood of z_0. Consider now the set M' of all points in M where M is locally contained in A. M' is clearly open in M and it is easily seen to be also closed thus $M \subseteq A$. $\qquad\square$

Definition 4.4 Let N be a **K**-manifold. If $A \subseteq N$ is an arbitrary definable set then $Reg_{\mathbf{K}} A$ is the set of all points $z \in A$ such that in some neighborhood of a, the set A is a **K**-submanifold of N. We let $Sing_{\mathbf{K}} A = A \setminus Reg_{\mathbf{K}} A$.

Later on we will prove that $Sing_{\mathbf{K}} A$ is itself a **K**-analytic set. At this stage we can prove the following approximation.

Lemma 4.5 *If N is a $\mathbf{K}$-manifold and A is a $\mathbf{K}$-analytic subset of N then $Reg_{\mathbf{K}} A$ is dense in A and*

$$\dim_{\mathbf{R}}(Sing_{\mathbf{K}} A) \leqslant \dim_{\mathbf{R}} A - 2.$$

Proof. We prove the theorem by induction on $n = \dim_{\mathbf{K}} N$. We may assume that N is an open definable subset of $\mathbf{K}^n$, which we call U. Since we prove the theorem for an arbitrary such U the density of $Reg_{\mathbf{K}} A$ will follow from the dimension inequality in the lemma.

Take $a \in A$ and assume that $g_1, \ldots, g_r : W \to K$ are $\mathbf{K}$-holomorphic functions on a definably connected open $W \subseteq U$, $a \in W$, such that $A \cap W$ is the intersection of the zero-sets of the g_i's. It is sufficient to show that $\dim_{\mathbf{R}}(Sing_{\mathbf{K}}(A \cap W)) \leqslant \dim_{\mathbf{R}}(A \cap W) - 2$, thus we may assume that $A \subseteq W$. Notice that $A \cap W$ remains $\mathbf{K}$-analytic in elementary extensions thus we may use generic points. We assume that A and the g_i's are definable over $\emptyset$, and that $A \neq W$.

Let z_0 be a generic point in $Sing_{\mathbf{K}} A$. We claim that there is a 0-definable $\mathbf{K}$-holomorphic function $f : W \to \mathbf{K}$ which vanishes on A, such that one of its partial first derivatives $\partial f / \partial z_i$ is not identically zero on any relatively open subset of A containing z_0.

Indeed, we may take f to be a partial derivative of sufficiently large order of one of the g_i's. If no such f satisfies the above property then in particular the partial derivatives of all g_i's, of any order, vanish at a, and then (see Theorem 2.13 in [14]) all the g_i's are zero functions, thus $A = W$.

Let $B = \{z \in W : \partial f / \partial z_i(z) = 0\}$, $N' = Z(f) \setminus B$ and $A' = A \setminus B$. By the implicit function theorem, N' is a submanifold of W of dimension $n - 1$, and A' is a $\mathbf{K}$-analytic subset of N'. By our assumptions on f, z_0 is in the closure of A'. Notice that because B is a closed set,

$$Reg_{\mathbf{K}} A' = Reg_{\mathbf{K}}(A \setminus B) = Reg_{\mathbf{K}}(A) \setminus B$$

and

$$Sing_{\mathbf{K}} A' = Sing_{\mathbf{K}}(A) \setminus B.$$

By induction, $Reg_{\mathbf{K}} A'$ is dense in A' and $\dim_{\mathbf{R}}(Sing_{\mathbf{K}} A') \leqslant \dim_{\mathbf{R}} A' - 2$.

Case 1 $z_0 \notin B$.

In this case z_0 is in $Sing_{\mathbf{K}} A'$, whence $\dim_{\mathbf{R}}(z_0/\emptyset) \leqslant \dim_{\mathbf{R}}(Sing_{\mathbf{K}} A')$ (since A' is $\emptyset$-definable). Because z_0 was taken generic in $Sing_{\mathbf{K}} A$ we have that

$$\dim_{\mathbf{R}}(Sing_{\mathbf{K}} A) \leqslant \dim_{\mathbf{R}}(Sing_K A') \leqslant \dim_{\mathbf{R}} A' - 2 \leqslant \dim_{\mathbf{R}} A - 2.$$

Case 2 $z_0 \in B$.

Because $Reg_{\mathbf{K}} A'$ is dense in A', we have $z_0 \in B \cap Cl_U(Reg_{\mathbf{K}} A')$ and hence $z_0 \in Cl_U(B \cap (Reg_{\mathbf{K}}(A) \setminus B))$. We may now apply Lemma 4.3 as follows:

Denote the submanifold $Reg_{\mathbf{K}}(A) \setminus B$ by M'. We claim that $\dim_{\mathbf{R}}(B \cap Cl_U(M')) \leqslant \dim_{\mathbf{R}} A - 2$.

Indeed, if not then

$$\dim_{\mathbf{R}}(B \cap Cl_U(M')) \geqslant \dim_{\mathbf{R}} A - 1 \geqslant \dim_{\mathbf{R}} M' - 1,$$

and, by Lemma 4.3, there is some (nonempty) definably connected component of M' which is contained in B. This is absurd because M' is disjoint from B.

We thus showed that $\dim_{\mathbf{R}}(B \cap Cl_U(M')) \leqslant \dim_{\mathbf{R}} A - 2$. Since z_0 belongs to this last intersection and is generic in $Sing_{\mathbf{K}} A$ we have $\dim_{\mathbf{R}}(Sing A) \leqslant \dim_{\mathbf{R}} A - 2$. $\qquad \square$

Lemma 4.6 *If A is a $\mathbf{K}$-analytic subset of a $\mathbf{K}$-manifold N and $Reg_{\mathbf{K}} A$ is definably connected then A is irreducible. In particular, the number of irreducible components of A is finite.*

Proof. This follows from the density of $Reg_{\mathbf{K}} A$ in A, together with the fact that a $\mathbf{K}$-holomorphic function which vanishes on some open subset of its domain must vanish on a whole definably connected component of the domain (we identify for this purpose the submanifold $Reg_{\mathbf{K}} A$ with an open subset of some $\mathbf{K}^d$). $\qquad \square$

Our main technical result is the following.

Lemma 4.7 *Let U be a definable open subset of $\mathbf{K}^n$, A a definable relatively closed subset of U whose $\mathbf{R}$-dimension is $2d$. Assume that $Reg_{\mathbf{K}} A$ is definably connected, dense in A, and that $\dim_{\mathbf{R}}(Sing_{\mathbf{K}} A) \leqslant 2d - 2$. Assume also that the projection map onto the first d coordinates, $\pi : A \to \mathbf{K}^d$, is finite-to-one and definably proper over its image.*

Then, $\pi(A)$ is open in $\mathbf{K}^d$ and there is an $r \in \mathbb{N}$ and a $\mathbf{K}$-holomorphic map $\Psi : \pi(A) \times \mathbf{K}^{n-d} \to \mathbf{K}^r$ such that A is the zero set of Ψ.

Moreover, the map Ψ can be definably recovered in the structure $\langle \mathbf{R}, <, +, \cdot, A \rangle$.

Proof. We first prove that $\pi(A)$ is open in $\mathbf{K}^d$.

We assume that A and U are definable over the empty set. Since A is a locally closed, definably connected set then, by Lemma 2.6 (ii), $\pi(A)$ is relatively closed in some definably open set W. Since $\dim_{\mathbf{R}} \pi(A) = 2d$, then either $\pi(A) = W$ or the boundary of $\pi(A)$ in W has **R**-dimension $2d - 1$ (see for example Proposition 2 in [10]). If the latter holds then there is a point y on the boundary of $\pi(A)$ in W, such that $\dim_{\mathbf{R}}(y/\emptyset) = 2d - 1$. But, since $\pi(A)$ is relatively closed in W, there is $x \in A$, $\dim_{\mathbf{R}}(x/\emptyset) \geqslant 2d - 1$, such that $\pi(x) = y$. Because of the assumption on $Sing_{\mathbf{K}}A$, x must be in $Reg_{\mathbf{K}}A$ and furthermore, by Lemma 3.5, $\pi|Reg_{\mathbf{K}}A$ is a local **K**-biholomorphism near x, which implies that y is in the interior of $\pi(A)$. Contradiction. This implies that $\pi(A)$ is open.

Let $S_1 = \pi(Sing_{\mathbf{K}}A)$. By assumption, $\dim_{\mathbf{R}}(Cl(S_1)) \leqslant 2n - 2$, and since $\pi|A$ is finite-to-one, we also have $\dim_{\mathbf{R}}(\pi^{-1}(Cl(S_1)) \leqslant 2n - 2$.

Let $M = A \cap (\mathbf{K}^n \setminus \pi^{-1}(Cl(S_1)))$. Notice that M is a dense, relatively open subset of $Reg_{\mathbf{K}}A$ and that $\dim_{\mathbf{R}}(A \setminus M) \leqslant 2n - 2$. Moreover, $\pi|M$ is definably proper over its image.

Let $S \subseteq \mathbf{K}^d$ be the set from Lemma 3.6 (3) and let $V = \pi(M) \setminus S$. Namely, every $x \in V$ has precisely m preimages in M.

We choose $\phi_1(x), \ldots, \phi_m(x)$ definable (but not necessarily continuous) maps from V into $\mathbf{K}^{n-d}$, which for every $x \in V$ give all the y's in $\mathbf{K}^{n-d}$ such that $(x, y) \in M$. We denote $(\phi_1, \ldots, \phi_m)$ by Φ.

We use the following fact, which appears in different forms in any basic book on complex analytic or algebraic geometry[1]:

Fact. There is a polynomial map $F(\bar{y}_1, \ldots, \bar{y}_m, \bar{x})$, from $\mathbf{K}^{(n-d)(m+1)}$ into $\mathbf{K}^r$, for some r, such that
(i) F is symmetric under any permutation of $\{\bar{y}_1, \ldots, \bar{y}_m\}$.
(ii) For every $(\bar{b}_1, \ldots, \bar{b}_m, \bar{c}) \in \mathbf{K}^{(n-d)(m+1)}$,

$$(1) \qquad F(\bar{b}_1, \ldots, \bar{b}_m, \bar{c}) = 0 \Leftrightarrow \bigvee_{i=1}^m \bar{c} = \bar{b}_i.$$

Notice that each coordinate function of F can be viewed as a polynomial in $\bar{x}$ whose coefficients, which we denote by $a_i(\bar{y}_1, \ldots, \bar{y}_m)$, are symmetric functions of the $\bar{y}_i$'s.

Consider the map, defined on $V \times \mathbf{K}^{n-d}$,

$$\Psi(\bar{x}', \bar{x}) = F(\phi_1(\bar{x}'), \ldots, \phi_m(x'), \bar{x})$$

The ϕ_i's are **K**-holomorphic on V outside a set S' of **R**-dimension $2d - 1$ (see the proof of Lemma 3.6 (2)). Each coordinate function of Ψ is a polynomial in $\bar{x}$ whose coefficients, which we write as $a_i(\Phi)$,

---

1 One way to obtain F is as follows ([6], p.44 for a similar argument): Choose $\bar{u}_1, \ldots, \bar{u}_{(n-d)m}$ to be a sequence of vectors in $\mathbf{K}^{n-d}$, each $n - d$ of them are linearly independent over **K**, and for any $i = 1, \ldots, (n-d)m$, consider the function $f_i(\bar{Y}, \bar{x}) = \Pi_{j=1}^m \langle \bar{u}_i, (\bar{x} - \bar{y}_j) \rangle$. Now $F = (f_1, \ldots, f_{(n-d)m})$ will do the job.

are symmetric functions of $(\phi_1, \ldots, \phi_m)$. It follows that the $a_i(\Phi)$'s are **K**-holomorphic outside S' and invariant under permutations of the ϕ_i's. We claim that these coefficients are continuous on V and hence **K**-holomorphic .

Indeed, since $\pi|M$ is definably proper over V and a local homeomorphism on M, we can, for every $\bar{x}' \in S'$, re-define the ϕ_i's in a neighborhood of $\bar{x}'$ so they become continuous there, without changing the value that the a_i's take there. Therefore, each $a_i(\Phi)$ is continuous on V.

It follows from the theorem on the removal of singularities (see [14]) that the $\bar{a}_i$'s are **K**-holomorphic on V, and hence Ψ is **K**-holomorphic on $V \times \mathbf{K}^{n-d}$. By our choice of F, for every $(\bar{a}', \bar{a}) \in V \times \mathbf{K}^{n-d}$, $\Psi(\bar{a}', \bar{a}) = 0$, if and only if $\bar{a} = \phi_i(\bar{a}')$ for some $i = 1, \ldots m$ (see the proof of Claim 2.25 in [14] for a similar argument).

We now extend Ψ to $W \times \mathbf{K}^{n-d}$, where $W = \pi(Cl_U(M)) = \pi(A)$. We only need to show that each $a_i(\Phi)$, as functions on V, can be extended to W. Since $\dim_{\mathbf{R}}(W \setminus V) \leqslant 2d - 2$, it is enough to show that they are bounded at all points of W (see Theorem 2.15 in [14]).

Take $\bar{x}' \in W$ then, since $\pi|A$ is definably proper, the functions ϕ_i, $i = 1, \ldots, m$, are bounded in a neighborhood of $\bar{x}'$, and therefore $a_i(\Phi)$ is bounded there as well. It follows that Ψ can be extended **K**-holomorphically to $W \times \mathbf{K}^{n-d}$.

It is left to show that $\Psi^{-1}(0) = A$.

By our choice of Ψ, the set $\Psi^{-1}(0)$ is a **K**-analytic subset of $W \times \mathbf{K}^{n-d}$, containing A. We need to show that the opposite inclusion is true as well.

Take $(\bar{a}', \bar{a}) \in \Psi^{-1}(0) \cap U \subseteq W \times \mathbf{K}^{n-d}$. Because $\bar{a}'$ is in the closure of V there is a curve $\gamma : (0, 1) \to V$, which approaches $\bar{a}'$ as t tends to 0. Consider

$$\Psi(\gamma(t), \bar{a}) = F(\phi_1(\gamma(t)), \ldots, \phi_m(\gamma(t)), \bar{a}).$$

By the continuity of F, as $\gamma(t)$ tends to $\bar{a}'$ the function $\Psi(\gamma(t), \bar{a})$ tends to

$$F(lim_{t \to 0}\phi_1(\gamma(t)), \ldots, lim_{t \to 0}\phi_m(\gamma(t)), \bar{a}).$$

(these limits exist since $\pi|A$ is definably proper over $\bar{a}'$).

But Ψ itself is continuous hence,

$$F(lim_{t \to 0}\phi_1(\gamma(t)), \ldots, lim_{t \to 0}\phi_m(\gamma(t)), \bar{a}) = 0.$$

However, our choice of F now implies that for some $i = 1, \ldots, m$, $\bar{a} = \lim_{t \to 0} \phi_i(\gamma(t))$ and therefore $(\bar{a}', \bar{a}) = lim_{t \to 0}(\gamma(t), \phi_i(\gamma(t)))$.

Since for every t we have $(\gamma(t), \phi_1(t), \ldots, \phi_m(t)) \in M$, it follows that (a', a) is in $Cl(M \cap V \times \mathbf{K}^{n-d})$. But this last set is clearly contained in A, and hence we showed that $\Psi^{-1}(0) = A$. $\square$

Remark 4.8

1. We will later need the following observation regarding the above proof (we stick to the same notation):

If $z = (\bar{x}', \bar{x}) \in M$ and $\bar{x}' \in V$, then $Rank(D\Psi)_z = n - d$, where V is as above and $D\Psi$ is the differential with respect to $\mathbf{K}$.

To see that, first notice that if $\bar{x}' \in V$ then, as in the proof above, we may assume that $\phi_1, \ldots, \phi_m$ are $\mathbf{K}$-holomorphic in a neighborhood of $\bar{x}'$.

Take F as in the footnote of the previous page. We now can show that the matrix $(\partial\Psi_i/\partial x_j), j = 1, \ldots, n - d$, has rank $n - d$ at z over $\mathbf{K}$, as required (similar proofs can be found in most literature on complex analytic geometry).

2. The proof of the last lemma implies that, under the assumptions of the lemma, there is a definable set $S \subseteq \pi(A)$, with $\dim_{\mathbf{R}} S \leqslant 2d - 2$, such that every $y \in \pi(A) \setminus S$ has exactly m preimages in A under π. Consider now an arbitrary $y_0 \in \pi(A)$, and let $z_1, \ldots, z_t \in A$ be its preimages under π. By definable properness, there are pairwise-disjoint neighborhoods $U_1, \ldots, U_t$ of $z_1, \ldots, z_t$, respectively, and a neighborhood $G \subseteq \pi(A)$ of y_0 such that

$$\pi^{-1}(G) \cap A = \bigcup_i U_i \cap A.$$

By applying the last lemma to each $U_i \cap A$ (and possibly shrinking G) we may assume $\pi(U_i \cap A) = G$ for every $i = 1, \ldots, t$. By considering $\pi^{-1}(y) \cap A$ for a generic $y \in G$, it is easy to deduce that $t \leqslant m$. We thus showed that under the assumptions of the last lemma, a generic fiber in A has maximal number of elements.

The following theorem, in the classical setting, is a corollary of Shiffman's Theorem (see [6]). A definable set A is called of *pure dimension* d if the dimension of every nonempty relatively open subset of A is d.

Corollary 4.9 *Let M be a $\mathbf{K}$-manifold, $F \subseteq M$ a definable closed set and A a $\mathbf{K}$-analytic subset of $M \setminus F$ of pure dimension d. If $\dim_{\mathbf{R}} F \leqslant 2d - 2$ then $Cl(A)$ is a finitely $\mathbf{K}$-analytic subset of M.*

Proof. Consider each definably connected component of $Reg_{\mathbf{K}} A$. It is sufficient to show that the closure in M of each such component is $\mathbf{K}$-analytic in M. Let A' be such a component and let A'' be the closure of A' in M. Notice that the set of $\mathbf{K}$-singular points of A'' is contained in $Sing_{\mathbf{K}}(A) \cup F$ and therefore, by Lemma 4.5, its $\mathbf{R}$-dimension is at most $2d - 2 = \dim_{\mathbf{R}} A - 2$.

After partitioning M further, working in charts, and using Theorem 2.14, we may assume that A'' satisfies the assumptions of Theorem 4.7 thus concluding that A'' (and hence also A) is a finitely $\mathbf{K}$-analytic subset of M. $\square$

One advantage of the o-minimal setting over the general one is our ability to handle well the intersection of one $\mathbf{K}$-analytic set with the closure of another. For that we need the following lemma, which is an analogue of Theorem 3.2 from [16].

Corollary 4.10 *Let M be a $\mathbf{K}$-manifold and let $A_1 \subseteq M$ be a definable, locally $\mathbf{K}$-analytic subset of M such that $Reg_{\mathbf{K}} A_1$ is definably connected and $\dim_{\mathbf{K}} A_1 = d$. Assume that $A_2 \subseteq M$ is another definable, locally $\mathbf{K}$-analytic subset of M. Then either $A_1 \subseteq A_2$ or $\dim_{\mathbf{R}}(Cl(A_1) \cap A_2) \leqslant 2d - 2$.*

Proof. It is sufficient to prove the result locally, in a neighborhood of every point in M. Thus we may assume that A_1 and A_2 are $\mathbf{K}$-analytic subset of M. We apply Lemma 4.3 with A_2 playing the role of A and $Reg_{\mathbf{K}} A_1$ playing the role of M in that lemma (notice that $Reg_{\mathbf{K}} A_1$ is dense in A_1). $\square$

We can now prove a generalization of of Theorem 4.9, with the purity assumption omitted. As was pointed out in [16], the result is not true outside the o-minimal setting.

Theorem 4.11 *Let M be a $\mathbf{K}$-manifold and A a definable closed subset of M. Assume that for every open $U \subseteq M$, $\dim_{\mathbf{R}}(Sing_{\mathbf{K}}(A \cap U)) \leqslant \dim_{\mathbf{R}}(A \cap U) - 2$. Then A is a finitely $\mathbf{K}$-analytic subset of M.*

In particular, every $\mathbf{K}$-analytic subset of M is finitely $\mathbf{K}$-analytic.

Proof. This is an analogue of Corollary 4.2, from [16], which is itself based on Theorem 4.1 there. The proof of 4.1 goes through in our setting, with Theorem 3.2 there replaced in the current paper by Corollary 4.10. The finiteness result is obtained since in the very last step, where originally

Shiffman's Theorem was used, we are now using Corollary 4.9 to obtain a finitely **K**-analytic set. □

Actually, the proof of Theorem 4.1 in [16] shows in particular that the closure in M of every definably connected component of $Reg_{\mathbf{K}} A$ is itself a **K**-analytic subset of M. Thus, just as in the classical case, we have:

Lemma 4.12 *If A is a **K**-analytic subset of M then its irreducible components are exactly the closures of the definably connected components of $Reg_{\mathbf{K}} A$.*

The following strong version of the Remmert-Stein Theorem is a direct analogue of Theorem 4.4 from [16]. The proof there works here as well, using Lemma 3.7 and Theorem 4.11.

Theorem 4.13 *Let M be a **K**-manifold and $E \subseteq M$ a **K**-analytic subset of M (of arbitrary dimension). If A is a **K**-analytic subset of $M \setminus E$ then its closure in M is a **K**-analytic subset of M.*

To sum-up the main results so far, we have:

Corollary 4.14 *If M is a **K**-manifold and A a closed definable subset of M then the following are equivalent:*
(i) For every open $W \in \mathbf{K}^n$, $\dim_{\mathbf{R}}(Sing_{\mathbf{K}}(A \cap W)) \leqslant \dim_{\mathbf{R}}(A \cap W) - 2$.
*(ii) A is **K**-analytic subset of M.*
*(iii) A is finitely **K**-analytic subset of M.*
  *Moreover, in (iii) the open sets and **K**-holomorphic functions which carve A in each of them can be chosen be definable in the structure $\langle R, <, +, \cdot, A \rangle$.*

Proof. The proof of (i) ⇒ (iii) follows from Theorem 4.11. (ii) ⇒ (i) is Lemma 4.5. We only need to note why the definability clause is true. This follows from 4.7. □

Note that both Clause (i) and Clause (iii) guarantee that **K**-analytic sets remain **K**-analytic in every elementary extension.

5 K-analytic subsets of $\mathbf{K}^n$; Chow's Theorem

We now present an o-minimal proof of a strong version of Chow's Theorem. Note that since every analytic subset of $\mathbb{P}(\mathbb{C})$ is definable in the o-minimal structure $\mathbb{R}_{an}$, the classical Chow's Theorem is an immediate corollary.

Theorem 5.1 *Let A be a definable $\mathbf{K}$-analytic subset of $\mathbf{K}^n$. Then A is an algebraic set over $\mathbf{K}$.*

Proof. We may assume that A is irreducible of $\mathbf{K}$-dimension d. We may also assume that A is definable over $\emptyset$. Consider a generic (over $\emptyset$) orthogonal projection π of $\mathbf{K}^n$ onto a d-dimensional $\mathbf{K}$-linear subspace L.

Since A is a closed subset of $\mathbf{K}^n$, it follows from Lemma 2.13 that $\pi|A$ is definably proper over the whole of L, and in particular, $\pi(A)$ is a closed subset of L. By Lemma 4.5 and Lemma 4.7, (this last lemma applied to $M = Reg_{\mathbf{K}}A$), $\pi(A)$ is also open and therefore $\pi(A) = L$.

To simplify notation we will assume now that $L = \mathbf{K}^d$ and π is the projection onto the first d coordinates. By Lemma 4.7, there is a $\mathbf{K}$-holomorphic map $\Psi : \mathbf{K}^n \to \mathbf{K}^m$ such that $A = \Psi^{-1}(0)$. By Theorem 2.17 of [14], every $\mathbf{K}$-holomorphic function on $\mathbf{K}^n$ is a polynomial, therefore Ψ is a polynomial map, and A must be algebraic. $\qquad \square$

The following corollary is a very important ingredient in the development of the theory of analytic sets.

Corollary 5.2 *Every bounded $\mathbf{K}$-analytic subset of $\mathbf{K}^n$ is finite.*

Proof. One can prove the result directly but instead, we may use our version of Chow's theorem and then transfer this fact for algebraic sets from the complex and real fields to $\mathbf{K}^n$. $\qquad \square$

6 The set of singular points

As in the classical case, we are now ready to prove a strong version of Lemma 4.5.

Theorem 6.1 *Let M be a $\mathbf{K}$-manifold and let A be a $\mathbf{K}$-analytic subset of M. Then $Sing_{\mathbf{K}}A$ is a $\mathbf{K}$-analytic subset of M.*

Proof. Let $A = \bigcup A_i$ be the decomposition of A into its irreducible components. Notice that $z \in Sing_{\mathbf{K}}A$ if and only if $z \in A_i \cap A_j$ for $i \neq j$, or $z \in Sing_{\mathbf{K}}A_i$ for some i. Therefore we may assume that A is irreducible of dimension d. Also, we may assume that $M = U$ a definable open subset of $\mathbf{K}^n$.

Since $Sing_{\mathbf{K}} A$ is closed in A, it is sufficient to prove that it is locally $\mathbf{K}$-analytic. Fix $a \in Sing_{\mathbf{K}} A$ and let $L_1, \ldots, L_k$ be a sequence of independent and generic over a, d-dimensional $\mathbf{K}$-linear subspaces of $\mathbf{K}^n$ and let $U_1, \ldots, U_k$ be the corresponding open sets as in Theorem 2.14. Because the L_i's are generic over a, the point a belongs to the intersection of the U_i's. By Lemma 4.7 for every $i = 1, \ldots, k$ there is a $\mathbf{K}$-holomorphic map $\Psi_i : U_i \to \mathbf{K}^{d_i}$ whose zero set is $A \cap U_i$. Moreover, each L_i has a definable open subset V_i such that the following properties hold, for each $i = 1, \ldots, k$:

(1) $\pi_i^{-1}(V_i) \subseteq Reg_{\mathbf{K}} A$.
(2) For all $z \in A \cap \pi_i^{-1}(V_i)$, $Rank_{\mathbf{K}}(D\Psi_i)_z = n - d$.
(3) For every $z \in Reg_{\mathbf{K}} A$ there is an $i = 1, \ldots, k$ such that $z \in \pi^{-1}(V_i)$.

Indeed, the existence of V_i satisfying (1) and (2) above can be read-off from the proof of Lemma 4.7 and Remark 4.8 which follows it (V_i is the set of points $z' \in L_i$ such that A is a $\mathbf{K}$-submanifold at every point of the set $X = \pi^{-1}(z')$ and $\pi_i|A$ is a local $\mathbf{K}$-biholomorphism at every point of X). As for (3), as was pointed out in the proof of the lemma, for each $z \in Reg_{\mathbf{K}} A$ there is an i such that π_i is generic over z. Let H_i be the orthogonal complement to L_i. Then, by Lemma 2.12, $z + H_i$ intersects A transversally and in particular, every point of intersection is in $Reg_{\mathbf{K}} A$. It follows that $z \in \pi^{-1}(V_i)$.

For every $i = 1, \ldots, k$, let

$$Z_i = \{z \in U_i : Rank_{\mathbf{K}}(D\Psi_i)_z < n - d\}.$$

Claim $(\bigcap_{i=1}^{k} U_i) \cap Sing_{\mathbf{K}} A \doteq \bigcap_{i=1}^{k} Z_I$.

Proof Assume that $z \in (\bigcap_i U_i) \cap Reg_{\mathbf{K}} A$. Then, by (3) above, there is an i such that $z \in V_i$. By (2), $Rank_{\mathbf{K}}(D\Psi_i)_z = n - d$, and so $z \notin Z_i$.

For the opposite inclusion, assume that $Rank_{\mathbf{K}}(D\Psi_i)_z = n - d$ for some $i = 1, \ldots, n$. Then, by the Implicit Functions Theorem, there is a d-dimensional $\mathbf{K}$-submanifold M containing z which is contained in A. Since $\dim_{\mathbf{K}} A = d$, it easily follows that $M = A$ in some neighborhood of z, and thus $z \in Reg_{\mathbf{K}} A$. $\square$

7 Dimension, rank and Remmert's Theorem

Once we know that every bounded analytic subset of $\mathbf{K}^n$ is finite (see Corollary 5.2), the following three results on dimension are standard. We include their proofs for the sake of completeness.

Theorem 7.1 *Let $A \subseteq U \subseteq \mathbf{K}^n$ be an $\mathbf{K}$-analytic subset of a definable open set, $a \in A$. Assume that H is a p-dimensional affine $\mathbf{K}$-subspace of $\mathbf{K}^n$ such that a is an isolated point of $A \cap H$, and let $H^\perp$ be the orthogonal complement of H. Then*
(i) the projection π of A onto $H^\perp$ is at most finite-to-one, in some neighborhood of a.
(ii) $\dim_a A \leqslant n - p$.

Proof. (i) By continuity of π, there is a neighborhood $V \subseteq \mathbf{K}^n$ of a such that for every $y \in \pi(V)$, $\pi^{-1}(y) \cap A \cap V$ is closed and bounded near a. Since each such fiber is $\mathbf{K}$-analytic in V it is also $\mathbf{K}$-analytic in $\mathbf{K}^n$ and so, by Corollary 5.2, it is either empty or finite. (ii) follows immediately from (i). □

Theorem 7.2 *Assume that $A \subseteq U \subseteq \mathbf{K}^n$ is a $\mathbf{K}$-analytic set and $f : U \to \mathbf{K}^d$ is $\mathbf{K}$-holomorphic . Then the map $x \mapsto \dim_{\mathbf{K}}(A \cap f^{-1}(f(x)))_x$ is upper semicontinuous on A.*

Proof. We need to show that for every natural number $p \geqslant 0$, the set $A_0 = \{x \in A : \dim_{\mathbf{K}}(A \cap f^{-1}(f(x)))_x \geqslant p\}$ is closed in U.

Take $x_0 \in Cl(A_0)$, and denote $A \cap f^{-1}(f(x_0))$ by B. Assume that $\dim_{\mathbf{K}} B_{x_0} = q$ and let H be a generic $\mathbf{K}$-space of K-dimension $n - q$. By Lemma 2.10, x_0 is an isolated point in $B \cap x_0 + H$. Let $V \subseteq \mathbf{K}^n$ be a small ball around x_0 such that $V \cap B \cap x_0 + H = \{x_0\}$. By the continuity of f and since A is closed in U, there is a neighborhood W of x_0 such that for every $x \in A \cap W$, the set $A \cap f^{-1}(f(x)) \cap x + H$ is closed in $\mathbf{K}^n$ and contained in V. By Theorem 7.1, it must be also finite. It follows that every $x \in A \cap W$, is an isolated point of $A \cap f^{-1}(f(x)) \cap x + H$. By 7.1, $\dim_{\mathbf{K}}(A \cap f^{-1}(f(x)))_x \leqslant q$ for all $x \in W$. Since $x_0 \in Cl(A_0)$, it follows that $p \leqslant q$. □

Theorem 7.3 *(The Dimension Theorem) Let M be an n-dimensional $\mathbf{K}$-manifold, X, Y definable, irreducible $\mathbf{K}$-analytic subsets of M. Then every irreducible component of $X \cap Y$ has $\mathbf{K}$-dimension not less than $\dim_{\mathbf{K}} X + \dim_{\mathbf{K}} Y - n$.*

Proof. We may clearly assume that $M = \mathbf{K}^n$ and A is 0-definable. Since $X \cap Y$ is $\mathbf{K}$-biholomorphic with $X \times Y \cap \Delta$, where Δ is the diagonal in $\mathbf{K}^n \times \mathbf{K}^n$ it is sufficient to prove the following:

Given an irreducible d-dimensional $\mathbf{K}$-analytic set $A \subseteq \mathbf{K}^n$ and given $a \in A$, if H_0 is an $n - 1$-dimensional $\mathbf{K}$-linear subspace of $\mathbf{K}^n$, then

$$\dim_{\mathbf{K}}(A \cap a + H_0)_a \geqslant d - 1.$$

Assume to the contrary that $\dim_{\mathbf{K}}(A \cap a + H_0)_a \leqslant d - 2$. By 2.10, there is a $\mathbf{K}$-linear space $H \subseteq H_0$, with $\dim_{\mathbf{K}}(H) = n - 1 - (d - 2)$ such that a is an isolated point of $A \cap a + H$. We now apply Theorem 7.1 to A, H and $\mathbf{K}^n$, and conclude that $\dim_{\mathbf{K}}(A)_a \leqslant n - (n + 1 - d) = d - 1$. Contradiction. $\qquad \square$

We can now prove the following version of Remmert's proper mapping theorem.

Theorem 7.4 *Let $f : M \to N$ be a $\mathbf{K}$-holomorphic map between definable $\mathbf{K}$-manifolds. Let $A \subseteq M$ be a $\mathbf{K}$-analytic subset. Assume that $f(A)$ is a closed subset of N. Then $f(A)$ is a $\mathbf{K}$-analytic subset of N.*

Proof. The proof is a much simplified variant of the proof of Theorem 6.1 in [16]. Namely, instead of A and $f(A)$ being only locally definable we now have both of these sets definable. The rest of the argument is identical, using Theorem 4.11. The upper semi-continuity of the function $x \mapsto \dim_{\mathbf{K}} f^{-1}(f(x))_x$ is given by Theorem 7.2. $\qquad \square$

Note that if $f : M \to N$ is a $\mathbf{K}$-holomorphic map which is definably proper over N then for every $\mathbf{K}$-analytic $A \subseteq M$, $f(A)$ is closed in N and therefore, by the above, it is $\mathbf{K}$-analytic in N. This is more or less the precise content of Remmert's original theorem.

Example 7.5 Let A be the following analytic subset of $\mathbb{C}^3$:

$$A = \mathbb{C} \times \{(0,0)\} \cup \{(0,0)\} \times \mathbb{C} \cup \bigcup_{n=1}^{\infty} \{(z, 1/nz, n) : z \in \mathbb{C}\}.$$

Let π be the projection of A on the first two coordinates. Then

$$\pi(A) = \mathbb{C} \times \{0\} \cup \bigcup_{i=1}^{\infty} \{(z, 1/nz) : z \in \mathbb{C}\},$$

which is a closed set but not analytic.

Although A above has infinitely many irreducible components we expect that a similar example can be obtained with A irreducible.

8 K-manifolds as Zariski structures

The initial motivation to consider, model theoretically, compact complex manifolds and their analytic subsets is due to the following theorem:

Theorem 8.1 [21] *Consider the structure whose universe is a compact complex manifold, and whose atomic relations are all the complex analytic subsets of M and of its cartesian products. Then the structure eliminates quantifiers, it is stable of finite Morley Rank and moreover, it is a Zariski structure (with the closed sets taken as the **K**-analytic ones).*

We are almost ready to prove a generalization of that theorem. But first we need one more definability result.

Lemma 8.2 *Let $f : M \to N$ be a **K**-holomorphic map between **K**-manifolds and let A be a **K**-analytic subset of M. If $f|A$ is definably proper over N then for every $k \in \mathbb{N}$, the set $B = \{y \in N : \dim_{\mathbf{K}} f^{-1}(y) \geqslant k\}$ is a **K**-analytic subset of N.*

Proof. Consider the set

$$A(k) = \{z \in A : \dim_z f^{-1} f(z) \geqslant k\}.$$

The upper semi-continuity of dimension implies that this set is closed, but moreover it is a **K**-analytic subset of M. The argument for the latter statement is described in details in Whitney's book, [19], in the proof of Theorem 9F on page 240 there. (The two theorems used in this argument, from Chapters 2 and 4, are easily seen to hold in our setting as well. Notice that we only need the analyticity result there).

Since $f|A$ is definably proper, the image of $A(k)$ under f is closed in N and therefore, by Theorem 7.4, it is **K**-analytic in N. This is precisely the set B. $\qquad\square$

Together with the dimension Theorem, Remmert mapping theorem, and the decomposition of **K**-analytic sets into finitely many irreducible components we obtain, exactly as in theorem 8.1:

Theorem 8.3 *Let M be a definably compact $\mathbf{K}$-manifold, equipped with all $\mathbf{K}$-analytic subsets of cartesian products of M. Then M eliminates quantifiers, it is stable of finite Morley rank and a (complete) Zariski structure (with the closed sets taken as the $\mathbf{K}$-analytic ones).*

Remarks

1. Notice that not every $\mathbf{K}$-manifold in an o-minimal structure, when equipped with all its $\mathbf{K}$-analytic subsets, is stable. Indeed, work in the field of real numbers and take for example M to be the open unit disc in $\mathbf{K}$, equipped with all semialgebraic analytic subsets of its cartesian products. One of the analytic subsets of $M \times M$ is the graph of the function $z \mapsto \frac{1}{2}z$ but its image in M is an open disc of radius $1/2$. This is easily seen to contradict stability.

2. It is not necessary for a $\mathbf{K}$-manifold to be compact, or definably compact, in order for the induced structure to be stable. E.g., by Theorem 5.1, if we take $M = \mathbf{K}^n$, (in any o-minimal structure) then the structure induced by all the $\mathbf{K}$-analytic sets is just that of an algebraically closed field.

Question Let M be a $\mathbf{K}$-manifold and assume that M, when equipped with all $\mathbf{K}$-analytic subsets of its cartesian products, is stable. Is it possible to definably "compactify" M? I.e., is M a Zariski open subset of a definably compact $\mathbf{K}$-manifold (either in the same structure or in some o-minimal expansion)?

In [14] we discussed several examples of definably compact $\mathbf{K}$-manifolds in this nonstandard o-minimal setting and showed how one obtains in this manner new objects, which do not arise in nonstandard models of the theory of compact complex spaces. We now return to one of these examples.

8.1 Locally modular elliptic curves

In [14] and in [15] we showed how to view the family of all 1-dimensional complex tori as definable in the structure $\langle \mathbb{R}, +, \cdot \rangle$ with the upper half plane $\mathbb{H} \subseteq \mathbf{K}$ as the parameter set. We briefly recall this definition here:

Given τ in $\mathbb{H}$, we consider the half-open parallelogram E_τ in $\mathbb{H}$ whose sides are determined by $1, \tau \in \mathbf{K}$. By covering $Cl(E_\tau)$ with finitely many definable open sets we can endow E_τ with a $\mathbf{K}$-manifold structure which in fact "glues" the two opposite sides of the parallelogram.

We thus obtain in every o-minimal expansion of a real closed field a definably compact one dimensional **K**-manifold, call it $\mathcal{E}_\tau$, for every $\tau \in \mathbb{H}(\mathbf{K})$. Moreover, the natural group structure on $\mathcal{E}_\tau$ is a definable **K**-holomorphic map, making $\mathcal{E}_\tau$ into a **K**-group. Several basic results on such **K**-groups are proved in [15], section 5.1.

Assume now that our structure is an o-minimal expansion of a structure which is itself elementarily equivalent to $\mathbb{R}_{an,exp}$. In [15] we gave a characterization of all those $\tau \in \mathbb{H}(\mathbf{K})$ such that $\mathcal{E}_\tau$ is definably **K**-biholomorphic with a nonsingular algebraic projective cubic curve. Let us call $\mathcal{E}_\tau$ a *nonstandard elliptic curve* if it is not such a curve. For example, if the real part of τ, as an element of **K**, is infinitely large, then $\mathcal{E}_\tau$ is nonstandard. Or, if the imaginary part of τ is infinitesimally small (and positive) and the real part is infinitesimally close to an irrational number then $\mathcal{E}_\tau$ is nonstandard.

Consider now $\mathcal{E}_\tau$ with the induced **K**-analytic structure, as above. Since it is definably compact the structure we get is a Zariski structure of finite Morley Rank. Moreover, since it is definably connected of **K**-dimension one, it is also strongly minimal. The proof of the following theorem will appear elsewhere.

Theorem 8.4 *Let $\mathcal{E}_\tau$ be a nonstandard elliptic curve as above. Then, when equipped with all **K**-analytic subsets of the cartesian products, $\mathcal{E}_\tau$ is a locally modular, strongly minimal structure.*

9 Meromorphic maps

Let M, N be **K**-manifolds, U a definable open subset of M whose complement is a **K**-analytic subset of M. We say that a **K**-holomorphic map $f : U \to N$ is **K**-*meromorphic* if the closure of its graph in $M \times N$ is a **K**-analytic subset of $M \times N$. (Although we will not use it here, the above definition generalizes Definition 2.29 from [14] of a *definably meromorphic function* from a **K**-manifold M into **K**. Namely, f is a **K**-meromorphic map, in our sense, from M into **K** if and only if it can be written locally, at every point of M, as the quotient of two **K**-holomorphic functions).

As the following theorem shows, the requirement about the closure of the graph of f in the above definition is obtained for free in the o-minimal setting.

Corollary 9.1 *Let* M, N *be* **K**-*manifolds,* $S \subseteq M$ *an irreducible* **K**-*analytic subset of* M, *and assume that* L *is a closed definable subset of* S *which contains* $\mathrm{Sing}_{\mathbf{K}} S$. *If* $f : S \setminus L \to N$ *is a* **K**-*holomorphic map and* $\dim_{\mathbf{R}} L \leqslant \dim_{\mathbf{R}} S - 2$ *then the closure of the graph of* f *in* $M \times N$ *is a* **K**-*analytic subset of* $M \times N$.

Proof. This is an analogue of Theorem 6.4 from [16]. With the obvious adjustments the proof goes through almost in full in our setting. One exception however is Proposition 6.5 from that paper which we use in the proof. There we originally quoted a result of Kurdyka and Parusinski, but it is not difficult to see that a corresponding result can be proved in our setting as well. $\qquad\square$

10 Campana-Fujiki

One of the applications of the theory developed thus far, in the classical case, is a generalization of a theorem by Campana and Fujiki (see [5], [9]) which is discussed in [16] (see Theorem 9.2 there). The theorem of Campana and Fujiki has attracted the attention of people in model theory when it was noticed that in some cases one could replace the "Zariski structure machinery" with the geometric tool provided by the theorem, in order to establish connections between certain definable objects and algebraic varieties (see [18], [12], [17]). In [16] we gave a slightly different proof of the theorem and generalized the result from compact complex manifolds to arbitrary definable complex manifolds. Here we will just formulate the generalization to the nonstandard setting and refer to the proof in [16].

We first need the definition of a **K**-holomorphic -vector-bundle (we did not choose the most general one but the one sufficient for our purposes):

Definition 10.1 Let M be a **K**-manifold, *a* **K**-*holomorphic vector-bundle over* M, *of dimension* d consists of a **K**-manifold X, a **K**-holomorphic map $\rho : X \to M$, such that:

There is a finite cover of M by definable open sets $M = \bigcup_{i=1}^{k} V_i$, and for each $i = 1, \ldots, k$ there is a **K**-biholomorphism $\phi_i : V_i \times \mathbf{K}^d \to \rho^{-1}(V_i)$ such that $\rho \phi_i(x, y) = x$. Furthermore, if $x \in V_i \cap V_j \neq \emptyset$ then $\phi_i^{-1} \phi_j$ induces a **K**-linear automorphism $g_{i,j}(x)$ of the vector space $\mathbf{K}^d$, in the fiber above x. The map $x \mapsto g_{i,j}(x)$ is **K**-holomorphic from $V_i \cap V_j$ into $GL(d, \mathbf{K})$.

In the theorem below, we use the term *a Zariski open subset* of a **K**-analytic set S, to mean the complement of another **K**-analytic set in S. For $S \subseteq M \times N$, we let π_M denote the projection onto M.

Theorem 10.2 *Assume that N, M are* **K**- *manifolds, and S is an irreducible* **K**-*analytic subset of $N \times M$. Then there is a* **K**-*holomorphic vector-bundle* $\pi : V \to M$, *a* **K**-*meromorphic map* $\sigma : S \to \mathbb{P}(V)$, *and a Zariski open subset S^0 of S such that for all $(b, a), (b', a) \in S^0$, $\sigma(b, a) = \sigma(b', a)$ if and only if $S_b = S_{b'}$ near a, and the following diagram of* **K**-*meromorphic maps is commutative (we view π as a* **K**-*meromorphic map from the projectivization* $P(V)$ *of V into M)*

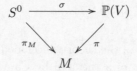

11 A finite version of the coherence theorem

The coherence of analytic sheaves, due to Cartan and Oka, is one of the most important results in the theory of complex analytic spaces. Our goal in this section is to formulate the essential ingredients of the theorem in our setting, and show how o-minimality yields here once again a strong finiteness result.

Notation Let M be a **K**-manifold of dimension n. For $U \subseteq M$ a definable open set, we let $\mathcal{O}(U)$ denote the ring of **K**-holomorphic functions on U.

For $p \in M$, we will denote by $\mathcal{O}_p$ the ring of germs at p of **K**-holomorphic functions. When $M = \mathbf{K}^n$, we sometimes write $\mathcal{O}_{n,p}$. If A is a **K**-analytic subset of M, and $p \in M$ then we denote by $\mathcal{I}(A)_p$ the ideal in $\mathcal{O}_p$ of all germs at p which vanish on A in some neighborhood of p. For a definable open $V \subseteq M$, we let $\mathcal{I}_V(A)$ be the set of all **K**-holomorphic functions on V which vanish on $A \cap V$; this is a module over $\mathcal{O}(V)$. When A is a **K**-analytic subset of V we just write $\mathcal{I}(A)$ for $\mathcal{I}(A)_V$.

A function $f : A \to \mathbf{K}$ is called **K**-holomorphic if there is an open definable set V, $A \subseteq V \subseteq U$, and a **K**-holomorphic function on V which extends f. For p in A, we denote by $\mathcal{O}(A)_p$ the ring of of germs at p of

K-holomorphic functions on $A \cap W$ for some open neighborhood W of p. The ring $\mathcal{O}(A)_p$ can be identified with the quotient ring $\mathcal{O}_p/\mathcal{I}(A)_p$.

Our formulation of the Coherence Theorem is as follows:

Theorem 11.1 *Let M be a **K**-manifold and $A \subseteq M$ a **K**-analytic subset of M. Then there are finitely many open sets $V_1, \ldots, V_k$ whose union covers M and for each $i = 1 \ldots, k$ there are finitely many **K**-holomorphic functions $f_{i,1}, \ldots, f_{i,m_i}$ in $\mathcal{I}_{V_i}(A)$, such that for every $p \in V_i$ the functions $f_{i,1}, \ldots, f_{i,m_i}$ generate the ideal $\mathcal{I}(A)_p$ in $\mathcal{O}_p$.*

Moreover, the V_i's and the $f_{i,j}$'s are all definable over the same parameters defining M and A.

Remarks

1. The fact that at each point $p \in A$ the ideal $\mathcal{I}(A)_p$ is finitely generated was already established in [14]. The content of the Coherence Theorem as well as Theorem 11.1 is of course much stronger than that.

2. If we eliminate in the theorem the requirement for finitely many V_i's then what we have is the coherence of a certain sheaf of **K**-holomorphic functions, just as in the classical case.

3. If M is a compact complex manifold then the theorem above follows immediately from the classical Coherence Theorem, together with the compactness of M. Once again, o-minimality plays a similar role to compactness and yields a finiteness result even when the M is not compact (or even definably compact).

4. Because of the "moreover" clause, it is sufficient to prove the theorem in a sufficiently saturated elementary extension of the original structure, and the same V_i's and $f_{i,j}$'s will work for all elementarily equivalent structures. Indeed, assume that we already proved the theorem for a saturated structure, with the V_i's and the $f_{i,j}$'s all $\emptyset$-definable, and fix V_i. It follows that for every definable family of functions $\{h_{a,p} : p \in A \cap V_i, a \in S\}$, where each $h_{a,p}$ is in $\mathcal{I}(A)_p$, there exist m_i definable families

$$\{\{g_{a,p,j} : p \in A \cap V_i, a \in S\} : i = 1, \ldots, m_i\},$$

each $g_{a,p,j}$ in $\mathcal{O}_p$, such that for all $a \in S$ and $p \in A \cap V_i$,

$$h_{a,p} = \Sigma_{j=1}^{m_i} g_{a,p,j} f_{i,j},$$

in $\mathcal{O}_p$. When quantifying over a and p this last equality carries over to any elementarily equivalent structure.

The proof of Theorem 11.1 follows closely ideas from Sections 8,9 in Chapter 8 of [19]. The only novelty is the fact that we are trying to obtain a finite covering of M. The proof requires the following two subtheorems.

We first state the two subtheorems, and before proving them we will show how indeed the two imply Theorem 11.1.

Theorem 11.2 *Assume that $U \subseteq \mathbf{K}^n$ is a definable open set and $A \subseteq U$ an irreducible $\mathbf{K}$-analytic subset of U of dimension d. Assume also:*

(i) The projection π of A on the first d coordinates is definably proper over its image, and $\pi(A)$ is open in $\mathbf{K}^d$.

(ii) There is a definable set $S \subseteq \mathbf{K}^d$, of $\mathbf{R}$-dimension $\leqslant 2d - 2$ and a natural number m, such that $\pi|A$ is m-to-1 outside the set $A \cap \pi^{-1}(S)$, π is a local homeomorphism outside of the set $\pi^{-1}(S)$, and $A \setminus \pi^{-1}(S)$ is dense in A.

(iii) The coordinate function $z \mapsto z_{d+1}$ is injective on $A \cap \pi^{-1}(x')$ for every $\mathbf{R}$-generic $x' \in \pi(A)$. Namely, for all $z, w \in \pi^{-1}(x')$, if $z_{d+1} = w_{d+1}$ then $z = w$.

Then, there is a definable open set $U' \subseteq U$ containing A, a natural number s and $\mathbf{K}$-holomorphic functions $G_1, \ldots G_r, D : U' \to \mathbf{K}$, such that for every $p \in A$ and $f \in \mathcal{O}_p$, if we let $g_1, \ldots, g_r, \delta$ be the germs at p of $G_1, \ldots, G_r, D$, respectively, then:

$$f \in \mathcal{I}(A)_p \Leftrightarrow \exists f_1 \ldots, f_r \in \mathcal{O}_p \ (\delta^s f = f_1 g_1 + \cdots + f_r g_r).$$

For the next theorem we need one more definition. Let $p \in \mathbf{K}^n$ and $g_1, \ldots, g_t$ germs at p of $\mathbf{K}$-holomorphic maps into $\mathbf{K}^N$. The *module of relations associated to* $g_1, \ldots, g_t$ at p is the module over $\mathcal{O}_p$ defined as

$$R_p(g_1, \ldots, g_t) = \{(f_1, \ldots, f_t) \in \mathcal{O}_p^t : f_1 g_1 + \cdots + f_t g_t = 0\}.$$

Theorem 11.3 *Assume that A is a $\mathbf{K}$-analytic subset of $U \subseteq \mathbf{K}^n$ and assume that $G_1, \ldots, G_t$ are $\mathbf{K}$-holomorphic maps from A into $\mathbf{K}^N$. Then we can write A as a union of finitely many relatively open sets $A_1, \ldots, A_m$ such that on each A_i the following holds:*

For some $k = k(i)$, there are finitely many tuples of $\mathbf{K}$-holomorphic functions on A_i, $\{(H_{j,1}, \ldots, H_{j,t}) : j = 1, \ldots, k\}$, with the property that for every $p \in A_i$, the module $R_p(g_1, \ldots, g_t)$ equals its submodule generated by $\{(h_{j,1}, \ldots, h_{j,t}) : j = 1, \ldots, k\}$ over $\mathcal{O}_p$ (where g_i and $h_{i,j}$ are the germs of G_i and $H_{i,j}$ at p, respectively).

Let us see first how the two subtheorems above, taken together, imply Theorem 11.1:

Our intention, of course, is to use Theorem 2.14, and Theorem 3.6. For that, we need to treat each of the irreducible components of A separately. The missing ingredient is the following:

Claim Assume that $A = A_1 \cup \cdots \cup A_r$, where each of the A_i's is **K**-analytic in U. If Theorem 11.1 is true for each of the A_i's then it holds for A as well.

To prove the claim, we basically repeat the argument of Theorem 8C, p.279 from [19]: We may assume that $r = 2$.

By our assumptions on each of the A_i's we may assume, after replacing U by an open subset, that there are **K**-holomorphic Φ_i, $i = 1, \ldots, s$ in $\mathcal{I}(A_1)$ which at every point $p_1 \in A_1$ generate $\mathcal{I}(A_1)_{p_1}$ in $\mathcal{O}_{p_1}$. Similarly, there are Ψ_j, $j = 1, \ldots, t$ in $I(A_2)$.

Consider the relation module associated to the tuple

$$(\Phi, \Psi) = (\Phi_1, \ldots, \Phi_s, \Psi_1, \ldots, \Psi_t),$$

at every point $p \in U$. By Theorem 11.3, we may assume, possibly after replacing U by a smaller open subset, that there are tuples $(\xi^i, \eta^i) = (\xi_1^i, \ldots, \xi_s^i, \eta_1^i, \ldots, \eta_t^i)$, $i = 1, \ldots, r$, which, at every point $p \in U$, generate the relation module $R_p(\phi, \psi)$. Let

$$\Theta^i = \Sigma_{j=1}^s \xi_j^i \phi_j, \quad i = 1, \ldots, r.$$

As is shown in [19], the germs of these functions generate $\mathcal{I}(A_1 \cup A_2)_p$ in $\mathcal{O}_p$ at every $p \in A_1 \cap A_2$. All other points of $A_1 \cup A_2$ are handled by Φ and Ψ separately. We thus finished proving the claim.

Using the claim, we may assume that A is of pure **K**-dimension d. As in the proof or Theorem 2.14, we consider a sequence of generic independent **K**-subspaces of $\mathbf{K}^n$ of dimension d, $L_1, \ldots, L_{2n+1}$. For each i choose a generic orthogonal basis $B = \{v_1, \ldots, v_n\}$ for $\mathbf{K}^n$ such that $v_1, \ldots, v_d$ form a basis for L_i. By Fact 2.8, every m-subset of B generate a generic m-dimensional subspace. In particular, if $m = d + 1$ and L is such a $d + 1$-dimensional subspace then for every $x \in A$ which is generic over the parameters defining the L_i's, the space L is generic over x. Hence, by Theorem 2.10 (ii), $x + L^\perp$ intersects A at exactly x.

We may now replace U by each one of the U_i's as in the proof of Theorem 2.14 and after performing a suitable linear transformation, we may assume that the orthogonal projection onto L_i is actually the projection of $\mathbf{K}^d$ onto the first d coordinates. Therefore, assumptions (i), (ii) and (iii), from Theorem 11.2 now hold, because of Theorems 2.14, 3.6 and the above discussion. We may therefore apply Theorem 11.2 and obtain $\mathbf{K}$-holomorphic functions on $A, G_1, \ldots, G_r$ and D which satisfy the statement in that theorem.

Take at each point in A the relation module associated to the tuple $(G_1, \ldots, G_r, D^s)$, as given by 11.2, with $N = 1$. Next, replace A by an A_i as given by Theorem 11.3 and let $\{(H_{j,1}, \ldots, H_{j,r+1}) : j = 1, \ldots, k\}$ be as in that theorem. We claim that $H_{1,r+1}, \ldots, H_{k,r+1}$ generate, at each $p \in A_i$, the ideal $\mathcal{I}(A_i)_p$ in $\mathcal{O}_p$.

If $f \in \mathcal{I}(A_i)_p$ then, by Theorem 11.2, there exist $f_1, \ldots, f_r \in \mathcal{O}_p$ such that $F = (f_1, \ldots, f_r, f) \in R_p(g_1, \ldots, g_r, \delta^s)$. By our assumptions, F is in the submodule of $\mathcal{O}_p^{r+1}$ generated by the germs at p of $\{(H_{j,1}, \ldots, H_{j,r+1}) : j = 1, \ldots, k\}$ over $\mathcal{O}_p$ and in particular, f is in the ideal generated by the germs at p of $\{H_{j,r+1} : j = 1, \ldots, k\}$. As for the opposite inclusion, it follows from Theorem 11.3 that this last ideal is contained in $\mathcal{I}(A_i)_p$.

We therefore showed how Theorem 11.1 follows from Theorems 11.2 and 11.3.

Proof of Theorem 11.2

We follow the proof of Theorem 9B on p. 280 of [19].

For $\ell \leqslant n$, we will denote by $\mathbf{K}^\ell$ the subspace $z_{\ell+1} = \ldots = z_n = 0$, and by $\pi_\ell : \mathbf{K}^n \to \mathbf{K}^\ell$ the natural projection. For $p \in \mathbf{K}^n$, $\ell \leqslant n$ and $p' = \pi_\ell(p)$, the ring $\mathcal{O}_{\ell,p'}$ has a natural embedding into $\mathcal{O}_{n,p}$, and hence we consider it as a subring of $\mathcal{O}_{n,p}$ and denote it also by $\mathcal{O}_{\ell,p}$. In other words, $\mathcal{O}_{\ell,p}$ is the subring of $\mathcal{O}_{n,p}$ consisting of functions depending on variables $z_1, \ldots, z_\ell$ only. We denote by $\mathcal{O}(A)_{\ell,p}$ the ring $\mathcal{O}(A)_p \cap \mathcal{O}_{\ell,p}$.

By working in charts, we may assume that $M = U$ is a definable open subset of $\mathbf{K}^n$ and $A \subseteq U$ is $\mathbf{K}$-analytic in U. For $p \in \mathbf{K}^n$, we denote by p' the projection $\pi(p)$. We also denote by z the tuple of variables $(z_1, \ldots, z_n)$ and by z' the tuple $(z_1, \ldots, z_d)$.

Claim 11.4 *Let f be a $\mathbf{K}$-holomorphic function on A. Then there is a monic polynomial $P_f(z', u) \in \mathcal{O}(V)[u]$ of degree m such that $P_f(z', f(z))$ vanishes on A (P might be a reducible polynomial).*

Proof. For every $z' \in V \setminus S$ (S as in Clause (ii) of the theorem), let

$\phi_1(z'), \ldots, \phi_m(z')$, be all points in $\mathbf{K}^{n-d}$ such that $(z', \phi_i(z')) \in A$ and write $\phi_i = (\phi_{i,d+1}, \ldots, \phi_{i,n})$. As we saw, the ϕ_i's are $\mathbf{K}$-holomorphic outside a set of $\mathbf{R}$-dimension $2d - 1$.

Now, take an arbitrary $f \in \mathcal{O}(A)$, and consider the polynomial $P(X_1, \ldots, X_m, Y) = \Pi_{i=1}^m (Y - X_i)$. Since P is symmetric in $X_1, \ldots, X_m$, it follows that the function

$$G(z', Y) = \Pi_{i=1}^m (Y - f(z', \phi_i(z')))$$

is $\mathbf{K}$-holomorphic in $V \times \mathbf{K}$ and can be written as a polynomial in Y over the ring $\mathcal{O}(V)$ (see Lemma 4.7 for a similar argument). We claim that $G(z', f(z)) = 0$ for every $z = (z', z'') \in A$.

Indeed, for every $z' \in \mathbf{K}^d \setminus S$ and $z = (z', z'') \in A$, we have $z'' = \phi_i(z')$ for some $i = 1, \ldots, m$ and hence $G(z', f(z)) = 0$. But $\pi^{-1}(\mathbf{K}^d \setminus S) \cap A$ is dense in A therefore $G(z', f(z)) = 0$. $\qquad\square$

For each $i = d+1, \ldots, n$, we denote by Z_i the coordinate function from $\mathbf{K}^n$ into $\mathbf{K}$ that assigns to each $a \in \mathbf{K}^n$ its i-th coordinate. We denote by $P_i(z', u)$ the polynomial $P_{Z_i}(z', u)$ from Claim 11.4. Each $P_i(z', u)$ is a monic polynomial in u, of degree m, over $\mathcal{O}(V)$, and $P_i(z', z_i)$ vanishes on A.

We denote by $D(z')$ the discriminant of the polynomial $P_{d+1}(z', -)$. Namely,

$$D(z') = \Pi_{1 \leqslant i < j \leqslant m} (\phi_{i,d+1}(z') - \phi_{j,d+1}(z')).$$

By the symmetric nature of D, it is $\mathbf{K}$-holomorphic on V.

Claim 11.5 *The zero set of $D(z')$ is nowhere dense in V.*

Proof. From Assumptions (ii) and (iii) it follows that there is a definable open dense subset $V_0 \subseteq V$ such that for every $a' \in V_0$ the preimage $\pi^{-1}(a')$ contains exactly m points and the $d+1$-coordinate of these points are all distinct. Since $P_{d+1}(a', u)$ has degree m in the u-variable and $P_{d+1}(a', z_{d+1})$ vanishes on A we obtain that for all $a' \in V_0$, the function Z_{d+1} is a bijection between $\pi^{-1}(a')$ and the zero set of $P_{d+1}(a', u)$. Thus, for $a' \in V_0$, $P_{d+1}(a', z_{d+1})$ has only simple roots and $D(a') \neq 0$. Since $D(z')$ does not have zeroes in V_0, the zero set of $D(z')$ is nowhere dense in V. $\qquad\square$

The main ingredient in the proof of the theorem is the following:

Claim 11.6 *For each $i = d+2, \ldots, n$ there is a polynomial $R_i(z', u) \in \mathcal{O}(V)[u]$ such that $D(z')Z_i - R_i(z', z_{d+1})$ vanishes on A.*

Proof. See Theorem 4A, p.84 in [19], and take P_i to be g_i from that theorem. The proof there works in our setting as well. $\square$

Let $G_1, \ldots, G_r$ be a listing of the functions $P_{d+1}, \ldots, P_n$ and $D(z')z_i - R_i(z', z_{d+1})$.

We fix $p \in A$, and for $i = d+1, \ldots, n$ we denote by $p_i(z', z_i)$ the germs of the functions $P_i(z', z_i)$ in the ring $\mathcal{O}(A)_p$. Notice that all $p_i(z', z_i)$ are polynomials in z_i over $\mathcal{O}(A)_{d,p}$ of degree at most m. We denote by δ the germ of $D(z')$ in $\mathcal{O}(A)_p$, and, for $i = d+2, \ldots, n$, we denote by $r_i(z', z_{d+1})$ the germ of the function $R_i(z', z_{d+1})$ in the ring $\mathcal{O}(A)_p$. Again we have that each $r_i(z', z_{d+1})$ is a polynomial in z_{d+1} over $\mathcal{O}(A)_{d,p}$.

Let J_p be the ideal of $\mathcal{O}_p$ generated by the germs of $G_1, \ldots, G_r$ at p and let $s = (m-1)(n - (d+1))$. In order to prove the theorem we need to prove:

For every $f \in \mathcal{O}_{n,p}$, $f \in \mathcal{I}(A)_p$ if and only if $\delta^s f \in J_p$.

Assume first that $\delta^s f \in J_p$. By Claim 11.4 and Claim 11.6, $J_p \subseteq \mathcal{I}(A)_p$, hence $\delta^s f \in \mathcal{I}(A)_p$. Since $\delta^s f$ vanishes on A near p, the germ f vanishes at all points of A near p where $\delta \neq 0$. It follows from Claim 11.5 that the latter set is dense in A and hence $f \in \mathcal{I}(A)_p$.

The opposite inclusion will again be proved via a sequence of claims.

Claim 11.7 *For every $h(z', z_{d+1}) \in \mathcal{O}_{d+1,p}$, if $h \in \mathcal{I}(A)_p$ then h is divisible by $p_{d+1}(z', z_{d+1})$.*

Proof. Consider a small open definable set $U_1 = V_1 \times W_1 \subseteq \mathbf{K}^d \times \mathbf{K}^{n-d}$ containing p such that h is defined on U and such that $A \cap (V_1 \times \partial W_1) = \emptyset$. By our choice of p_{d+1}, for every $z' \in V_1$, every zero of $p_{d+1}(z', -)$ (in the $d+1$ variable) is also a zero of $h(z', -)$, of the same multiplicity. It follows that p_{d+1} divides $f(z', z_{d+1})$ in $\pi_{d+1}(U_1)$ (see p.17 in [14] for a similar argument). $\square$

Claim 11.8 *For every $g(z) \in \mathcal{O}_{n,p}$ there is $h(z', u_{d+1}, \ldots, u_n) \in \mathcal{O}_{d,p}[\bar{u}]$ of degree less than m in each variable $u_{d+1}, \ldots, u_n$ such that $g(z)$ is equivalent to $h(z', z_{d+1}, \ldots, z_n)$ modulo J_p.*

Proof. Given $g(z)$, we first divide g by $p_n(z_n)$, using Weierstrass Division Theorem (see Theorem 2.23 in [14]). The germ $g(z)$ is then equivalent, modulo J_p, to a polynomial $r(z_n)$ over $\mathcal{O}_{n-1,p}$ whose degree is smaller than $m(= deg(p_n))$. We next consider each of the coefficients of $r(z_n)$, and replace it, after dividing by $p_{n-1}(z_{n-1})$, with a polynomial in z_{n-1}, over $\mathcal{O}_{n-2,p}$, of degree at most m. We continue until we get, modulo J_p, a polynomial in $z_{d+1},\ldots,z_n$, over $\mathcal{O}_{d,p}$. $\qquad\square$

Claim 11.9 *For every* $g(z) \in \mathcal{O}_{n,p}$ *there is* $q(z',u) \in \mathcal{O}_{d,p}[u]$ *such that* $\delta^s g(z)$ *is equivalent modulo* J_p *to* $q(z',z_{d+1})$.

Proof. Let $h(z',u_{d+1},\ldots,u_n) \in \mathcal{O}_{d,p}[\bar{u}]$ be as in Claim 11.8. Each monomial appearing in h has total degree at most s. Hence

$$\delta^s h(z) = h_1(z',z_{d+1},\delta z_{d+2},\ldots,\delta z_n),$$

where $h_1(z',u_{d+1},\ldots,u_n) \in \mathcal{O}_{d,p}[\bar{u}]$. Since, for $i = d+2,\ldots,d_n$, each δz_i is equivalent (modulo J_p) to $r_i(z',z_{d+1})$, we obtain that $\delta^s h$ is equivalent to a polynomial over $\mathcal{O}_{d,p}$ in one variable z_{d+1}. $\qquad\square$

We can now finish the proof of Theorem 11.2. If $f \in \mathcal{I}(A)_p$ then, by Claim 11.9, $\delta^s f$ is equivalent (modulo J_p) to some $q(z',z_{d+1})$, where $q(z',u) \in \mathcal{O}_{d,p}[u]$. Since $J_p \subseteq \mathcal{I}(A)_p$, we have $q(z',z_{d+1}) \in \mathcal{I}(A)_p$. By Claim 11.7, $q(z',z_{d+1})$ is divisible by $p_{d+1}(z',z_{d+1})$ and therefore it belongs to J_p. $\qquad\square$

Proof of Theorem 11.3.

The proof we use here is almost identical to the proof of the classical corresponding theorem in [19] (see p. 275 Theorem 8B in Chapter 8).

We first use induction on n, the dimension of the domain space. The case $n = 0$ is just about vector spaces and is easy to verify (see [19]).

Take $n > 0$ and assume first that the dimension of the target space is $N = 1$. Take $f = (f_1,\ldots,f_s)$ a **K**-holomorphic map on A.

Claim 1 Let $V = \pi(U) \subseteq \mathbf{K}^{n-1}$ be the projection of U onto the first $n - 1$ coordinates. We may assume that V is open and that $f_1,\ldots,f_s$ are polynomials in the n-th variable with coefficients in $\mathcal{O}(V)$.

We first take a sequence of $2n + 1$ generic and independent $n - 1$-dimensional **K**-subspaces and cover U by finitely many open sets, as in Theorem 2.14, with respect, simultaneously, to the sets $Z(f_1),\ldots,Z(f_s)$.

We therefore may assume that the projection map onto the first $n - 1$ coordinates, when restricted to the zero set for each f_i is definably proper over its image, finite-to-one and its image is open in $\mathbf{K}^{n-1}$. We now want to use the following strong form of the Weierstrass Preparation Theorem:

Under the above assumptions, for each i, there is a Weierstrass polynomial $\omega_i(z', z_n)$ in the variable z_n, whose coefficients are $\mathbf{K}$-holomorphic functions on $V = \pi(U)$, and there is a $\mathbf{K}$-holomorphic nonvanishing function $u_i(z', z_n)$ on $V \times \mathbf{K}$ such that

$$\forall (z', z_n) \in V \times \mathbf{K} \;\; f_i(z', z_n) = u_i(z', z_n)\omega_i(z', z_n).$$

This is a modified, "global" version of Theorem 2.20 in [14]. The assumption on the zero set of $f_i(z', z_n)$ allows for this modification (notice that $V \times \mathbf{K}$ here satisfies the assumptions on the open set W, at the top of p.16, in [14]).

It is now easy to verify that it is sufficient to prove the theorem for the relation module associated to the germs of $\omega_1, \ldots, \omega_s$.

Let $\omega = (\omega_1, \ldots, \omega_s)$ be polynomials in z_n over $\mathcal{O}(V)$, where V open in $\mathbf{K}^{n-1}$. Assume that the degree of all of these polynomials is at most m. Let $\pi = \pi_{n-1}$ be the projection onto the first $n - 1$ coordinates, and for $z \in \mathbf{K}^n$, let $z' = \pi(z)$. The following local claim will allow us to treat only tuples of polynomials in z_n, $P = (P_1, \ldots, P_s)$, in $R_p(\omega)$.

Claim 2 For $p \in U$, let $\phi = (\phi_1, \ldots, \phi_s)$ be a tuple of $\mathbf{K}$-holomorphic functions, $\phi \in R_p(\omega)$. Then there are tuples $\psi^i = (\psi_1^i, \ldots, \psi_s^i)$, for $i = 1, \ldots, t$, where each ψ_j^i is a polynomial in z_n of degree $\leqslant m$ over $\mathcal{O}_{p'}$, such that ϕ is in the module generated by $\psi^1, \ldots, \psi^t$ over $\mathcal{O}_p$.

We omit the proof of Claim 2 since the one given in Whitney's book goes through in our setting as well (see the proof of statement (b) on p. 276, from [19]).

The remainder of the argument is very close to Whitney's proof. However, because of our special formulation of the theorem and because Whitney's proof contains a small error (involving indices, see bottom of p. 277) of we repeat the proof in almost full details.

Claim 2 allows us to consider, for each $p \in U$, only tuples $\phi = (\phi_1, \ldots, \phi_s)$ in $R_p(\omega)$ where each ϕ_i is polynomial in z_n of degree $\leqslant m$ over some neighborhood of $\pi(p)$.

For each $i = 1, \ldots, s$, we write $\omega_i = \Sigma_{j=0}^m A_{i,j} x_n^j$. Notice that if

$P = (P_1, \ldots, P_s)$ is a tuple of **K**-holomorphic polynomials in the variable z_n, all over $\mathcal{O}_{p'}$, and $P_i = \Sigma_{i=0}^{m} B_{i,j} z_n^j$ then

$$P \in R_p(\omega) \Leftrightarrow \text{ for every } k = 0, \ldots, 2m, \ \Sigma_{i=1}^{s} \Sigma_{j=0}^{m} B_{i,j} A_{i,k-j} = 0$$

(where $A_{i,k-j}$ is taken to be zero when $k - j < 0$ or $k - j > m$).

We want to translate this last condition to another relation module, of **K**-holomorphic maps over open subsets of $\mathbf{K}^{n-1}$. We do it as follows: For each $k = 0, \ldots, 2m$ and for each $i = 1, \ldots, s$, we consider the tuple $A_i^k = (a_{i,j})_{j=0,\ldots,m}$, defined by

$$a_{i,j} = \begin{cases} A_{i,k-j} & \text{if } 0 \leqslant k - j \leqslant m \\ 0 & \text{otherwise.} \end{cases}$$

We then let, for $k = 0, \ldots, 2m$, $G^k = (G_0^k, \ldots, G_{s(m+1)}^k)$ be the $s(m+1)$-tuple obtained by the concatenation $A_1^k {}^\frown \cdots {}^\frown A_s^k$.

Finally, we consider an $s(m+1)$-tuple $H = (H_0, \ldots, H_{s(m+1)})$ of **K**-holomorphic maps from $V = \pi(U) \subseteq \mathbf{K}^{n-1}$ into $\mathbf{K}^{2m+1}$, defined by

$$H_i = \begin{pmatrix} G_i^0 \\ \vdots \\ G_i^{2m} \end{pmatrix}.$$

By induction on n, the theorem holds for the relation modules of the form $R_q(H_1, \ldots, H_{s(m+1)})$, for $q \in V$. Therefore, after replacing V by an open subset, we may assume that there is a sequence of **K**-holomorphic maps $B^1, \ldots, B^t$, where each B^j is an $s(m+1)$-tuple of **K**-holomorphic functions on V, such that at every point $q \in V$, the relation module $R_q(H)$ is generated by the germs of the B^j's over $\mathcal{O}_q$. Now, each B^j stores the coefficients of s polynomials $P^j = (P_1^j, \ldots, P_s^j)$ of degree $\leqslant m$, in the variable z_n. We claim that $P^1, \ldots, P^t$ generate, at every point $p \in U$, the relation module $R_p(\omega)$.

By Claim 2, it is sufficient to check $(\phi_1, \ldots, \phi_s) \in R_p(\omega)$, where the ϕ_i's are polynomial maps of degree $\leqslant m$ in z_n, over $\mathcal{O}_{\pi(p)}$. Our construction implies the result for such polynomials.

We now finished the proof for the case $N = 1$. For the general case, see the argument on p.278 in [19]. $\qquad \square$

The proof of Theorem 11.1 is thus finished, except for the "moreover" clause. For that, notice that the only external parameters which were used in the argument were the ones needed for the linear subspaces and projections from Theorem 2.14. It is true that in the proof we referred to generic subspaces, thus requiring possibly external parameters, but the properties satisfied by these subspaces (e.g. clauses (i), (ii) and (iii) of Theorem 11.3) can all be expressed in a first order way and hence, by "definable choice" in o-minimal structures, these subspaces can be chosen to be definable over the original parameters.   □

12 Appendix

We are going to prove Theorem 4.2. We start with several lemmas, interesting on their own right.

For $W \subseteq \mathbf{K}^n$ a definable open set and $f : W \to \mathbf{K}$, we denote by $W_{\mathbf{R}}$ the set W, when viewed as a subset of $\mathbf{R}^{2n}$, and by $f_{\mathbf{R}}(x, y)$ the associated $\mathbf{R}$-function from $W_{\mathbf{R}}$ into $\mathbf{R}^2$. Using the real and imaginary parts of f, we write, for every $x + iy \in W$,

$$f_{\mathbf{R}}(x, y) = f(x + iy) = u_f(x, y) + iv_f(x, y).$$

Recall that $\mathbf{R}^n$ is viewed as a subset of $\mathbf{K}^n$ by identifying it with the set $\{(x, 0) \in \mathbf{R}^{2n}\}$.

Lemma 12.1 *Assume that $W \subseteq \mathbf{K}^m$ is a definably connected open set and $G \subseteq W \cap \mathbf{R}^m$ is a definable, nonempty, relatively open subset of $\mathbf{R}^m$. Assume that $f, g : W \to \mathbf{K}$ are $\mathbf{K}$-holomorphic functions such that $f|G = g|G$. Then $f(z) = g(z)$ for all $z \in W$.*

Proof. Because W is definably connected it is sufficient to find an open subset of W where the two functions agree. Equivalently, (see Theorem 2.13(2) in [14]) it is sufficient to find a point in W where f and g have the same partial $\mathbf{K}$-derivatives, of every order.

For any point $a \in G \subseteq W$, the partial $\mathbf{K}$-derivatives of f and g at a of first order, can be calculated along $W \cap \mathbf{R}^m$. Since f and g agree on G it follows that the partial $\mathbf{K}$-derivatives of f and g, of first order, agree on G. We may proceed, and thus show that all the $\mathbf{K}$-derivatives of f and g, of every order, agree at every point in G. It follows that f and g agree on all of W.   □

In the language of the complex numbers the theorem below says that if f is a definable (in some o-minimal structure) holomorphic function of n-variables then $Re(f)$ and $Im(f)$ can be extended to definable holomorphic functions of $2n$-variables.

Lemma 12.2 *Let $f : G \to \mathbf{K}$ be a $\mathbf{K}$-holomorphic function on some definable open $G \subseteq \mathbf{K}^n$. Then the functions $u_f(x, y)$ and $v_f(x, y)$, of $2n$*

R-*variables, can be extended to* **K**-*holomorphic (in particular definable) functions of* $2n$ **K**-*variables*, U_f *and* V_f, *respectively.*

U_f *and* V_f *are unique in the sense that any other two such extensions,* U', V' *must agree with* U_f *and* V_f, *respectively, in some open neighborhood of* $G_\mathbf{R}$ *in* $\mathbf{K}^{2n}$.

Proof. The uniqueness of U_f and V_f follows from Lemma 12.1. Thus, if we can, for every $(x, y) \in G_\mathbf{R}$, find an open neighborhood $W_{f,(x,y)} \subseteq \mathbf{K}^{2n}$ and definable $U_{f,(x,y)}, V_{f,(x,y)} : W_{f,(x,y)} \to \mathbf{K}$ as needed, and if furthermore $U_{f,(x,y)}$, $V_{f,(x,y)}$ and $W_{f,(x,y)}$ are definable uniformly in (x, y), then, by the uniqueness statement, the union of the $U_{f,(x,y)}$'s gives a function U_f on some open set in $\mathbf{K}^{2n}$ which contains $G_\mathbf{R}$, and the same is true for the union of the $V_{f,(x,y)}$'s.

Consider $a = x + iy \in G$ (so $(x, y) \in G_\mathbf{R}$). Without loss of generality, $a = 0$, for if not, replace f with $f(z + a)$. We now define $U_f(z, w)$ in an open neighborhood of $(0, 0) \in \mathbf{K}^{2n}$ as follows (this is a variation on the so-called "Halmos trick"):

For $z = x + iy \in \mathbf{K}$, we let $\bar{z} = x - iy$ be the **K**-conjugate of z, and for $z = (z_1, \ldots, z_n) \in \mathbf{K}^n$ we let $\bar{z} = (\bar{z_1}, \ldots, \bar{z_n})$ be the n-tuple of **K**-conjugates. For $(z, w) \in \mathbf{K}^{2n}$ near $(0, 0)$, let

$$U_f(z, w) = 1/2(f(z + iw) + \overline{f(\bar{z} + i\bar{w})}).$$

It is immediate to see that the restriction of U_f to $\mathbf{R}^{2n}$ equals u (because $Re(z) = 1/2(z + \bar{z})$. To check that U_f is **K**-holomorphic in each variable one has to consider the corresponding limits and verify that they indeed exist. The continuity and therefore the **K**-holomorphicity of U_f in all $2n$ variables is immediate (notice that the definition of U_f makes sense only for $(z, w) \in \mathbf{K}^{2n}$ such that $z + iw$ and $\bar{z} + i\bar{w}$ belong to W).

To see that $V_f(z, w)$ is also definable just note that V_f is the real part of $-if$. The uniformity of U_f and V_f in (x, y) is easy to verify. $\square$

The idea of the following lemma is taken from [3] (Theorem 4), but since converging power series are not available in our nonstandard setting we replace them by definable functions.

Lemma 12.3 *Let f be a* **K**-*holomorphic function on some open subset of* $\mathbf{K}^n$ *and write* $f(x + iy) = u_f(x, y) + iv_f(x, y)$. *Assume that either the function* u_f *or the function* v_f *(as functions of $2n$ real variables) are* **R**-*algebraic in some neighborhood of* $(a, b) \in \mathbf{R}^{2n}$. *Then the function f is* **K**-*algebraic in some open neighborhood of* $a + ib$.

Proof. We assume that u_f is **R**-algebraic. Therefore, there is a polynomial $P(x, y, w)$ over **R**, in $2n + 1$ variables, such that in some neighborhood $G \subseteq \mathbf{R}^{2n}$ of (a, b) we have $P(x, y, u_f(x, y)) = 0$. We now view P as a polynomial over **K**, take $U_f(z, w)$ to be as in Lemma 12.2 and consider the function $h(z, w) = P(z, w, U_f(z, w))$. This function is defined in some open set $W \subseteq \mathbf{K}^{2n}$ containing G but since its restriction to G is zero, it follows from Lemma 12.1 that it vanishes everywhere, and hence $U_f(z, w)$ is **K**-algebraic as well.

By replacing f with $f(z + a + ib)$ we may assume that $(a, b) = (0, 0)$. It follows from the definition of $U_f(z, w)$ in the proof of Lemma 12.2 that for z near 0 we have

$$U_f(z/2, z/2i) = f(z) + \overline{f(0)}.$$

Since $U_f(z, w)$ is **K**-algebraic it follows that $f(z)$ is also algebraic.

The case where $v_f(x, y)$ is **R**-algebraic is treated similarly, using the fact that v is the real part of $-if$.   □

We now restrict ourselves to the classical setting, where our o-minimal structure is assumed to be an expansion of the real field $\langle \mathbb{R}, <, +, \cdot \rangle$.

The following theorem is taken from [4] (see Theorem 6 on p. 202). Its proof makes use of the Baire Category Theorem and we do not know whether it is true for o-minimal structures in general (with the notion of "holomorphic" replaced by "**K**-holomorphic ").

Theorem 12.4 *Assume that $f(z, w)$ is a holomorphic function defined on $G \times W \subseteq \mathbb{C}^{n+m}$. Assume that for each $z \in G$, the function $f(z, w)$ is complex-algebraic in w and for each $w \in W$, $f(z, w)$ is complex-algebraic in z. Then $f(z, w)$ is complex-algebraic as a function of (z, w).*

We are now ready to prove the main theorem of this appendix.

Theorem 12.5 *Assume that $\mathcal{M}$ is an o-minimal structure expanding the field of real numbers such that every definable holomorphic function of 1-variable is locally semialgebraic.*

Let $G \subseteq \mathbb{C}^n$ be a definable open set and X a definable irreducible complex analytic subset of G. Then there is a complex algebraic set $A \subseteq \mathbb{C}^n$ such that X is one of the irreducible components of $A \cap G$.

Proof. We first prove the theorem for $n = 2$. Fix $a \in Reg_{\mathbb{C}}X$. After a change of coordinates we may assume that near a the set X is the graph of a definable holomorphic map h from some open subset of $\mathbb{C}$ into $\mathbb{C}$. By our assumption, h is locally semialgebraic and therefore there is an open set $W \subseteq \mathbb{C}$ such that $h|W$ is real algebraic and in particular its real and imaginary parts are real algebraic. By Lemma 12.3, h is an algebraic function on some open subset of W containing a.

It follows that there is an algebraic set $A \subseteq \mathbb{C}^2$ and an open set $W \subseteq \mathbb{C}^2$ such that $X \cap W = A \cap W$. If we now define

$$B = \{z \in Reg_{\mathbb{C}}X : \exists \text{ open } W \ (X \cap W = A \cap W)\},$$

then B is non-empty, open and closed in $Reg_{\mathbb{C}}X$. Because $Reg_{\mathbb{C}}X$ is connected, it follows that $B = Reg_{\mathbb{C}}X$ and hence $Reg_{\mathbb{C}}X \subseteq A \cap G$. Since $Reg_{\mathbb{C}}X$ is dense in X, we have $X \subseteq A \cap G$. By dimension considerations, X is an irreducible component of $A \cap G$.

Consider now the general case $X \subseteq G \subseteq \mathbb{C}^n$, $\dim_{\mathbb{C}} X = d$, and again take $a \in Reg_{\mathbb{C}}X$. After a change in coordinates we may assume that near a, the set X is the graph of a holomorphic map Φ from an open set in $\mathbb{C}^d$ into $\mathbb{C}^{n-d}$. Let h be one of the coordinate functions of this map. Namely, $h(z_1, \ldots, z_d)$ is a definable holomorphic function from an open subset of $\mathbb{C}^d$ into $\mathbb{C}$. By the earlier paragraph, if we fix any of the $d - 1$ variables then the function we obtain in the remaining variable is algebraic, in the sense that it satisfies a nontrivial algebraic equation. It follows from Theorem 12.4 that $h(z_1, \ldots, z_d)$ satisfies an algebraic equation. Since this is true for any one of the coordinate functions of Φ, there is some algebraic set $A \subseteq \mathbb{C}^n$ which agrees with X on some open set. Just as before, X is an irreducible component of $A \cap G$. $\square$

The following corollary answers a question of Chris Miller which was posed in [3]. (The original conjecture implies that X below must definable in $\langle \mathbb{C}, +, \cdot \rangle$ but this is clearly false because G may not be definable there).

Corollary 12.6 *Let X be an analytic subset of an open set $G \subseteq \mathbb{C}^n$. Assume that G and X are definable in $\mathbb{R}_{exp}$. Then there is a complex algebraic set $A \subseteq \mathbb{C}^n$ such that X is an irreducible component of $A \cap G$.*

Proof. This is an immediate corollary of Bianconi's theorem (see [3]) that every $\mathbb{R}_{exp}$-definable holomorphic function of 1-variable is semialgebraic, together with Theorem 12.5 above. $\square$

We do not know whether the same theorem remains true in structures which are elementarily equivalent to $\mathbb{R}_{exp}$.

References

[1] Alessandro Berarducci, Margarita Otero, o-minimal fundamental group, homology and manifolds, *J. London Math. Soc.* (2), **65**, (2002), Nr 2, 257–270.

[2] Jacek Bochnak, Michel Coste and Marie-Françoise Roy, *Real algebraic geometry*, Ergebnisse der Mathematik und ihrer Grenzgebiete (3), **36**, (Translated from the 1987 French original; Revised by the authors), Springer-Verlag, Berlin 1998.

[3] Ricardo Bianconi, Undefinability results in o-minimal expansions of the real numbers, *Ann. Pure Appl. Logic* **134**, (2005), 43–51.

[4] Salomon Bochner and William Ted Martin, *Several Complex Variables*, Princeton Mathematical Series, vol. 10, Princeton University Press, Princeton, N. J., 1948.

[5] F. Campana, Algébricité et compacité dans l'espace des cycles d'un espace analytique complexe (French), *Math. Ann.* **251**, 1980, Nr 1, 7–18.

[6] E. M. Chirka, *Complex analytic sets*, Mathematics and its Applications (Soviet Series) **46**, Kluwer Academic Publishers Group, Dordrecht, 1989.

[7] Lou van den Dries, *Tame topology and o-minimal structures*, London Mathematical Society Lecture Note Series **248**, Cambridge University Press, Cambridge, 1998.

[8] Lou van den Dries and Chris Miller, Geometric categories and o-minimal structures, *Duke Math. J.* **84**, 1996, Nr2, 497–540.

[9] Akira Fujiki, On the Douady space of a compact complex space in the category C, *Nagoya Math. J.* **85**, 1982, 189–211.

[10] Joseph Johns, An open mapping theorem for o-minimal structures, *J. Symbolic Logic* **66**, 2001, Nr4, 1817–1820.

[11] K. Kurdyka, and A. Parusinski, Quasi-convex Decomposition in o-Minimal Structures. Application to the Gradient Conjecture, University of Angers preprint no. 138, October 2001.

[12] Rahim Moosa, Jet spaces in complex analytic geometry: an exposition, Notes.

[13] Ya'acov Peterzil and Sergei Starchenko, Expansions of algebraically closed fields in o-minimal structures, *Selecta Math. (N.S.)* **7**, 2001, Nr 3, 409–445.

[14] Ya'acov Peterzil and Sergei Starchenko, Expansions of algebraically closed fields. II. Functions of several variables, *J. Math. Log.* **3** 2003, Nr 1, 1–35.

[15] Ya'acov Peterzil and Sergei Starchenko, Uniform definability of the Weierstrass ℘-functions and generalized tori of dimension one, To appear in *Selecta Math. (N.S.)*.

[16] Ya'acov Peterzil and Sergei Starchenko, Complex analytic geometry and analytic geometric categories, Submitted.

[17] Anand Pillay, Model-theoretic consequences of a theorem of Campana and Fujiki, *Fund. Math.* **174**, 2002, Nr 2, 187–192.

[18] Anand Pillay and Martin Ziegler, Jet spaces of varieties over differential and difference fields, *Selecta Math. (N.S.)* **9**, 2003, Nr 4, 579–599.

[19] Hassler Whitney, *Complex analytic varieties*, Addison-Wesley Publishing Co., Reading, Mass.-London-Don Mills, Ont., 1972.

[20] A. J. Wilkie, Model completeness results for expansions of the ordered field of real numbers by restricted Pfaffian functions and the exponential function, *J. Amer. Math. Soc.* **9**, 1996, Nr 4, 1051–1094.

[21] Boris Zil'ber, Model theory and algebraic geometry, in *Proceedings of the Tenth Easter Conference on Model Theory*, Seminarberichte **93**, Martin Weese, Helmut Wolter eds., Humboldt Universität Fachbereich Mathematik, Berlin, 1993, 202-222.

Model theory and Kähler geometry

Rahim Moosa[†]
University of Waterloo

Anand Pillay[‡]
University of Leeds

Summary

We survey and explain some recent work at the intersection of model theory and bimeromorphic geometry (classification of compact complex manifolds). Included here are the essential saturation of the many-sorted structure $\mathcal{C}$ of Kähler manifolds, the conjectural role of hyperkähler manifolds in the description of strongly minimal sets in $\mathcal{C}$, and Campana's work on the isotriviality of hyperkähler families and its connection with the nonmultidimensionality conjecture.

1 Introduction

The aim of this paper is to discuss in some detail the relationship between ideas from model theory (classification theory, geometric stability theory) and those from bimeromorphic geometry (classification of compact complex manifolds), with reference to current research. Earlier work along these lines is in [17], [18], [21], [22], [23], [14], [15], [16], [1]. We will also take the opportunity here to describe for model-theorists some of the basic tools of complex differential geometry, as well as summarise important notions, facts and theorems such as the Hodge decomposition, and local Torelli.

Zilber [26] observed some time ago that if a compact complex manifold M is considered naturally as a first order structure (with predicates for analytic subsets of M, $M \times M$, etc.) then $\mathrm{Th}(M)$ has finite Morley rank. The same holds if we consider the category $\mathcal{A}$ of compact complex (possibly singular) spaces as a many-sorted first order structure. This

† Supported by an NSERC grant
‡ Supported by a Marie Curie chair

observation of Zilber was closely related, historically, to the work on Zariski structures and geometries by Hrushovski and Zilber [11].

There is a rich general theory of theories of finite Morley rank, encompassing both Shelah's work on classification theory (classifying first order theories and their models) as well as the more self-consciously geometric theory of 1-basedness (modularity), definable groups, definable automorphism groups, etc ... It turns out that $\text{Th}(\mathcal{A})$ witnesses most of the richness of this theory. Among the main points of the current article is that notions belonging to the Shelah theory such as nonorthogonality and nonmultidimensionality, have a very clear geometric content, and are connected with things such as "variation of Hodge structure".

The class of compact Kähler manifolds has been identified as an important rather well-behaved class of compact complex manifolds, where there is a better chance of classification. The first author [16] observed that such manifolds can to all intents and purposes be treated as saturated structures (inside which one can apply the compactness theorem). We give some more details in section 3, explaining the role of the Kähler condition.

The category of compact Kähler manifolds (or rather compact complex analytic spaces that are holomorphic images of compact Kähler manifolds) is a "full reduct" of the many-sorted structure $\mathcal{A}$. We call it $\mathcal{C}$. In [21] it was pointed out how, from work of Lieberman, one can see that $\text{Th}(\mathcal{A})$ is about as complicated as it can be from the point of view of Shelah's theory (it has the DOP). We have conjectured on the other hand that $\text{Th}(\mathcal{C})$ is rather tame. $\text{Th}(\mathcal{C})$ could not be uncountably categorical (unidimensional), but we believe it to be the next best thing, nonmultidimensional. The description of U-rank 1 types (equivalently *simple* compact complex manifolds) in $\text{Th}(\mathcal{C})$ which are *trivial* is still open, and it is conjectured that they are closely related to so-called irreducible hyperkähler manifolds. As we explain in section 5, an isotriviality result for families of hyperkähler manifolds in $\mathcal{C}$, due to Campana, represents some confirmation of the nonmultidimensionality of $\text{Th}(\mathcal{C})$.

We now give a brief survey of the model theory of compact complex manifolds, continuing in a sense [19]. There are several published survey-type articles, such as [15] and [17], to which the interested reader is referred for more details. We assume familiarity with the notion of a complex manifold M. An *analytic* subset X of M is a subset X such that for each $a \in M$ there is an an open neighbourhood U of a in M such that $X \cap U$ is the common zero-set of a finite set of holomorphic functions on U. A compact complex manifold M is viewed as a first order structure by equipping it with predicates for all the analytic subsets of M and its cartesian powers. The fundamental fact observed by Zilber is that the theory of this first order structure has quantifier elimination and has finite Morley rank. We can of course consider the collection of *all* compact complex manifolds (up to biholomorphism) and view it is a many-sorted first order structure (predicates for all analytic subsets of cartesian products of sorts) which again has QE and finite Morley rank (sort by sort). The same holds for the larger class of *compact complex analytic spaces*. A compact complex analytic space is (a compact topological space) locally modelled on zero-sets of finitely many holomorphic functions on open domains in $\mathbb{C}^n$ with of course biholomorphic transition maps. We have the notion of an analytic subset of X and its cartesian powers, and we obtain thereby a first order structure as before. As above we let $\mathcal{A}$ denote the many-sorted structure of compact complex analytic spaces, and we let $\mathcal{L}$ denote its language. (A complex analytic space is usually presented as a ringed space, where the rings may be nonreduced. We refer to [15] for more discussion of this. In any case by a *complex variety* we will mean a reduced and irreducible complex analytic space.)

If X is a compact complex variety and $a \in X$, then $\{a\}$ is an analytic subset of X, and hence is essentially named by a constant. So really $\mathcal{A}$ has names for all elements (of all sorts). Let $\mathcal{A}'$ be a saturated elementary extension of $\mathcal{A}$. If S is a sort in $\mathcal{A}$, we let S' be the corresponding sort in $\mathcal{A}'$. For example, $(\mathbb{P}_1)'$ denotes the projective line over a suitable elementary extension $\mathbb{C}'$ of $\mathbb{C}$.

Among the basic facts connecting definability and geometry are:

(i) For X, Y sorts in $\mathcal{A}$ the definable maps from X to Y are precisely the piecewise meromorphic maps.

(ii) If $p(x)$ is a complete type of $\text{Th}(\mathcal{A})$ then p is the generic type (over $\mathcal{A}$) of a unique compact complex variety X. That is, $p(x)$ is axiomatised by "$x \in X'$ but $x \notin Y'$ for any proper analytic subset $Y \subset X$".

(iii) If a, b are tuples from $\mathcal{A}'$ with $tp(a)$ the generic type of X and $tp(b)$ the generic type of Y, then $dcl(a) = dcl(b)$ iff X and Y are bimeromorphic.

(iv) Suppose a, b are tuples from $\mathcal{A}'$, and $tp(a/b)$ is stationary. Then there are compact complex varieties X, Y and a meromorphic dominant map $f : X \to Y$ whose fibres over a non-empty Zariski open subset of Y are irreducible, and such that: ab is a generic point (realizes the generic type) of X, b is a generic point of Y, and (in $\mathcal{A}'$) $f(ab) = b$. So $tp(a/b)$ is the "generic type" of the "generic fibre" X_b of $f : X \to Y$. We consider definable sets such as X_b as "nonstandard" analytic subsets of X".

Algebraic geometry lives in $\mathcal{A}$ on the sort $\mathbb{P}_1$. Any irreducible complex quasi-projective *algebraic* variety V has a compactification $\bar{V}$ which will be a compact complex variety living as a sort in $\mathcal{A}$, and biholomorphic with a closed subvariety of $\mathbb{P}_1^n$ for some $n > 0$. The variety V will be a Zariski open, hence definable, subset of $\bar{V}$.

A compact complex variety X is said to be *Moishezon* if X is bimeromorphic with a complex projective algebraic variety. This is equivalent to X being internal to the sort $\mathbb{P}_1$, and also equivalent to a generic point a of X being in the definable closure of elements from $(\mathbb{P}_1)'$. The expression "algebraic" is sometimes used in place of Moishezon. We extend naturally this notion to nonstandard analytic sets as well as definable sets and stationary types in $\mathcal{A}'$. The "strong conjecture" from [19] then holds in $\mathcal{A}'$ in the more explicit form: if $(Y_z : z \in Z)$ is a normalized family of definable subsets of a definable set X, then for $a \in X$, $Z_a = \{z \in Z : a \text{ is generic on } Y_z\}$ is Moishezon (namely generically internal to $(\mathbb{P}_1)'$). This result was derived in [18] from a theorem due independently to Campana and Fujiki.

In [21] it was shown that any strongly minimal modular group definable in $\mathcal{A}$ is definably isomorphic to a complex torus. An appropriate generalization to strongly minimal modular groups in $\mathcal{A}'$ was obtained in [1].

More details on the classification of strongly minimal sets (or more generally types of U-rank 1) in $\mathrm{Th}(\mathcal{C})$ will appear in section 5.

2 Preliminaries on complex forms

We give in this section a brief review of the basic theory of complex-valued differential forms. The reader may consult [8] or [25] for a more detailed treatment of this material.

Suppose X is an n-dimensional complex manifold. By a *coordinate system* (z, U) on X we mean an open set $U \subset X$ and a homeomorphism z from U to an open ball in $\mathbb{C}^n$. Composing with the coordinate projections we obtain *complex coordinates* $z_i : U \to \mathbb{C}$ for $i = 1, \ldots, n$, which we decompose into real and imaginary parts as $z_i = x_i + iy_i$.

Fix a coordinate system (z, U) on X and a point $x \in U$. Let $T_{X,x}$ denote the (real) tangent space of X at x. Viewed as the space of $\mathbb{R}$-linear derivations on real-valued smooth functions at x, we have that

$$\left\{ \frac{\partial}{\partial x_1}, \ldots, \frac{\partial}{\partial x_n}, \frac{\partial}{\partial y_1}, \ldots, \frac{\partial}{\partial y_n} \right\}$$

forms an $\mathbb{R}$-basis for $T_{X,x}$. But the complex manifold structure on X gives $T_{X,x}$ also an n-dimensional *complex* vector space structure, which can be described as follows. Let $T_{X,\mathbb{C},x} := T_{X,x} \otimes \mathbb{C}$ denote the complexification of the real tangent space. We have a decomposition $T_{X,\mathbb{C},x} = T_{X,x}^{1,0} \oplus T_{X,x}^{0,1}$ where $T_{X,x}^{1,0}$ is the complex subspace generated by

$$\left\{ \frac{\partial}{\partial z_i} := \frac{\partial}{\partial x_i} - i\frac{\partial}{\partial y_i} \;\middle|\; i = 1, \ldots, n \right\}$$

and $T_{X,x}^{0,1}$ is generated by

$$\left\{ \frac{\partial}{\partial \bar{z}_i} := \frac{\partial}{\partial x_i} + i\frac{\partial}{\partial y_i} \;\middle|\; i = 1, \ldots, n \right\}.$$

If we view $T_{X,\mathbb{C},x}$ as the space of $\mathbb{C}$-linear derivations on *complex*-valued smooth functions at x, then $T_{X,x}^{1,0}$ corresponds to those that vanish on all the anti-holomorphic functions (functions whose complex conjugates are holomorphic at x). In any case, $T_{X,x}^{1,0}$ is called the *holomorphic tangent space* of X at x. The natural inclusion $T_{X,x} \subset T_{X,\mathbb{C},x}$ followed by the projection $T_{X,\mathbb{C},x} \to T_{X,x}^{1,0}$ produces an $\mathbb{R}$-linear isomorphism between the real tangent space and the holomorphic tangent space. This isomorphism makes $T_{X,x}$ canonically into a complex vector space.

Despite our presentation, the above constructions do not depend on the coordinates and extend globally: We have the complexification of the (real) tangent bundle $T_{X,\mathbb{C}} := T_X \otimes \mathbb{C}$, and a decomposition into complex vector subbundles, $T_{X,\mathbb{C}} = T_X^{1,0} \oplus T_X^{0,1}$, whereby the holomorphic tangent bundle $T_X^{1,0}$ is naturally isomorphic as a real vector bundle with T_X. It is with respect to this isomorphism that we treat T_X as a complex vector bundle.

A *complex-valued differential k-form* (or just a *k-form*) at $x \in X$ is an alternating k-ary $\mathbb{R}$-multilinear map $\phi : T_{X,x} \times \cdots \times T_{X,x} \to \mathbb{C}$. The complex vector space of all k-forms at x is denoted by $F_{X,\mathbb{C},x}^k$. The *real* differential k-forms, $F_{X,\mathbb{R},x}^k$, are exactly those forms in $F_{X,\mathbb{C},x}^k$ that are real-valued. So $F_{X,\mathbb{C},x}^k = F_{X,\mathbb{R},x}^k \otimes \mathbb{C}$. In particular, $F_{X,\mathbb{C},x}^1 = \operatorname{Hom}_{\mathbb{R}}(T_{X,x}, \mathbb{R}) \otimes \mathbb{C}$ is the complexification of the real cotangent space at x. Hence, in a coordinate system (z, U) about x, if we let

$$\{dx_1, \ldots, dx_n, dy_1, \ldots, dy_n\}$$

be the dual basis to $\left\{ \frac{\partial}{\partial x_1}, \ldots, \frac{\partial}{\partial x_n}, \frac{\partial}{\partial y_1}, \ldots, \frac{\partial}{\partial y_n} \right\}$ for $\operatorname{Hom}_{\mathbb{R}}(T_{X,x}, \mathbb{R})$, then

$$dz_i := dx_i + idy_i,$$

$$d\overline{z}_i := dx_i - idy_i$$

for $i = 1, \ldots, n$, form a $\mathbb{C}$-basis for $F^1_{X,\mathbb{C},x}$.

Now $F^k_{X,\mathbb{C},x}$ is the kth exterior power of $F^1_{X,\mathbb{C},x}$. Given $I = (i_1, \ldots, i_p)$ an increasing sequence of numbers between 1 and n, let dz_I be the p-form $dz_{i_1} \wedge \cdots \wedge dz_{i_p}$. Similarly, let $d\overline{z_I} := d\overline{z_{i_1}} \wedge \cdots \wedge d\overline{z_{i_p}}$. Then

$$\{dz_I \wedge d\overline{z_J} \mid I = (i_1, \ldots, i_p), J = (j_1, \ldots, j_q), p + q = k\}$$

is a $\mathbb{C}$-basis for $F^k_{X,\mathbb{C},x}$. This gives us a natural decomposition

$$F^k_{X,\mathbb{C},x} = \bigoplus_{p+q=k} F^{p,q}_{X,x}$$

where $F^{p,q}_{X,x}$ is generated by the forms $dz_I \wedge d\overline{z_J}$ where $I = (i_1, \ldots, i_p)$ and $J = (j_1, \ldots, j_q)$ are increasing sequences of numbers between 1 and n. The complex vector subspaces $F^{p,q}_{X,x}$ can also be more intrinsically described as made up of those k-forms ϕ such that

$$\phi(cv_1, \ldots, cv_n) = c^p \overline{c}^q \phi(v_1, \ldots, v_n)$$

for all $v_1, \ldots, v_k \in T_{X,x}$ and $c \in \mathbb{C}$. Such forms are said to be of *type* (p, q).

Once again, these constructions extend globally to X and we have complex vector bundles $F^k_{X,\mathbb{C}} = \bigoplus_{p+q=k} F^{p,q}_{X,\mathbb{C}}$. For $U \subseteq X$ an open set, by a complex k form *on* U we mean a smooth section to the bundle $F^k_{X,\mathbb{C}}$ over the set U. Similarly for forms of *type* (p, q) *on* U. We denote by $\mathcal{A}^k$ and $\mathcal{A}^{p,q}$ the sheaves on X of smooth sections to $F^k_{X,\mathbb{C}}$ and $F^{p,q}_{X,\mathbb{C}}$ respectively. So $\mathcal{A}^k(U)$ is the space of all complex k-forms on U while $\mathcal{A}^{p,q}(U)$ is the space of all complex forms of type (p, q) on U. Given a coordinate system (z, U) on X, a k-form $\omega \in \mathcal{A}^k(U)$ can be expressed as $\omega = \sum_{|I|+|J|=k} f_{IJ} dz_I \wedge d\overline{z_J}$ where $f_{IJ} : U \to \mathbb{C}$ are smooth. Note that dz_i and $d\overline{z_i}$ are being viewed here as 1-forms on U. By convention, $\mathcal{A}^0$ is the sheaf of $\mathbb{C}$-valued smooth functions.

The *exterior derivative* map $d : \mathcal{A}^k(U) \to \mathcal{A}^{k+1}(U)$ is defined by

$$d\left(\sum_{|I|+|J|=k} f_{IJ} dz_I \wedge d\overline{z_J} \right) = \sum_{|I|+|J|=k} df_{IJ} \wedge dz_I \wedge d\overline{z_J}$$

where for any smooth function $f : U \to \mathbb{C}$, $df \in \mathcal{A}^1(U)$ is given by

$$df := \sum_{i=1}^{n} \frac{\partial f}{\partial z_i} dz_i + \sum_{i=1}^{n} \frac{\partial f}{\partial \overline{z_i}} d\overline{z_i}.$$

If we define $\partial f := \sum_{i=1}^n \frac{\partial f}{\partial z_i} dz_i$ and $\overline{\partial} f = \sum_{i=1}^n \frac{\partial f}{\partial \overline{z}_i} d\overline{z}_i$, and then extend these maps so that $\partial : \mathcal{A}^{p,q}(U) \to \mathcal{A}^{p+1,q}(U)$ is given by

$$\partial \left(\sum f_{IJ} dz_I \wedge d\overline{z_J} \right) = \sum \partial f_{IJ} \wedge dz_I \wedge d\overline{z_J}$$

and $\overline{\partial} : \mathcal{A}^{p,q}(U) \to \mathcal{A}^{p,q+1}(U)$ is given by

$$\overline{\partial} \left(\sum f_{IJ} dz_I \wedge d\overline{z_J} \right) = \sum \overline{\partial} f_{IJ} \wedge dz_I \wedge d\overline{z_J};$$

then we see that $d = \partial + \overline{\partial}$.

One can show that d, ∂, and $\overline{\partial}$ are all independent of the coordinate system and extend to sheaf maps on $\mathcal{A}^k$ and $\mathcal{A}^{p,q}$ as the case may be. Moreover,

- $d, \partial, \overline{\partial}$ are $\mathbb{C}$-linear;
- $d \circ d = 0$, $\partial \circ \partial = 0$, and $\overline{\partial} \circ \overline{\partial} = 0$; and,
- $d(\overline{\phi}) = \overline{d\phi}$, $\partial(\overline{\phi}) = \overline{\overline{\partial}\phi}$, and $\overline{\partial}(\overline{\phi}) = \overline{\partial\phi}$.

We say that $\omega \in \mathcal{A}^k(X)$ is *d-closed* if $d\omega = 0$, and *d-exact* if $\omega = d\phi$ for some $\phi \in \mathcal{A}^{k-1}(X)$. Since $d \circ d = 0$ the exact forms are closed. The *De Rham cohomology groups* are the complex vector spaces

$$H^k_{\mathrm{DR}}(X) := \frac{\{d\text{-closed } k\text{-forms}\}}{\{d\text{-exact } k\text{-forms}\}}.$$

We can relate this cohomology to the classical singular cohomology (which we assume the reader is familiar with) by integration: Given a complex form $\omega \in \mathcal{A}^k(X)$ and a *k-simplex*

$$\phi : \Delta_k := \left\{ (t_1, \ldots, t_{k+1}) \in [0,1]^{k+1} \;\middle|\; \sum_{i=1}^{k+1} t_i = 1 \right\} \longrightarrow X,$$

it makes sense to consider $\int_\phi \omega := \int_{\Delta_k} \phi^* \omega \in \mathbb{C}$. Every complex k-form thereby determines a homomorphism from the free abelian group generated by the k-simplices (i.e., the group of *singular k-chains*) to the complex numbers. We restrict this homomorphism to the *singular k-cycles* (those chains whose boundary is zero), and denote it by

$$\int \omega : \{k\text{-cycles on } X\} \longrightarrow \mathbb{C}.$$

If ω is d-closed then $\int \omega$ vanishes on boundaries by Stokes' theorem, and hence $\int \omega$ induces a complex-valued homomorphism on the *singular homology* group $H_k(X) = \frac{k\text{-cycles in } X}{k\text{-boundaries in } X}$. Moreover, by Stokes' theorem again, $\int \omega = 0$ if ω is d-exact. So $\omega \mapsto \int \omega$ induces a homomorphism

$$H^k_{\mathrm{DR}}(X) \longrightarrow \mathrm{Hom}_{\mathbb{Z}}\left(H_k(U), \mathbb{C}\right) = H^k_{\mathrm{sing}}(X, \mathbb{C})$$

where $H^k_{\mathrm{sing}}(X, \mathbb{C})$ is the *singular cohomology* group with complex coefficients.

De Rham's Theorem (cf. Section 4.3.2 of [25]). *The above homomorphism is an isomorphism:* $H^k_{\mathrm{DR}}(X) \cong H^k_{\mathrm{sing}}(X, \mathbb{C})$.

3 Saturation and Kähler manifolds

One obstacle to the application of model-theoretic methods to compact complex manifolds is that the structure $\mathcal{A}$ is not saturated; for example, every element of every sort of $\mathcal{A}$ is $\emptyset$-definable. However, for some sorts this can be seen to be an accident of the language of analytic sets in which we are working: Suppose V is a complex projective algebraic variety viewed as a compact complex manifold and consider the structure

$$\mathcal{V} := (V; P_A \mid A \subseteq V^n \text{ is a subvariety over } \mathbb{Q}, \ n \geq 0)$$

where there is a predicate for every subvariety of every cartesian power of V *defined over the rationals*. It is not hard to see that $\mathcal{V}$ is saturated (it is ω_1-compact in a countable language). Moreover, Chow's theorem says that every complex analytic subset of projective space is complex algebraic, and hence a subset of a cartesian power of V is definable in $\mathcal{A}$ if and only if it is definable (with parameters) in $\mathcal{V}$. That is, with respect to the sort V, the lack of saturation in $\mathcal{A}$ is a result of working in too large (and redundant) a language. This property was formalised in [16] as follows.

Definition 3.1 A compact complex variety X is *essentially saturated* if there exists a countable sublanguage of the language of $\mathcal{A}$, $\mathcal{L}_0$, such that every subset of a cartesian power of X that is definable in $\mathcal{A}$ is already definable (with parameters) in the reduct of $\mathcal{A}$ to $\mathcal{L}_0$.

The structure induced on X by such a reduct will be saturated.

It turns out that essential saturation, while motivated by internal model-theoretic considerations, has significant geometric content. The purpose of this section is to describe this geometric significance and to show in particular that compact Kähler manifolds are essentially saturated.

We will make use of Barlet's construction of the space of compact cycles of a complex variety. For X any complex variety and n a natural number, a (*holomorphic*) *n-cycle* of X is a (formal) finite linear combination $Z = \sum_i n_i Z_i$ where the Z_i's are distinct n-dimensional irreducible compact analytic subsets of X, and each n_i is a positive integer called the *multiplicity* of the component Z_i.[1] By $|Z|$ we mean the underlying set or *support* of Z, namely $\bigcup_i Z_i$. We denote the set of all n-cycles of X by $\mathcal{B}_n(X)$, and the set of all cycles of X by $\mathcal{B}(X) := \bigcup_n \mathcal{B}_n(X)$. In [2] Barlet endowed $\mathcal{B}(X)$ with a natural structure of a reduced complex analytic space. If for $s \in \mathcal{B}_n(X)$ we let Z_s denote the cycle represented by s, then the set $\{(s,x) : s \in \mathcal{B}_n(X), x \in |Z_s|\}$ is an analytic subset of $\mathcal{B}_n(X) \times X$. Equipped with this complex structure, $\mathcal{B}(X)$ is called the *Barlet space of X*. When X is a projective variety the Barlet space coincides with the Chow scheme.

An cycle is called *irreducible* if it has only one component and that component is of multiplicity 1. In [6] it is shown that

$$\mathcal{B}^*(X) := \{s \in \mathcal{B}(X) : Z_s \text{ is irreducible}\}$$

is a Zariski open subset of $\mathcal{B}(X)$. An irreducible component of $\mathcal{B}(X)$ is *prime* if it has nonempty intersection with $\mathcal{B}^*(X)$. Suppose S is a prime component of the Barlet space and set

$$G_S := \{(s,x) : s \in S, x \in |Z_s|\}.$$

Then G_S is an irreducible analytic subset of $S \times X$ and, if $\pi : G_S \to S$ denotes the projection map, the general fibres of π are reduced and irreducible. We call G_S the *graph* of (the family of cycles parametrised) by S.

Fact 3.2 (cf. Theorem 3.3 of [16]). *A compact complex variety X is essentially saturated if and only if every prime component of $\mathcal{B}(X^m)$ is compact for all $m \geq 0$.*

One direction of 3.2 is straightforward: If every prime component of $\mathcal{B}(X^m)$ is compact, then they are all sorts in $\mathcal{A}$ and their graphs are definable in $\mathcal{A}$. Consider the sublanguage $\mathcal{L}_0$ of the language of $\mathcal{A}$ made up of predicates for the graphs G_S as S ranges over all prime components of $\mathcal{B}(X^m)$ for all $m \geq 0$. Then $\mathcal{L}_0$ is countable because the Barlet space has countably many irreducible components (this actually follows from

---

1 We hope the context will ensure that holomorphic n-cycles will not be confused with the singular n-cycles of singular homo logy discussed in the previous section.

Lieberman's Theorem 3.6 below). Every irreducible analytic subset of X^m, as it forms an irreducible cycle, is a fibre of G_S for some such S, and hence is $\mathcal{L}_0$-definable. By quantifier elimination for $\mathcal{A}$, it follows that every $\mathcal{A}$-definable subset of every cartesian power of X is $\mathcal{L}_0$-definable. So X is essentially saturated. The converse makes use of Hironaka's flattening theorem and the universal property of the Barlet space.[2]

It is in determining whether a given component of the Barlet space is compact that Kähler geometry intervenes. We review the fundamentals of this theory now, and suggest [25] for further details.

Suppose X is a complex manifold. A *hermitian metric* h on X assigns to each point $x \in X$ a positive definite *hermitian form* h_x on the tangent space $T_{X,x}$. That is, $h_x : T_{X,x} \times T_{X,x} \to \mathbb{C}$ satisfies:

(i) $h_x(-, w)$ is $\mathbb{C}$-linear for all $w \in T_{X,x}$,

(ii) $h_x(v, w) = \overline{h_x(w, v)}$ for all $v, w \in T_{X,x}$, and

(iii) $h_x(v, v) > 0$ for all nonzero $v \in T_{X,x}$.

Note that h_x is $\mathbb{R}$-bilinear and that $h_x(v, -)$ is $\mathbb{C}$-antilinear for all $v \in T_{X,x}$. Moreover this assignment should be smooth: Given a coordinate system (z, U) on X, a hermitian metric is represented on U by

$$h = \sum_{i,j=1}^{n} h_{ij} dz_i \otimes d\overline{z_j}$$

where $h_{ij} : U \to \mathbb{C}$ are smooth functions and $dz_i \otimes d\overline{z_j}$ is the map taking a pair of tangent vectors (v, w) to the complex number $dz_i(v)d\overline{z_j}(w)$.

A hermitian metric encodes both a riemannian and a symplectic structure on X. The real part of h, $\mathrm{Re}(h_x) : T_{X,x} \times T_{X,x} \to \mathbb{R}$, is positive definite, symmetric, and $\mathbb{R}$-bilinear. That is, $\mathrm{Re}(h)$ is a riemannian metric on X. On the other hand, the imaginary part, $\mathrm{Im}(h_x) : T_{X,x} \times T_{X,x} \to \mathbb{R}$, is an alternating $\mathbb{R}$-bilinear map. So $\mathrm{Im}(h)$ is a real 2-form on X. Moreover, if in a coordinate system (z, U) we have $h = \sum_{i,j=1}^{n} h_{ij} dz_i \otimes d\overline{z_j}$, then a straightforward calculation shows that $\mathrm{Im}(h) = -\frac{i}{2} \sum_{i,j=1}^{n} h_{ij} dz_i \wedge d\overline{z_j}$. So as a complex 2-form on X, $\mathrm{Im}(h)$ is of type $(1, 1)$.

The assignment $h \mapsto -\mathrm{Im}(h)$ is a bijection between hermitian metrics and positive real 2-forms of type $(1, 1)$ on X. We call $\omega := -\mathrm{Im}(h)$ the *Kähler form* associated to h. A hermitian metric is a *Kähler metric* if

---

2 Actually, this is done in [16] with restricted *Douady spaces* (the complex analytic analogue of the Hilbert scheme) rather than Barlet spaces. However, it is routine to see that the argument works for Barlet spaces as well.

its Kähler form is d-closed. A complex manifold is a *Kähler manifold* if it admits a Kähler metric.

Example 3.3 The *standard* Kähler metric on $\mathbb{C}^n$ is $\sum_{i=1}^{n} dz_i \otimes d\overline{z_i}$.

Example 3.4 Every complex manifold admits a hermitian metric (but not necessarily a Kähler one). Indeed, given any complex manifold X, a cover $\mathcal{U} = (z^\iota, U_\iota)_{\iota \in I}$ by coordinate systems, and a partition of unity $\rho = (\rho_\iota)_{\iota \in I}$ subordinate to $\mathcal{U}$, $h := \sum_{\iota \in I} \rho_\iota \left(\sum_{i=1}^{n} dz_i^\iota \otimes d\overline{z_i^\iota} \right)$ is a hermitian metric on X.

Example 3.5 (Fubini-Study) Let $[z_0; \ldots; z_n]$ be complex homogeneous coordinates for $\mathbb{P}_n$. For each $i = 0, \ldots, n$ let U_i be the affine open set defined by $z_i \neq 0$. Let $F_i : U_i \to \mathbb{R}$ be the smooth function given by $\log \left(\frac{|z_0|^2 + \cdots + |z_n|^2}{|z_i|^2} \right)$. Then $i \partial \overline{\partial} F_i \in F_X^{1,1}(U_i)$ is real-valued. Moreover, for all $j = 0, \ldots, n$, $i \partial \overline{\partial} F_i$ agrees with $i \partial \overline{\partial} F_j$ on $U_i \cap U_j$. Hence, the locally defined forms $i \partial \overline{\partial} F_0, \ldots, i \partial \overline{\partial} F_n$ patch together and determine a global complex 2-form on $\mathbb{P}_n$. It is real-valued, of type $(1,1)$, and d-closed. The associated Kähler metric is called the *Fubini-Study* metric on $\mathbb{P}_n$.

Suppose h is a hermitian metric on X. There is strong interaction between the Kähler form ω and the riemannian metric $\mathrm{Re}(h)$. This is encapsulated in Wirtinger's formula for the volume of compact submanifolds of X. When we speak of the *volume* of a submanifold of X with respect to h, denoted by vol_h, we actually mean the riemannian volume with respect to $\mathrm{Re}(h)$.

Wirtinger's Formula (cf. Section 3.1 of [25]). *If $Z \subset X$ is a compact complex submanifold of dimension k, then*

$$\mathrm{vol}_h(Z) = \int_Z \omega^k$$

where $\omega^k = \omega \wedge \cdots \wedge \omega$ is the kth exterior power of ω.

Note that Z is of real-dimension $2k$ and ω^k is a real $2k$-form on X, and hence it makes sense to integrate ω^k along Z. If $X = \mathbb{P}_n$ and h is the Fubini-Study metric of Example 3.5, then for any algebraic subvariety $Z \subseteq \mathbb{P}_n$, $\mathrm{vol}_h(Z)$ is the degree of Z.

For possibly singular complex analytic subsets $Z \subset X$ (irreducible, compact, dimension k), Wirtinger's formula can serve as the *definition* of volume; it agrees with the volume of the regular locus of Z. More generally, if $Z = \sum_i n_i Z_i$ is a k-cycle of X, then the volume of Z with respect to h is $\mathrm{vol}_h(Z) := \sum_i n_i \, \mathrm{vol}_h(Z_i)$.

Taking volumes of cycles induces a function $\mathrm{vol}_h : \mathcal{B}(X) \to \mathbb{R}$ given by

$$\mathrm{vol}_h(s) := \mathrm{vol}_h(Z_s).$$

The link between hermitian geometry and saturation comes from the following striking fact.

Theorem 3.6 (Lieberman [13]) *Suppose X is a compact complex manifold equipped with a hermitian metric h, and W is a subset of $\mathcal{B}_k(X)$. Then W is relatively compact in $\mathcal{B}_k(X)$ if and only if vol_h is bounded on W.*

Sketch of proof. Wirtinger's formula tells us that vol_h is computed by integrating ω^k over the fibres of a morphism.[3] It follows that vol_h is continuous on $\mathcal{B}_k(X)$ and hence is bounded on any relatively compact subset.

The converse relies on Barlet's original method of constructing the cycle space. We content ourselves with a sketch of the ideas involved. First, let $K(X)$ denote the space of closed subsets of X equipped with the Hausdorff metric topology. So given closed subsets $A, B \subset X$,

$$\mathrm{dist}(A, B) := \frac{1}{2} \left[\max\{\mathrm{dist}_h(A, b) : b \in B\} + \max\{\mathrm{dist}_h(a, B) : a \in A\} \right].$$

Now suppose $W \subset \mathcal{B}(X)$ is a subset on which vol_h is bounded. Given a sequence $(s_i : i \in \mathbb{N})$ of points in W we need to find a convergent subsequence. Consider the sequence $(|Z_{s_i}| : i \in \mathbb{N})$ of points in $K(X)$. Since X is compact, so is $K(X)$, and hence there exists a subsequence $(|Z_{s_i}| : i \in I)$ which converges in the Hausdorff metric topology to a closed set $A \subset X$. Since $\mathrm{vol}_h(|Z_{s_i}|)$ is bounded on this sequence, a theorem of Bishop's [5] implies that A is in fact complex analytic. Now, by Barlet's construction, the topology on $\mathcal{B}(X)$ is closely related to the Hausdorff topology on $K(X)$. In particular, it follows from the fact that $(|Z_{s_i}| : i \in I)$ converges in the Hausdorff topology to a complex analytic subset $A \subset X$, that some subsequence of $(s_i : i \in I)$ converges to a point $t \in \mathcal{B}(X)$ with $|Z_t| = A$. In particular, $(s_i : i \in \mathbb{N})$ has a convergent subsequence. Hence W is relatively compact. $\qquad\square$

---

3 To be more precise, given a component S of $\mathcal{B}_k(X)$, one considers the *universal cycle* Z_S on $S \times X$ whose fibre at $s \in S$ is the cycle Z_s. Then integrating ω^k over the fibres of $Z_S \to S$ gives us vol_h on S.

Corollary 3.7 (Lieberman [13]) *If X is a compact Kähler manifold then the prime components of $\mathcal{B}(X)$ are compact.*

Sketch of proof. Let h be a Kähler metric on X. The d-closedness of the Kähler form $\omega = -\operatorname{Im}(h)$ will imply by Wirtinger's formula that vol_h is constant on the components of the Barlet space. We sketch the argument for this here, following Proposition 4.1 of Fujiki [9]. Fix a prime component S of $\mathcal{B}_k(X)$. By continuity of vol_h, we need only show that for sufficiently general points $s, t \in S$, $\operatorname{vol}_h(Z_s) = \operatorname{vol}_h(Z_t)$. Let $G_S \subset S \times X$ be the graph of the cycles parametrised by S, and let $\pi_X : G_S \to X$ and $\pi_S : G_S \to S$ be the natural projections. We work with a prime component so that for general $s, t \in S$, the fibres of G_S over s and t are the reduced and irreducible complex analytic subsets Z_s and Z_t. Now let I be a piecewise real analytic curve in S connecting s and t. For the sake of convenience, let us assume that there is only one piece: so we have a real analytic embedding $h : [0,1] \to S$ with $h(0) = s$, $h(1) = t$ and $I = h([0,1])$. Consider the semianalytic set $R := \pi_S^{-1}(I) \subset G_S$. Given the appropriate orientation we see that the boundary ∂R of R in G_S is $\pi_S^{-1}(s) - \pi_S^{-1}(t)$. Note that π_X restricts to an isomorphism between $\pi_S^{-1}(s)$ and Z_s (and similarly for $\pi_S^{-1}(t)$ and Z_t). Also, if $\pi_X^*(\omega^k)$ is the pull-back of ω^k to G_S, then $d\pi_X^*(\omega^k) = 0$ since $d\omega = 0$. Using a semianalytic version of Stokes' theorem (see, for example, Herrera [10]), we compute

$$
\begin{aligned}
0 &= \int_R d\pi_X^*(\omega^k) \\
&= \int_{\partial R} \pi_X^*(\omega^k) \\
&= \int_{\pi_S^{-1}(s)} \pi_X^*(\omega^k) - \int_{\pi_S^{-1}(t)} \pi_X^*(\omega^k) \\
&= \int_{Z_s} \omega^k - \int_{Z_t} \omega^k \\
&= \operatorname{vol}_h(Z_s) - \operatorname{vol}_h(Z_t),
\end{aligned}
$$

We have shown that vol_h is constant on S, and hence S is compact by Theorem 3.6. $\qquad\square$

If X is Kähler then so is X^m for all $m > 0$. Hence, from Corollary 3.7 and Fact 3.2 we obtain:

Corollary 3.8 *Every compact Kähler manifold is essentially saturated.*

A complex variety is said to be of *Kähler-type* if it is the holomorphic image of a compact Kähler manifold. The class of all complex varieties of Kähler-type is denoted by $\mathcal{C}$, and was introduced by Fujiki [9]. This class is preserved under cartesian products and bimeromorphic equivalence. Many of the results for compact Kähler manifolds discussed above extend to complex varieties of Kähler-type. In particular, Kähler-type varieties are essentially saturated (see Lemma 2.5 of [16] for how this follows from Corollary 3.8 above), and their Barlet spaces have compact components which are themselves again of Kähler-type.

Model-theoretically we can therefore view $\mathcal{C}$ as a many-sorted structure in the language where there is a predicate for each G_S as S ranges over all prime components of the Barlet space of each cartesian product of sorts. We call this the *Barlet language*.[4] Note that every analytic subset of every cartesian product of Kähler-type varieties is definable (with parameters) in this language, and so we are really looking at the full induced structure on $\mathcal{C}$ from $\mathcal{A}$. Moreover, when studying the models of $\mathrm{Th}(\mathcal{C})$, we may treat $\mathcal{C}$ as a "universal domain" in the sense that we may restrict ourselves to definable sets and types in $\mathcal{C}$ itself. This is for the following reason: Fix some definable set F in an elementary extension $\mathcal{C}'$. So there will be some sort X of $\mathcal{C}$ such that F is a definable subset of the nonstandard X'. Essential saturation implies that there is a countable sublanguage $\mathcal{L}_0$ such that $X|_{\mathcal{L}_0}$ is saturated and every definable subset of X^n in $\mathcal{A}$ is already definable in $X|_{\mathcal{L}_0}$ (with parameters). In particular, F is definable in $X|_{\mathcal{L}_0}$ over some parameters, say b, in X'. The $\mathcal{L}_0$-type of b is realised in X, by say b_0. Let F_0 be defined in $X'|_{\mathcal{L}_0}$ over b_0 in the same way as F is defined over b. Then in $X'|_{\mathcal{L}_0}$ there is an automorphism taking F to F_0. So in so far as any structural properties of F are concerned we may assume it is $\mathcal{L}_0$-definable *over* X. But as $X|_{\mathcal{L}_0}$ is a saturated elementary substructure of $X'|_{\mathcal{L}_0}$, the first order properties of F are then witnessed by $F \cap X$. The latter is now a definable set in $\mathcal{C}$. The same kind of argument works also with types.

We can also work, somewhat more canonically, as follows: In any given situation we will be interested in at most countably many Kähler-type varieties, $(X_i : i \in \mathbb{N})$, at once. We then consider the smallest (countable) subcollection $\mathcal{X}$ of sorts from $\mathcal{C}$ containing the X_i's and closed under taking prime components of Barlet spaces of cartesian products of sorts in $\mathcal{X}$. We view $\mathcal{X}$ as a multi-sorted structure in the language where there is a predicate for each G_S as S ranges over all such prime

4 This is in analogy with the *Douady language* from Definition 4.3 of [16].

components of the Barlet spaces. We call this the *Barlet language of* $(X_i : i \in \mathbb{N})$. Then $\mathcal{X}$ is saturated (2^ω-saturated and of cardinality 2^ω), ω-stable, and every analytic subset of every cartesian product of sorts in $\mathcal{X}$ is definable in $\mathcal{X}$. Moreover, after possibly naming countably many constants, $\mathcal{X}$ has elimination of imaginaries (cf. Lemma 4.5 of [16]). When working with Kähler-type varieties we will in general pass to such countable reducts of $\mathcal{C}$ without saying so explicitly, and it is in this way that we treat $\mathcal{C}$ as a universal domain for $\mathrm{Th}(\mathcal{C})$.

4 Holomorphic forms on Kähler manifolds

In section 2 we defined the De Rham cohomology groups on any complex manifold X by

$$H_{\mathrm{DR}}^k(X) := \frac{\{d\text{-closed }k\text{-forms on }X\}}{\{d\text{-exact }k\text{-forms on }X\}}.$$

There are also cohomology groups of forms coming from the operators $\bar{\partial} : \mathcal{A}^{p,q}(X) \to \mathcal{A}^{p,q+1}(X)$. Since $\bar{\partial} \circ \bar{\partial} = 0$, the $\bar{\partial}$-exact forms are $\bar{\partial}$-closed. The *Dolbeaut cohomology groups* are the complex vector spaces

$$H^{p,q}(X) := \frac{\{\bar{\partial}\text{-closed forms of type }(p,q)\}}{\{\bar{\partial}\text{-exact forms of type }(p,q)\}}.$$

We denoted by $h^{p,q}(X)$ the dimension of $H^{p,q}(X)$.

A fundamental result about Kähler manifolds is the following fact:

Hodge decomposition (cf. Section 6.1 of [25]). *If X is a compact Kähler manifold then $H^{p,q}(X)$ is isomorphic to the subspace of $H_{\mathrm{DR}}^{p+q}(X)$ made up of those classes that are represented by d-closed forms of type (p,q). Moreover, under these isomorphisms,*

$$H_{\mathrm{DR}}^k(X) \cong \bigoplus_{p+q=k} H^{p,q}(X).$$

A consequence of Hodge decomposition is that complex conjugation, which takes forms of type (p,q) to forms of type (q,p), induces an isomorphism between $H^{p,q}(X)$ and $H^{q,p}(X)$. In particular, $h^{p,q}(X) = h^{q,p}(X)$. So for X compact Kähler and k odd, $\dim_{\mathbb{C}} H_{\mathrm{DR}}^k(X)$ – which is called the kth *Betti number* of X – is always even.

For any complex manifold X, given an open set $U \subset X$, a form $\omega \in \mathcal{A}^{p,0}(U)$ is called a *holomorphic p-form* on U if $\bar{\partial}\omega = 0$. The sheaf of holomorphic p-forms on X is denoted Ω^p. Since there can be no non-trivial $\bar{\partial}$-exact forms of type $(p, 0)$, the holomorphic p-forms make up the $(p, 0)$th Dolbeaut cohomology group – that is, $H^{p,0}(X) = \Omega^p(X)$. In terms of local coordinates a holomorphic p-form is just a form $\omega = \sum_{|I|=p} f_I dz_I$ where each $f_I : U \to \mathbb{C}$ is holomorphic. A straightforward calculation shows that holomorphic forms are also d-closed.

If X has dimension n, then Ω^n is locally of rank 1. The corresponding complex line bundle is called the *canonical bundle* of X, denoted by K_X. The triviality of K_X is then equivalent to the existence of a nowhere zero global holomorphic n-form on X. For X Kähler, this is precisely the condition for X to be a *Calabi-Yau* manifold.

Holomorphic 2-forms will play an important role for us. Being of type $(2, 0)$, a holomorphic 2-form $\omega \in \Omega^2(X)$ determines a $\mathbb{C}$-bilinear map $\omega_x : T_{X,x} \times T_{X,x} \to \mathbb{C}$ for each $x \in X$. Hence it induces a $\mathbb{C}$-linear map from $T_{X,x}$ to $\operatorname{Hom}_{\mathbb{C}}(T_{X,x}, \mathbb{C}) = \Omega^1_x$. To say that ω is *non-degenerate* at x is to say that this map is an isomorphism.

Definition 4.1 An *irreducible hyperkähler* manifold (also called *irreducible symplectic*) is a compact Kähler manifold X such that (i) X is simply connected and (ii) $\Omega^2(X)$ is spanned by an everywhere non-degenerate holomorphic 2-form.

The basic properties of irreducible hyperkähler manifolds can be found in Section 1 of [12]. Such properties include: $\dim(X)$ is even, $h^{2,0}(X) = h^{0,2}(X) = 1$, and K_X is trivial. (The latter is because, if ϕ is a holomorphic 2-form on X witnessing the hyperkähler condition, and $\dim(X) = 2r$ then ϕ^r is an everywhere nonzero holomorphic $2r$-form on X.) For surfaces, condition (ii) in Definition 4.1 is equivalent to the triviality of K_X. The so-called $K3$ *surfaces* are precisely the irreducible hyperkähler manifolds which have dimension 2. $K3$ surfaces have been widely studied since their introduction by Weil. A considerable amount of information on them can be found in [3]. Irreducible hyperkähler manifolds are now widely studied as higher-dimensional generalizations of $K3$ surfaces. We will see in the next section the (conjectured) role of $K3$ surfaces and higher dimensional irreducible hyperkählers in the model theory of Kähler manifolds.

Given a complex manifold X, a cohomology class $[\omega] \in H^k_{\mathrm{DR}}(X)$ is called *integral* if under the identification

$$H^k_{\mathrm{DR}}(X) = H^k_{\mathrm{sing}}(X, \mathbb{C}) = H^k_{\mathrm{sing}}(X, \mathbb{Z}) \otimes \mathbb{C},$$

$[\omega]$ is contained in $H^k_{\mathrm{sing}}(X, \mathbb{Z}) \otimes 1$. Equivalently, the map $\gamma \mapsto \int_\gamma \omega$ on real k-cycles is integer-valued. Similarly, $[\omega]$ is *rational* if it is contained in $H^k_{\mathrm{sing}}(X, \mathbb{Z}) \otimes \mathbb{Q}$ under the above identification.[5]

Definition 4.2 A *Hodge manifold* is a compact complex manifold which admits a hermitian metric h whose associated Kähler form ω – which recall is a real 2-form of type $(1,1)$ – is d-closed and $[\omega]$ is integral.

In particular, a Hodge manifold is Kähler.

For example, $\mathbb{P}_n(\mathbb{C})$ is a Hodge manifold. Indeed, if $\omega \in \mathcal{A}^{1,1}(\mathbb{P}_n)$ is the Kähler form associated to the Fubini-Study metric (see Example 3.5), and we view $\mathbb{P}_1$ as a real 2-cycle in $\mathbb{P}_n$, then $\int_{\mathbb{P}_1} \omega = \pi$. Since the (class of) $\mathbb{P}_1$ generates $H_2(\mathbb{P}_n)$, $[\frac{1}{\pi}\omega] \in H^2_{\mathrm{DR}}(\mathbb{P}_n)$ is integral. It follows that every projective algebraic manifold is Hodge. A famous theorem of Kodaira (sometimes called Kodaira's embedding theorem) says the converse:

Kodaira's Embedding Theorem. *Every Hodge manifold is (biholomorphic to) a projective algebraic manifold.*

A consequence of Kodaira's theorem relevant for us is:

Corollary 4.3 *Any compact Kähler manifold with no nonzero global holomorphic 2-forms is projective.*

Sketch of proof. If $0 = \Omega^2(X) = H^{2,0}(X)$ then also $H^{0,2}(X) = 0$. By Hodge decomposition it follows that $H^2_{\mathrm{DR}}(X) = H^{1,1}(X)$. Now let

$$H^{1,1}(X, \mathbb{R}) := \{[\omega] : \omega \text{ is real, } d\text{-closed, type}(1,1)\}.$$

Then the set

$$C := \{[\omega] \in H^{1,1}(X, \mathbb{R}) : \omega \text{ corresponds to a Kähler metric}\}$$

is *open* in $H^{1,1}(X, \mathbb{R})$ – the argument being that a small deformation of a Kähler metric is Kähler. As X is Kähler, $C \neq \emptyset$. On the other hand, as $H^{1,1}(X) = H^2_{\mathrm{DR}}(X)$, we have that $H^{1,1}(X, \mathbb{R}) = H^2_{\mathrm{sing}}(X, \mathbb{Z}) \otimes \mathbb{R}$, and hence C must contain an element of $H^2_{\mathrm{sing}}(X, \mathbb{Z}) \otimes \mathbb{Q}$. Taking a suitable

---

5 We have chosen not to go through the definitions of *sheaf cohomology*, but for those familiar with it, $H^k_{\mathrm{sing}}(X, \mathbb{Z})$ coincides with $H^k(X, \mathbb{Z})$, the kth sheaf cohomology group of X with coefficients in the constant sheaf $\mathbb{Z}$. Likewise for $\mathbb{Q}$, $\mathbb{R}$, or $\mathbb{C}$ in place of $\mathbb{Z}$. In fact for Kähler manifolds, the Dolbeaut cohomology group $H^{p,q}(X)$ coincides with $H^q(X, \Omega^p)$. In any case, the integral classes can be described as those in $H^k(X, \mathbb{Z})$ and the rational ones as those in $H^k(X, \mathbb{Q})$.

integral multiple, we obtain an integral class in C. Thus X is a Hodge manifold, and so projective by Kodaira's embedding theorem. $\square$

In fact we will require rather a *relative* version, proved in a similar fashion, and attributed in [7] to Claire Voisin:

Corollary 4.4 *Suppose that* $f : X \to S$ *is a fibration in* C, *and the generic fibre of* f *is not projective, then there exists a global holomorphic 2-form* $\omega \in \Omega^2(X)$ *whose restriction* ω_a *to a generic fibre* X_a *is a nonzero global holomorphic 2-form on* X_a.

5 Stability theory and Kähler manifolds

In this section we will discuss some outstanding problems concerning the model theory (or rather stability theory) of $\text{Th}(C)$. One concerns identifying (up to nonorthogonality, or even some finer equivalence relation), the trivial U-rank 1 types. The second is the conjecture that $\text{Th}(C)$ is *nonmultidimensional*. As we shall see the problems are closely related.

Because of the results in section 3, and as discussed at the end of that section, we may treat C as a universal domain for $\text{Th}(C)$. The main use of this is the existence of *generic* points: given countably many parameters A from C, and a Kähler-type variety X, there exist points in X that are not contained in any proper analytic subsets of X *defined over* A in the Barlet language of X. The fact that we need not pass to elementary extensions in order to find such generic points makes the model-theoretic study of $\text{Th}(C)$ much more accessible than that of $\text{Th}(A)$.

In [19] strongly minimal sets were discussed as "building blocks" for structures of finite Morley rank. In fact one needs a slightly more general notion, that of a *stationary type of* U-*rank* 1, sometimes also called a *minimal type*. Let us assume for now that T is a stable theory, and we work in a saturated model $\bar{M}$. A complete (nonalgebraic) type $p(x) \in S(A)$ is *minimal* or *stationary of* U-*rank* 1 if for any $B \supseteq A$, p has a unique extension to a nonalgebraic complete type over B. Equivalently any (relatively) definable subset of the set of realizations of p is finite or cofinite. If X is a strongly minimal set defined over A, and $p(x) \in S(A)$ is the "generic" type of X, then $p(x)$ is minimal. However not every minimal type comes from a strongly minimal set. For example take T to be the theory with infinitely many disjoint infinite unary predicates P_i (and nothing else), and take p to be the complete type over $\emptyset$ axiomatized by $\{\neg P_i(x) : i < \omega\}$. We discussed the notion of *modularity* of a definable set X in [19]. The original definition was that X is modular if for all tuples a, b of elements of X, a is independent from b over $\mathrm{acl}(a) \cap \mathrm{acl}(b)$ (together with a fixed set of parameters over which X is defined), where acl is computed in $\bar{M}^{\mathrm{eq}}$. The same definition makes sense with a type-definable set (such as the set of realizations of a complete type) in place of X. So we obtain in particular the notion of a modular minimal type. The minimal type $p(x) \in S(A)$ is said to be *trivial* if whenever $a, b_1, \ldots, b_n$ are realizations of p and $a \in \mathrm{acl}(A, b_1, \ldots, b_n)$ then $a \in \mathrm{acl}(A, b_i)$ for some i. Triviality implies modularity (for minimal types). On the other hand, if p is a modular nontrivial type then $p(x)$ is nonorthogonal (see below or [19]) to a minimal type q which is the generic type of a definable group G. Assuming the ambient theory to be totally transcendental, G will be strongly minimal, so p will also come from a strongly minimal set. Under the same assumption (ambient theory is totally transcendental), any nonmodular minimal type will come from a strongly minimal set. So a divergence between strongly minimal sets and minimal types is only possible for trivial types.

Let us apply these notions to $\mathrm{Th}(\mathcal{C})$. Those compact complex varieties in $\mathcal{C}$ whose generic type is minimal are precisely the so-called *simple* complex varieties. The formal definition is that a compact complex variety X in $\mathcal{C}$ is *simple* if it is irreducible and if a is a generic point of X (over some set of definition) then there is no analytic subvariety Y of X containing a with $0 < \dim(Y) < \dim(X)$. (There is an appropriate definition not mentioning generic points, and hence also applicable to all compact complex varieties.) We are allowing the possibility that $\dim(X) = 1$, although sometimes this case is formally excluded in the definition of simplicity. In fact, all compact complex curves are simple. Moreover, a projective algebraic variety is simple if and only if it is of dimension 1. If X is simple we may sometimes say "X is modular, trivial, etc." if its generic type has that property.

Example 5.1 Given a $2n$-dimensional lattice $\Lambda \leq \mathbb{C}^n$, the quotient $T = \mathbb{C}^n/\Lambda$ inherits the structure of an n-dimensional compact Kähler manifold. Such manifolds are called *complex tori*. The additive group structure on $\mathbb{C}^n$ induces a compact complex Lie group structure on T. If the lattice is chosen sufficiently generally – namely the real and imaginary parts of a $\mathbb{Z}$-basis for Λ form an algebraically independent set over $\mathbb{Q}$ – then it is a fact that T has no proper infinite complex analytic subsets, and hence is strongly minimal.

A complex torus which is *algebraic* (bimeromorphic with an algebraic variety) is a certain kind of complex algebraic group: an abelian variety. So the only strongly minimal algebraic complex tori are the elliptic curves, that is the 1-dimensional abelian varieties.

If p and q are the generic types of X and Y respectively, then p is nonorthogonal to q (we might say X is nonorthogonal to Y) if and only if there is a proper analytic subvariety $Z \subset X \times Y$ projecting onto both X and Y. Note that if p and q are minimal then Z must be a *correspondence*: the projections $Z \to X$ and $Z \to Y$ are both generically finite-to-one.

Fact 5.2 *Let $p(x)$ be a minimal type over C. Let X be the compact complex variety whose generic type is p. Then either:*

(i) p is nonmodular in which case X is an algebraic curve,

(ii) p is modular, nontrivial, in which case X is nonorthogonal to (i.e. in correspondence with) a strongly minimal complex nonalgebraic torus (necessarily of dimension > 1), or

(iii) p is trivial, and $\dim(X) > 1$.

Proof. This is proved in [22] for the more general case of $\mathcal{A}$. We give a slightly different argument here.

From the truth of the *strong conjecture* (cf. [19]) for $\mathcal{A}$ one deduces that if p is nonmodular then X is nonorthogonal to a simple algebraic variety Y. Y has to be of dimension 1. Simplicity implies that X is also of dimension 1, and so, by the Riemann existence theorem, an algebraic curve.

If p is modular and nontrivial, then as remarked above, up to nonorthogonality p is the generic type of a strongly minimal (modular) group G. It is proved in [22] that any such group is definably isomorphic to a (strongly minimal) complex torus T. If T had dimension 1 then by the Riemann existence theorem it would be algebraic, so not modular.

Likewise in the trivial case, X could not be an algebraic curve so has dimension > 1.   □

So the classification or description of simple trivial compact complex varieties in C remains. Various model-theoretic conjectures have been made in earlier papers: for example that they are strongly minimal, or even that they must be ω-categorical when equipped with their canonical Barlet language (see section 3).

To understand the simple trivial compact *surfaces* we look to the classification of compact complex surfaces carried out by Kodaira in a series of papers in the 1960's, extending the Enriques classification of algebraic surfaces. An account of Kodaira's work appears in [3]. In particular Table 10 in Chapter VI there is rather useful. From it we can deduce:

Proposition 5.3 *Let X be a simple trivial compact complex variety of dimension 2 which is in the class $\mathcal{C}$. Then X is bimeromorphic to a $K3$ surface. Expressed otherwise, a stationary trivial minimal type in $\mathcal{C}$ of dimension 2 is, up to interdefinability, the generic type of a $K3$ surface.*

Proof. The classification of Kodaira gives a certain finite collection of (abstractly defined) classes, such that every compact surface has a "minimal model" in exactly one of the classes, in particular is bimeromorphic to something in one of the classes. Suppose X is a simple trivial surface in $\mathcal{C}$. Then X has algebraic dimension 0 (namely X does not map holomorphically onto any algebraic variety of dimension > 0), X is not a complex torus, and X has first Betti number even. Moreover these properties also hold of any Y bimeromorphic to X. By looking at Table 10, Chapter VI of [3], the only possibility for a minimal model of X is to be a $K3$ surface. □

Among $K3$ surfaces are (i) smooth surfaces of degree 4 in $\mathbb{P}_3$, and (ii) *Kummer surfaces*. A Kummer surface is something obtained from a 2-dimensional complex torus by first quotienting by the map $x \mapsto -x$ and then taking a minimal resolution. See Chapter VIII of [3] for more details. In particular there are algebraic $K3$ surfaces, and there are simple $K3$ surfaces which are not trivial. However there do exist $K3$ surfaces of algebraic dimension 0 (that is, which do not map onto any algebraic variety) and which are not Kummer, and these *will be* simple and trivial (see [17]). On the other hand all $K3$ surfaces are *diffeomorphic* (that is, isomorphic as real differentiable manifolds), and in fact they were first defined by Weil precisely as compact complex analytic surfaces diffeomorphic to a smooth quartic surface in $\mathbb{P}_3$.

It is conceivable, and consistent with the examples, that the natural analogue of Proposition 5.1 holds for higher dimensions:

Conjecture I. Any simple trivial compact complex variety in $\mathcal{C}$ is bimeromorphic to (or at least in correspondence with) an irreducible hyperkähler manifold. Equivalently any trivial minimal type in $\mathcal{C}$ is nonorthogonal to the generic type of some irreducible hyperkähler manifold.

Note that any irreducible hyperkähler manifold has even dimension. Also, as with the special case of $K3$ surfaces, there are (irreducible) hyperkählers of any even dimension which are algebraic (and hence not trivial).

Let us now pass to the stability-theoretic notion of *nonmultidimensionality*. We start with an arbitrary complete (possibly many-sorted) stable theory T, and work in a saturated model $\bar{M}$ of T. Let $p(x) \in S(A)$, $q(y) \in S(B)$ be stationary types (over small subsets A, B of $\bar{M}$). Then p is said to be *nonorthogonal* to q if there is $C \supseteq A \cup B$ and realizations a of p, and b of q, such that: (i) a is independent from C over A, and b is independent from C over B, and (ii) a forks with b over C.

A stationary type $p(x) \in S(A)$ is said to be nonorthogonal to a *set* of parameters B if p is nonorthogonal to some complete type over acl(B). The theory T is said to be *nonmultidimensional* if every stationary nonalgebraic type $p(x) \in S(A)$, is nonorthogonal to $\emptyset$. An equivalent characterization is:

($\star$) Whenever $p(x, a)$ is a stationary nonalgebraic type (with domain enumerated by the possibly infinite tuple a), and $stp(a') = stp(a)$, then $p(x, a)$ is nonorthogonal to $p(x, a')$.

Remark 5.4 *If T happens to be superstable, then it suffices that* ($\star$) *holds for $p(x, a)$ regular, and moreover we may assume that a is a finite tuple. If moreover T has finite rank (meaning every finitary type has finite U-rank), then it suffices for* ($\star$) *to hold for types $p(x, a)$ of U-rank 1.*

A stronger condition than nonmultidimensionality is *unidimensionality* which says that any two stationary nonalgebraic types are nonorthogonal. This is equivalent to T having exactly one model of cardinality κ for all $\kappa > |T|$. Nonmultidimensionality was also introduced by Shelah [24] in connection with classifying and counting models. For totally transcendental T (namely every formula has ordinal valued Morley rank), T is nonmultidimensional if and only if there is some fixed cardinal μ_0 (which will be at most $|T|$) such that *essentially* the models of T are naturally in one-one correspondence with sequences $(\kappa_\alpha : \alpha < \mu_0)$

of cardinals. When μ_0 is finite, T is called *finite-dimensional*. Alternatively (for T superstable of finite rank) this means that there are only finitely many stationary U-rank 1 types up to nonorthogonality.

Remark 5.5 (cf. [20]) *Suppose that T is superstable of finite rank and nonmultidimensional. Suppose moreover that T is one-sorted and that every stationary type of U-rank 1 is nonorthogonal to a type of Morley rank 1. Then T is finite-dimensional.*

The following conjecture was formulated (by Thomas Scanlon and the second author) around 2000-2001. They also pointed out (in [21]) that it fails for $\text{Th}(\mathcal{A})$.

Conjecture II. $\text{Th}(\mathcal{C})$ is nonmultidimensional.

Let $p(x, a)$ be a stationary type in $\mathcal{C}$ realized by b say (where a is a finite tuple). Then $stp(a, b)$ is the generic type of a compact complex variety X, $stp(a)$ is the generic type of a compact complex variety S and the map $(x, y) \to x$ gives a dominant meromorphic map f from X to S, and $tp(a/b)$ is the generic type of the irreducible fibre X_a. Without changing $p(x, a)$ we may assume that X and S are manifolds and that f is a holomorphic submersion (so that the generic fibre of f is a manifold also). The requirement that for another realization a' of $stp(a)$, $p(x, a)$ and $p(x, a')$ are nonorthogonal, becomes: for a' another generic point of S, there is some proper analytic subset Z of $X_a \times X_{a'}$ which projects onto both X_a and $X_{a'}$. So by $(\star)$ above, we obtain the following reasonably geometric account or interpretation of the nonmultidimensionality of $\text{Th}(\mathcal{C})$: for any fibration $f : X \to S$ in $\mathcal{C}$, any two generic fibres have the feature that there is a proper analytic subset of their product, projecting onto each factor. By Remark 5.4, we may restrict to the case where the generic fibre X_a is *simple*. So we obtain:

Remark 5.6 *Conjecture II is equivalent to:* Whenever $f : X \to S$ is a fibration in $\mathcal{C}$ with generic fibre a simple compact complex manifold, then f is *weakly isotrivial* in the sense that for any generic fibres X_s, $X_{s'}$, there is a *correspondence* between X_s and $X_{s'}$.

Of course there are other stronger conditions than weak isotriviality which a fibration $f : X \to S$ may satisfy, for example that any two generic fibres are bimeromorphic or even that any two generic fibres are biholomorphic. If the latter is satisfied we will call the fibration *isotrivial*.

Let us begin a discussion of Conjecture II. Let $f : X \to S$ be a fibration in $\mathcal{C}$ with simple generic fibre X_s. By 5.2, X_s is either (i) an algebraic curve, (ii) a simple nonalgebraic complex torus, or (iii) has trivial generic type. In case (i), we obtain weak isotriviality (as any two algebraic curves project generically finite-to-one onto $\mathbb{P}_1$). So we are reduced to cases (ii) and (iii). Special cases of case (ii) are proved by Campana [7]. Assuming the truth of Conjecture I, case (iii) is also proved in [7]. An exposition of this work is one of the purposes of this paper and appears in the next section.

For now, we end this section with a few additional remarks on isotriviality.

Remark 5.7 *Let $f : X \to S$ be a fibration in $\mathcal{C}$. If f is locally trivial in the sense that for some nonempty open subset U of S, X_U is biholomorphic to $U \times Y$ over U for some compact complex variety Y, then f is isotrivial.*

Proof. By Baire category, we can find $s_1, s_2 \in U$ which are mutually generic. So X_{s_1} is isomorphic to X_{s_2} by assumption. But $tp(s_1, s_2)$ is uniquely determined by the mutually genericity of s_1, s_2. Hence for any mutually generic $s_1, s_2 \in S$, X_{s_1} is isomorphic to X_{s_2}. Now given generic $s_1, s_2 \in S$, choose $s \in S$ generic over $\{s_1, s_2\}$. So X_s is isomorphic to each of X_{s_1}, X_{s_2}. $\square$

Remark 5.8 *Suppose that $f : X \to S$ is a fibration in $\mathcal{C}$ whose generic fibre X_s is a simple nonalgebraic complex torus. Suppose moreover that any (some) two mutually generic fibres X_s, $X_{s'}$ are nonorthogonal. Then any two generic fibres are isomorphic (as complex tori).*

Proof. Fix two mutually generic fibres X_s and X'_s. These are both locally modular strongly minimal groups. Hence nonorthogonality implies that there is a strongly minimal subgroup C of $X_s \times X_{s'}$ projecting onto both factors, and this induces an isogeny from $X_{s'}$ onto X_s, and thus an isomorphism (of complex tori) between $X_{s'}/A_{s'}$ and X_s for some finite subgroup $A_{s'}$ of $X_{s'}$. Note that $A_{s'}$ is $\mathrm{acl}(s')$-definable. Now let s_1, s_2 be generic points of S. Let $s' \in S$ be generic over $\{s_1, s_2\}$. So there is an isomorphism f_1 between $X_{s'}/A_{s'}$ and X_{s_1} (for some finite, so $\mathrm{acl}(s')$-definable subgroup of $X_{s'}$) with X_{s_1}. As s_1 and s_2 have the same type over $\mathrm{acl}(s')$, we obtain an isomorphism f_2 between $X_{s'}/A_{s'}$ and X_{s_2}. Thus X_{s_1} and X_{s_2} are isomorphic. $\qquad\square$

6 Local Torelli and the isotriviality theorem

In this section we state and sketch the proof of a recent result of Campana [7], which was motivated by and partially resolves the nonmultidimensionality conjecture for $\mathcal{C}$ discussed in the previous section.

Suppose Y is a compact Kähler manifold and consider a *deformation* $f : X \to S$ – that is, f is a proper holomorphic submersion between complex manifolds X and S and there is a point $o \in S$ such that $X_o = Y$. For s near o, X_s will be a compact Kähler manifold (see Theorem 9.23 of [25]). Diffeomorphically, f is locally trivial: there exists an open neighbourhood $U \subseteq S$ of o such that X_U is diffeomorphic to $U \times Y$ over U. Letting u be this diffeomorphism we have the commuting diagram:

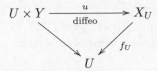

For each $s \in U$, the diffeomorphism $u_s : Y \to X_s$ induces an isomorphism of singular cohomology groups, and hence by De Rham's Theorem, of the De Rham cohomology groups. In particular, we obtain a group isomorphism, $\hat{u}_s : H^2_{\mathrm{DR}}(X_s) \to H^2_{\mathrm{DR}}(Y)$. From our discussion of De Rham's theorem in Section 2 it is not hard to see that, under the identification

$$H^2_{\mathrm{DR}}(Y) = H^2_{\mathrm{sing}}(Y, \mathbb{C}) = \mathrm{Hom}_{\mathbb{C}}\left(H_2(Y), \mathbb{C}\right),$$

the isomorphism $\hat{u}_s : H^2_{\mathrm{DR}}(X_s) \to H^2_{\mathrm{DR}}(Y)$ is given by

$$[\omega] \longmapsto \left([\gamma] \mapsto \int_\gamma u_s^* \omega\right)$$

where ω is a d-closed 2-form on X_s and and γ is a real 2-cycle on Y.

Since u may not be biholomorphic, $\hat{u}_s$ does not necessarily respect the Hodge decomposition of $H^2_{\mathrm{DR}}(X_s)$ and $H^2_{\mathrm{DR}}(Y)$. Indeed, one measure of how far u is from being a biholomorphic trivialisation is the *period map* of Y for holomorphic 2-forms (with respect to the deformation f):

$$p : U \to \mathrm{Grass}\left(H^2_{\mathrm{DR}}(Y)\right)$$

which assigns to each $s \in U$ the subspace $\hat{u}_s\left(H^{2,0}(X_s)\right)$. Recall that for any complex manifold M, $H^{2,0}(M)$ is just the space $\Omega^2(M)$ of global holomorphic 2-forms on M.

Note that if u is a biholomorphism then the period map is constant on U since for all $s \in U$ $\hat{u}_s(H^{2,0}(X_s)) = H^{2,0}(Y)$.

Definition 6.1 Suppose Y is a compact Kähler manifold. We say that Y satisfies *local Torelli for holomorphic 2-forms* if the following holds: given any deformation $f : X \to S$ of Y with a local diffeomorphic trivialisation $u : U \times Y \to X_U$, if the corresponding period map is constant on U then u is in fact a biholomorphic trivialisation.

Example 6.2 Complex tori and irreducible hyperkähler manifolds all satisfy local Torelli for holomorphic 2-forms. (See Theorem 5(b) of [4] for the case of irreducible hyperkähler manifolds.)

We can now state the isotriviality theorem we are interested in.

Theorem 6.3 (Campana [7]) *Suppose $f : X \to S$ is a fibration where X and S are compact Kähler manifolds. Assume that for $a \in S$ generic, (i) X_a is not projective, (ii) $\dim_{\mathbb{C}} \Omega^2(X_a) = 1$, and (iii) X_a satisfies local Torelli for holomorphic 2-forms. Then f is isotrivial.*

Sketch of proof. From condition (i) and Corollary 4.4 there exists a global holomorphic 2-form $\omega \in \Omega^2(X)$ whose restriction ω_a to the generic fibre X_a is a nonzero global holomorphic 2-form on X_a. Moreover by condition (ii), ω_a spans $\Omega^2(X_a)$.

Let U be an open neighbourhood of a such that there is a diffeomorphic trivialisation $u : U \times X_a \to X_U$ over U. We show that the corresponding period map $p : U \to \text{Grass}(H^2_{\text{DR}}(X_a))$ is constant on U. By local Torelli this will imply that u is biholomorphic and hence, by Remark 5.7, f is isotrivial.

For any $s \in U$ let $\hat{u}_s : H^2_{\text{DR}}(X_s) \to H^2_{\text{DR}}(X_a)$ be the isomorphism induced by u and discussed above. We need to show that $\hat{u}_s(H^{2,0}(X_s)) = \hat{u}_t(H^{2,0}(X_t))$ for all $s, t \in U$. But, shrinking U if necessary, $H^{2,0}(X_s) = \Omega^2(X_s)$ is spanned by the restriction ω_s of ω to X_s, for all $s \in U$. Hence it suffices to show that $\hat{u}_s(\omega_s) = \hat{u}_t(\omega_t)$ for all $s, t \in U$.

Now fix a real 2-cycle γ on X_a. Viewing $\hat{u}_s(\omega_s)$ and $\hat{u}_t(\omega_t)$ as elements of $\text{Hom}_{\mathbb{C}}(H_2(X_a), \mathbb{C}) = H^2_{\text{DR}}(X_a)$ we compute

$$\left(\hat{u}_s(\omega_s) - \hat{u}_t(\omega_t)\right)[\gamma] = \int_\gamma u_s^*\omega_s - \int_\gamma u_t^*\omega_t = \int_{u_s \circ \gamma} \omega_s - \int_{u_t \circ \gamma} \omega_t.$$

Here $u_s \circ \gamma$ and $u_t \circ \gamma$ are 2-cycles on X_s and X_t respectively. Viewed as 2-cycles on X we have

$$\int_{u_s \circ \gamma} \omega_s - \int_{u_t \circ \gamma} \omega_t = \int_{(u_s \circ \gamma - u_t \circ \gamma)} \omega.$$

But $(u_s \circ \gamma - u_t \circ \gamma)$ is the boundary of some 3-cycle λ on X. By Stokes', $\int_{(u_s \circ \gamma - u_t \circ \gamma)} \omega = \int_\lambda d\omega$. Since holomorphic forms are d-closed it follows that

$$\left(\hat{u}_s(\omega_s) - \hat{u}_t(\omega_t) \right)[\gamma] = 0$$

for all 2-cycles γ on X_a. That is, $\hat{u}_s(\omega_s) = \hat{u}_t(\omega_t)$ for all $s, t \in U$. So the period map is constant on U and f is isotrivial. $\qquad\square$

Remark 6.4 *The hypotheses of Theorem 6.3 are valid in the following cases:*
(a) The generic fibre X_a is irreducible hyperkähler and nonprojective,
(b) The generic fibre X_a is a simple complex torus of dimension 2.

Proof. We have already mentioned that complex tori and irreducible hyperkähler satisfy local Torelli for holomorphic 2-forms. Nonprojectivity is assumed in (a) and follows for (b) by the fact that the only simple projective varieties are curves. Finally, $\dim_{\mathbb{C}} \Omega^2(X_a) = 1$ is true of irreducible hyperkähler manifolds by definition, and true of simple complex tori of dimension 2 by the fact that dimension 2 forces $\dim_{\mathbb{C}} \Omega^2(X_a)$ to be at most 1 while nonprojectivity forces it to be at least 1.

$\qquad\square$

Let us return to the nonmultidimensionality conjecture (Conjecture II) from section 5, bearing in mind the equivalence stated in Remark 5.6.

Corollary 6.5 *The nonmultidimensionality conjecture holds in $\mathrm{Th}(\mathcal{C})$ for surfaces. In other words if $p(x)$ is a minimal type of dimension 1 or 2 over some model of $\mathrm{Th}(\mathcal{C})$ then p is nonorthogonal to $\emptyset$.*

Proof. As discussed at the end of section 3 we may work in $\mathcal{C}$ itself. Let $p(x) = tp(b/a)$ for a, b from $\mathcal{C}$ and b a generic point of an a-definable simple compact complex manifold X_a of dimension 1 or 2. We have already pointed out that in the case of dimension 1 (i.e. of projective curves), X_a is nonorthogonal to $X_{a'}$ whenever $stp(a) = stp(a')$. So assume X_a is a simple compact complex surface. It is then not projective.

By Fact 5.2 and Proposition 5.3, we may assume that X_a is either a 2-dimensional simple complex torus, or a nonprojective $K3$ surface. So by Theorem 6.3 and Remark 6.4, X_a is biholomorphic to $X_{a'}$ whenever $stp(a) = stp(a')$. Hence $tp(b/a)$ is nonorthogonal to $\emptyset$. $\square$

Condition (ii) of Theorem 6.3 seems rather strong, and indeed, Campana works with the following weaker condition: A *rational Hodge substructure* of $H^2_{DR}(X_a)$ is a $\mathbb{C}$-vector subspace V such that $V^{i,j} = \overline{V^{j,i}}$ where $V^{i,j} := V \cap H^{i,j}(X_a)$, and $V = V_{\mathbb{Q}} \otimes \mathbb{C}$ where

$$V_{\mathbb{Q}} := V \cap H^2_{\mathrm{sing}}(X_a, \mathbb{Q}).$$

Campana says that X_a is *irreducible in weight* 2 if for any rational Hodge substructure $V \subseteq H^2_{DR}(X_a)$, either $V^{2,0} = 0$ or $V^{2,0} = H^{2,0}(X_a)$. By a theorem of Deligne, the image of $H^2_{DR}(X)$ in $H^2_{DR}(X_a)$ under the restriction map is a rational Hodge substructure. Hence, the above proof of Theorem 6.3 works if condition (ii) is replaced by the irreducibility of X_a in weight 2. Campana proves that the "general" torus of dimension ≥ 3 is irreducible in weight 2. Apparently it is open whether any simple nonalgebraic torus is irreducible in weight 2. This together with Conjecture I (that any simple trivial compact Kähler manifold is nonorthogonal to an irreducible hyperkahler manifold) are the remaining obstacles to the nonmultidimensionality conjecture for $\mathcal{C}$.

References

[1] M. Aschenbrenner, R. Moosa, and T. Scanlon, Strongly minimal groups in the theory of compact complex spaces, *The Journal of Symbolic Logic*, 71(2):529–552, 2006.

[2] D. Barlet, Espace analytique réduit des cycles analytiques complexes compacts d'un espace analytique complexe de dimension finie, In *Fonctions de plusieurs variables complexes, II (Sém. François Norguet, 1974–1975)*, volume 482 of *Lecture Notes in Math.*, pages 1–158. Springer, 1975.

[3] W. Barth, C. Peters, and A. Van de Ven, *Compact complex surfaces*, Springer-Verlag, 1984.

[4] A. Beauville, Variétés Kähleriennes dont la première classe de Chern est nulle, *J. Differential Geometry* 18 (1983), 755-782.

[5] E. Bishop, Conditions for the analyticity of certain sets, *The Michigan Mathematical Journal*, 11:289–304, 1964.

[6] F. Campana, *Application de l'espace des cycles à la classification biméromorphe des espaces analytiques Kählériens compacts*, PhD thesis, Université Nancy 1, Prépublication de l'université Nancy 1, no. 2, Mai 1980, p. 1–163.

[7] F. Campana, Isotrivialité de certaines familles Kählériennes de variétés non projectives, *Math. Z.* 252 (2006), no. 1, 147–156.

[8] K. Fritzsche and H. Grauert, *From Holomorphic Functions to Complex Manifolds*, Springer, 2002.

[9] A. Fujiki, Closedness of the Douady spaces of compact Kähler spaces, *Publication of the Research Institute for Mathematical Sciences*, 14(1):1–52, 1978.

[10] M. Herrera, Integration on a semianalytic set, *Bulletin de la Société Mathématique de France*, 94:141–180, 1966.

[11] E. Hrushovski and B. Zilber, Zariski geometries, *Bulletin of the American Mathematical Society*, 28(2):315–322, 1993.

[12] D. Huybrechts, Compact hyper-Kähler manifolds: basic results, *Inventiones Mathematicae*, 135(1):63–113, 1999.

[13] D. Lieberman, Compactness of the Chow scheme: applications to automorphisms and deformations of Kähler manifolds, In *Fonctions de plusieurs variables complexes, III (Sém. François Norguet, 1975–1977)*, volume 670 of *Lecture Notes in Math.*, pages 140–186. Springer, 1978.

[14] R. Moosa, A nonstandard Riemann existence theorem, *Transactions of the American Mathematical Society*, 356(5):1781–1797, 2004.

[15] R. Moosa, The model theory of compact complex spaces, In *Logic Colloquium '01*, volume 20 of *Lect. Notes Log.*, pages 317–349. Assoc. Symbol. Logic, 2005.

[16] R. Moosa, On saturation and the model theory of compact Kähler manifolds, *Journal für die reine und angewandte Mathematik*, 586:1–20, 2005.

[17] A. Pillay, Some model theory of compact complex spaces, In *Hilbert's tenth problem: relations with arithmetic and algebraic geometry (Ghent, 1999)*, 323–338, Contemp. Math., 270, Amer. Math. Soc., Providence, RI, 2000.

[18] A. Pillay, Model-theoretic consequences of a theorem of Campana and Fujiki, *Fundamenta Mathematicae*, 174(2):187–192, 2002.

[19] A. Pillay, Model theory and stability theory, with applications in differential algebra and algebraic geometry, *This volume*.

[20] A. Pillay and W. Pong, On Lascar rank and Morley rank of definable groups in differentially closed fields, *The Journal of Symbolic Logic*, 67(3):1189–1196, 2002.

[21] A. Pillay and T. Scanlon, Compact complex manifolds with the DOP and other properties, *The Journal of Symbolic Logic*, 67(2):737–743, 2002.

[22] A. Pillay and T. Scanlon, Meromorphic groups, *Transactions of the American Mathematical Society*, 355(10):3843–3859, 2003.

[23] D. Radin, A definability result for compact complex spaces, *The Journal of Symbolic Logic*, 69(1):241–254, 2004.

[24] S. Shelah, *Classification theory and the number of nonisomorphic models*, volume 92 of *Studies in Logic and the Foundations of Mathematics*, North-Holland Publishing Co., Amsterdam, 1978.

[25] C. Voisin, *Hodge theory and complex algebraic geometry, I*, volume 76 of *Cambridge studies in advanced mathematics*, Cambridge University Press, 2002.

[26] B. Zilber, Model theory and algebraic geometry, In *Proceedings of the 10th Easter Conference on Model Theory*, Eds: M. Weese and H. Wolter, Humboldt Universität, Berlin 1993, 202–222.

Some local definability theory for holomorphic functions

A.J. Wilkie

The University of Manchester

Summary

Let $\mathcal{F}$ be a collection of holomorphic functions and let $\mathbb{R}(PR(\mathcal{F}))$ denote the reduct of the structure $\mathbb{R}_{an}$ to the ordered field operations together with the set of proper restrictions (see below) of the real and imaginary parts of all functions in $\mathcal{F}$. We ask the question: Which holomorphic functions are locally definable (i.e., have their real and imaginary parts locally definable) in the structure $\mathbb{R}(PR(\mathcal{F}))$? It is easy to see that the collection of all such functions is closed under composition, partial differentiation, implicit definability (via the Implicit Function Theorem in one dependent variable) and Schwarz Reflection. We conjecture that this exhausts the possibilities and we prove as much in the neighbourhood of generic points. More precisely, we show that these four operations determine the natural pregeometry associated with $\mathbb{R}(PR(\mathcal{F}))$-definable, holomorphic functions.

1 Introduction

In this paper a holomorphic function is always understood to have domain an open subset of $\mathbb{C}^n$ for some n. If Δ is an open box in $\mathbb{C}^n$ with rational data (i.e., $\Delta = D_1 \times \cdots \times D_n$ for some open rectangles $D_1, \ldots, D_n$ in $\mathbb{C}$ with Gaussian rational corners) and F is a holomorphic function whose domain contains the *closure* of Δ, $\overline{\Delta} \subseteq dom(F)$, then we say that Δ is *suitable* for F, or if a point $\mathbf{a} \in \Delta$ is given, suitable for F *around* $\mathbf{a}$. The holomorphic function $F|\Delta$ is then called a *proper restriction* of F.

Let $\mathcal{F}$ be a collection of holomorphic functions. We assume that $\mathcal{F}$ contains all polynomials with Gaussian rational coefficients (in any number of variables). Denote by $PR(\mathcal{F})$ the set of all proper restrictions of all functions in $\mathcal{F}$. Let $\mathbb{R}(PR(\mathcal{F}))$ be the expansion of the ordered field of real numbers by (the graphs of) the functions in $PR(\mathcal{F})$, where we identify $\mathbb{C}$ with $\mathbb{R}^2$ in the usual way. A holomorphic function F is called *locally definable* from $\mathcal{F}$ if all its proper restrictions are definable in the structure $\mathbb{R}(PR(\mathcal{F}))$, where "definable" always means first-order definable *without* parameters unless otherwise stated. Clearly a holomorphic function F is locally definable from $\mathcal{F}$ if and only if for all $\mathbf{w} \in dom(F)$ there is *some* Δ suitable for F around $\mathbf{w}$ such that $F|\Delta$ is definable in $\mathbb{R}(PR(\mathcal{F}))$.

1.1 Problem.

Given $\mathcal{F}$, characterize the class of functions locally definable from $\mathcal{F}$ in terms of *complex*-analytically natural closure conditions.

I cannot claim to solve 1.1 completely here. However, I do give a complex analytic characterization of the pregeometry arising from local definability, and hence answer 1.1 in neighbourhoods of *generic* points of $\mathbb{C}^n$.

Let us first make some simple observations concerning 1.1 which follow from the fact that in the structure $\mathbb{R}(PR(\mathcal{F}))$ we may definably separate out the real and imaginary parts of complex functions and apply definable constructions coming from *real* algebra and analysis to them (such as $\epsilon - \delta$ methods).

1.2 Differentiation.

If $F : U \to \mathbb{C}$ is locally definable from $\mathcal{F}$ (where $U \subseteq \mathbb{C}^n$) and $1 \leq i \leq n$, then so is the partial derivative $\frac{\partial F}{\partial z_i} : U \to \mathbb{C}$.

1.3 Schwarz Reflection.

If $F : U \to \mathbb{C}$ is locally definable from $\mathcal{F}$ then so is its *Schwarz Reflection* $F^{SR} : U' \to \mathbb{C}$, where $U' := \{\overline{\mathbf{z}} : \mathbf{z} \in U\}$ (the bar here denotes coordinatewise complex conjugation) and where $F^{SR}(\mathbf{z}) := \overline{F(\overline{\mathbf{z}})}$ for $\mathbf{z} \in U'$. (Note that Schwarz Reflection commutes with taking proper restrictions.)

Our results are most conveniently stated if we assume from the outset that $\mathcal{F}$ is closed under differentiation and under Schwarz Reflection (which, in view of 1.2 and 1.3 does not affect 1.1) and we fix such an $\mathcal{F}$ for the rest of this paper. Our conjectured answer to 1.1 is, roughly speaking, that a function is locally definable from $\mathcal{F}$ if and only if it can be obtained (locally) from $\mathcal{F}$ by finitely many applications of composition and extractions of *implicitly defined functions*:

Definition 1.4 Let $F : U \to \mathbb{C}$, $f : V \to \mathbb{C}$ be holomorphic functions where $U \subseteq \mathbb{C}^{n+1}$, and $V \subseteq \mathbb{C}^n$. Then we say that f is *implicitly defined* from F if for all $\mathbf{w} \in V$,

$$\langle \mathbf{w}, f(\mathbf{w}) \rangle \in U \quad \text{and} \quad F(\mathbf{w}, f(\mathbf{w})) = 0 \neq \frac{\partial F}{\partial z_{n+1}}(\mathbf{w}, f(\mathbf{w})).$$

1.5 Implicit Definability

Notice that if, in 1.4, F is locally definable from $\mathcal{F}$, then so is f. For if $\mathbf{a} \in V$ then by the Implicit Function Theorem we may choose a sufficiently small Δ, suitable for f around $\mathbf{a}$, and a rectangle D in $\mathbb{C}$ such that $\Delta \times D$ is suitable for F around $\langle \mathbf{a}, f(\mathbf{a}) \rangle$ and has the further property that for each $\mathbf{w} \in \Delta$ there is a *unique* $u \in D$ such that $F(\mathbf{w}, u) = 0$. Since this u is necessarily equal to $f(\mathbf{w})$ (for small enough Δ) and since the function $F|(\Delta \times D)$ is definable in the structure $\mathbb{R}(PR(\mathcal{F}))$, it follows that the function $f|\Delta$ is too. (No parameters are needed because Δ and D have rational data.)

1.6 Composition

I leave the reader to check that if $F : U \to \mathbb{C}$ (where $U \subseteq \mathbb{C}^n$) and $G_i : V_i \to \mathbb{C}$ (where $V_i \subseteq \mathbb{C}^m$ for $i = 1, \ldots, n$) are locally definable from $\mathcal{F}$, then so is their *composition*

$$F \circ \langle G_1, \ldots, G_n \rangle : \bigcap_{i=1}^{n} V_i \cap \langle G_1, \ldots, G_n \rangle^{-1}[U] \to \mathbb{C}.$$

(This is not an *immediate* consequence of the fact that definable functions are closed under composition: one does need to invoke the continuity of the G_i's.)

Definition 1.7 We denote by $\tilde{\mathcal{F}}$ the smallest class of functions containing $\mathcal{F}$ and closed under both composition and implicit definability.

Thus we have seen (1.5 and 1.6) that every function in $\tilde{\mathcal{F}}$ is locally definable from $\mathcal{F}$.

1.8 Conjecture. A function F is locally definable from $\mathcal{F}$ if and only if for all $\mathbf{a} \in dom(F)$ there exists a function $G \in \tilde{\mathcal{F}}$ with $\mathbf{a} \in dom(G)$, and some Δ suitable for both F and G around $\mathbf{a}$, such that $F|\Delta = G|\Delta$.

In order to be able to state what I can actually prove, I require the following

Definition 1.9 Let X be any subset of $\mathbb{C}$. Then $\tilde{D}(X)$ denotes the set of all complex numbers of the form $F(\mathbf{w})$ where $F \in \tilde{\mathcal{F}}$ and $\mathbf{w}$ is a tuple from X such that $\mathbf{w} \in dom(F)$. The set $LD(X)$ is defined similarly except that F is allowed to be any function locally definable from $\mathcal{F}$.

It follows immediately from the comment preceding 1.8 that $\tilde{D}(X) \subseteq LD(X)$ for all $X \subseteq \mathbb{C}$. The main result of this paper is the following

Theorem 1.10 *The operators LD and $\tilde{D}$ are both pregeometries on $\mathbb{C}$ and are identical. Further, the conjecture holds in neighbourhoods of generic points. In other words, if $a_1, \ldots, a_n$ are independent complex numbers (for either of the pregeometries) and F is a function locally definable from $\mathcal{F}$ with $\mathbf{a} = \langle a_1, \ldots, a_n \rangle \in dom(F)$, then there exists a function $G \in \tilde{\mathcal{F}}$ with $\mathbf{a} \in dom(G)$, and some Δ suitable for both F and G around $\mathbf{a}$, such that $F|\Delta = G|\Delta$.*

Presumably a (positive) solution to 1.8 would, in addition to 1.10, require some sort of resolution of singularities. However, for many model-theoretic purposes 1.10 is sufficient: non-generic points may be dealt with by a suitable inductive hypothesis.

The reader familiar with early work on o-minimal expansions of the real field may have noticed by now that $\mathbb{R}(PR(\mathcal{F}))$ is a structure to which Gabrielov's theorem on reducts of $\mathbb{R}_{an}$ applies (see [2]), and hence is model complete (and o-minimal). It is not hard to deduce from this that the analogue of 1.10 for *real* analytic functions, in particular where we replace $\mathcal{F}$ by the set of all real and imaginary parts of functions in $\mathcal{F}$, holds. That is, the real and imaginary parts of a function locally definable from $\mathcal{F}$ are (generically) locally equal to functions obtained from the real and imaginary parts of functions in $\mathcal{F}$ by finitely many applications of composition and extraction of implicitly defined (*real* analytic) functions. But it is by no means clear how to deduce from this that the (complex) locally defined function itself has (generically) such a characterization in terms of the (complex) functions themselves in $\mathcal{F}$. For example, in the algebraic case (where $\mathcal{F}$ is just the set of all polynomials with Gaussian rational coefficients) this amounts to showing that if $F(\mathbf{z}) = F(\mathbf{x} + \sqrt{-1}\mathbf{y}) = u(\mathbf{x}, \mathbf{y}) + \sqrt{-1}v(\mathbf{x}, \mathbf{y})$ is holomorphic for $\mathbf{z}$ in an open neighbourhood of a *generic* n-tuple $\mathbf{z}_0 = \mathbf{x}_0 + \sqrt{-1}\mathbf{y}_0$ (i.e., the coordinates of $\mathbf{z}_0$ are algebraically independent over $\mathbb{Q}$), and if u and v are definable functions in the ordered field of real numbers, then $F(\mathbf{z}_0)$ is algebraic over $\mathbb{Q}(\mathbf{z}_0)$ (and, indeed, that a polynomial relationship $P(\mathbf{z}_0, F(\mathbf{z}_0)) = 0$ extends to an open neighbourhood of $\mathbf{z}_0$). Even this special case of 1.10 does not seem obvious to me. In fact, this case arose out of a misinterpretation of mine of a question of Hrushovski, namely whether quantifier elimination for the complex field could be "easily deduced" from the deeper fact of quantifier elimination for the real ordered field. Neither of us can remember exactly what the precise formulation was (though Hrushovski now guesses that it probably had something to do with decidability) but, at any rate, I was motivated to revisit elimination procedures for the real field with a view to investigating to what extent they "preserve the Cauchy-Riemann equations", and this is really the issue here. Indeed, I suspect that 1.10 could be deduced using the methods of van den Dries from the important and influential paper [1], but my main point in this note is to connect real and complex definability via another pregeometry associated with *derivations* where the Cauchy-Riemann equations (and Schwarz Reflection) may be applied directly.

Another reason for looking at locally definable functions is that it might help in studying expansions of the complex field by certain entire functions, such as the exponential function, and thereby settling Zilber's conjecture: is every $\langle \mathbb{C}, +, \cdot, e^z \rangle$-definable (with parameters) subset of $\mathbb{C}$ either countable or co-countable? (See [5].) Here we would take $\mathcal{F}$ to be the collection of all polynomials in $z_1, z_2, \ldots, e^{z_1}, e^{z_2}, \ldots$ (with Gaussian rational coefficients)-note that this $\mathcal{F}$ is closed under differentiation and Schwarz Reflection-and the hope would be, as was successful in the real case (see [4]), that one could study the unrestricted complex exponential function *modulo* its restrictions, and 1.10 gives the necessary control on the latter. One should also remark here that because $\mathbb{R}(PR(\mathcal{F}))$ is an o-minimal structure, one has available the extensive theory, developed by Peterzil and Starchenko, of complex analysis over such structures (see their survey [3]).

Of course, we cannot fruitfully expand the real field by the *unrestricted* complex exponential function (i.e., by the unrestricted real exponential and sine functions) because this results in a highly wild structure (equivalent to second-order number theory). The point of 1.10 is that using first-order real definability methods in the study of *restricted* holomorphic functions does not take us outside the realms of complex geometry.

The plan of the proof of 1.10 is as follows. In the next section I shall show that both LD and $\tilde{D}$ are pregeometries. Then I shall introduce another pregeometry on $\mathbb{C}$, denoted DD, via the class of derivations on the field $\mathbb{C}$ that respect (i.e., satisfy the chain rule for) all the functions in $\mathcal{F}$. It is easy to show that $\tilde{D}$ and DD are identical, this being the analogue of the classical fact that if $k \subseteq K$ are fields of characteristic zero and $a \in K$, then a is algebraic over k if and only if every derivation on K that vanishes on k also vanishes at a. In the fourth section I set up the bijection between the class of derivations on $\mathbb{C}$ respecting the functions in $\mathcal{F}$ and the class of (pairs of) derivations on $\mathbb{R}$ respecting their real and imaginary parts, this being where the Cauchy-Riemann equations and Schwarz Reflection are used. Next I observe that the above pregeometric notions have analogues in the real analytic case (still over the structure $\mathbb{R}(PR(\mathcal{F}))$) and then derive from Gabrielov's theorem that the corresponding version of 1.10 holds in this case. (This might be new and, indeed, holds in general for expansions of the real field to which Gabrielov's theorem applies. But it *is* a very easy consequence of model completeness and I am not sure whether it is a particularly useful one in this form, since it only holds for archimedean models.) Finally, the required complex result, 1.10 itself, can now be read off from the bijective correspondence between the real and complex derivations.

2 The operators LD and $\tilde{D}$ are pregeometries

Let us show that local definability satisfies the axioms for a pregeometry.

Firstly, it follows immediately from 1.6 that $LD(LD(X)) \subseteq LD(X)$ for all $X \subseteq \mathbb{C}$, so it only remains to prove the Steinitz Exchange Principle, all the other axioms being trivially satisfied.

So suppose that $X \subseteq \mathbb{C}$, $a, b \in \mathbb{C}$ and that $a \in LD(X \cup \{b\})$. Then there exists a function F, locally definable from $\mathcal{F}$, and a tuple $\mathbf{w}$ from X with $\langle \mathbf{w}, b \rangle \in dom(F)$ such that $F(\mathbf{w}, b) = a$. Say $\mathbf{w}$ is an n-tuple and suppose first that $\frac{\partial^i F}{\partial z^i_{n+1}}$ vanishes at the point $\langle \mathbf{w}, b \rangle$ for all $i \geq 1$. Then the function $z_{n+1} \mapsto F(\mathbf{w}, z_{n+1})$ is (defined and) constant on an open neighbourhood of b, with value a. Let q be a Gaussian rational lying in this neighbourhood and define

$$G : \{\mathbf{z} : \langle \mathbf{z}, q \rangle \in dom(F)\} \to \mathbb{C}, \quad \mathbf{z} \mapsto F(\mathbf{z}, q).$$

Then clearly G is locally definable from $\mathcal{F}$ and $G(\mathbf{w}) = a$. So $a \in LD(X)$ in this case.

For the remaining case, let $i \geq 1$ be minimal such that $\frac{\partial^i F}{\partial z^i_{n+1}}$ does not vanish at $\langle \mathbf{w}, b \rangle$. Then we may suppose that $i = 1$, for if $i \geq 2$ then just replace F by $F + \frac{\partial^{i-1} F}{\partial z_{n+1}^{i-1}}$ (which is permissible by 1.2 and 1.6). Now define $H(\mathbf{z}, z_{n+1}, z_{n+2}) := F(\mathbf{z}, z_{n+2}) - z_{n+1}$ (for $\langle \mathbf{z}, z_{n+2} \rangle \in dom(F)$ and $z_{n+1} \in \mathbb{C}$) so that H is clearly locally definable from $\mathcal{F}$. Also $H(\mathbf{w}, a, b) = 0 \neq \frac{\partial H}{\partial z_{n+2}}(\mathbf{w}, a, b)$. It follows from the Implicit Function Theorem that there exists a holomorphic function g such that $g(\mathbf{w}, a) = b$ and, for all $\langle \mathbf{z}, z_{n+1} \rangle \in dom(g)$,

$$H(\mathbf{z}, z_{n+1}, g(\mathbf{z}, z_{n+1})) = 0 \neq \frac{\partial H}{\partial z_{n+2}}(\mathbf{z}, z_{n+1}, g(\mathbf{z}, z_{n+1})).$$

Then g is implicitly defined from H and hence, by 1.5, locally definable from $\mathcal{F}$. Thus $b \in LD(X \cup \{a\})$ in this case, and the proof of the Exchange Principle is complete.

Now notice that the argument above almost goes through for the operator $\tilde{D}$. The only thing missing is the fact (used in the second case) that the collection of functions under consideration be closed under differentiation.

Lemma 2.1   (i) *Let U be an open subset of $\mathbb{C} \setminus \{0\}$. Then the function $\iota : U \to \mathbb{C} : z_1 \mapsto z_1^{-1}$ lies in $\tilde{\mathcal{F}}$.*
(ii) *$\tilde{\mathcal{F}}$ is closed under differentiation.*

Proof For (i), one readily checks that the function ι is implicitly defined from the polynomial $z_1 \cdot z_2 - 1$ and so lies in $\tilde{\mathcal{F}}$. (I remind the reader that $\mathcal{F}$, and hence $\tilde{\mathcal{F}}$, contains all polynomials with Gaussian rational coefficients.)

For (ii), let $\mathcal{S}$ denote the set of all functions $F \in \tilde{\mathcal{F}}$ such that for all i, $\frac{\partial F}{\partial z_i} \in \tilde{\mathcal{F}}$. Certainly $\mathcal{F} \subseteq \mathcal{S}$ by our original assumption on $\mathcal{F}$, so we only need to show that $\mathcal{S}$ is closed under implicit definability and composition.

For the former, suppose that f, F are as in 1.4, that F lies in $\mathcal{S}$ and that $1 \leq i \leq n$. By differentiating the identity in 1.4 with respect to z_i we obtain, for $\mathbf{w} \in dom(f)$:

$$\frac{\partial f}{\partial z_i}(\mathbf{w}) = \frac{\partial F}{\partial z_i}(\mathbf{w}, f(\mathbf{w})) \cdot \left(\frac{\partial F}{\partial z_{n+1}}(\mathbf{w}, f(\mathbf{w})) \right)^{-1}.$$

It now follows from (i) and the fact that $\tilde{\mathcal{F}}$ is closed under composition that $\frac{\partial f}{\partial z_i} \in \tilde{\mathcal{F}}$. Since this also holds trivially for $i > n$ we see that $f \in \mathcal{S}$.

Now suppose that, in the notation of 1.6, the functions $F, G_1, \ldots, G_n$

all lie in $\mathcal{S}$. I leave the reader to apply the chain rule to the composite function $F \circ \langle G_1, \ldots, G_n \rangle$ (and invoke the closure of $\tilde{\mathcal{F}}$ under composition) to see that each of its first partial derivatives lies in $\tilde{\mathcal{F}}$, and hence that the composite function itself lies in $\mathcal{S}$, as required. $\square$

As remarked above, we have now established the following

Theorem 2.2 *The operators LD and $\tilde{D}$ are both pregeometries on $\mathbb{C}$.*

The following observation, which will be used repeatedly, captures the spirit of these pregeometries. Its proof may be extracted easily from the proof of 2.2.

Lemma 2.3 *Let $\mathbf{w} = \langle w_1, \ldots, w_n \rangle \in \mathbb{C}^n$. Then $w_1, \ldots, w_n$ are $\tilde{D}$-dependent (respectively, LD-dependent) if and only if there exists a function $F \in \tilde{\mathcal{F}}$ (respectively, a function F locally definable from $\mathcal{F}$) such that $\mathbf{w} \in dom(F)$ and $F(\mathbf{w}) = 0$, but such that F does not vanish identically on any open neighbourhood of $\mathbf{w}$ contained in $dom(F)$.* $\square$

3 Derivations

Let $K = \mathbb{R}$ or $\mathbb{C}$. By a *derivation* on K I shall simply mean a $\mathbb{Q}$-linear map from K to K which, in the case $K = \mathbb{C}$, is also $\mathbb{Q}(\sqrt{-1})$-linear.

Suppose that F is a holomorphic function (if $K = \mathbb{C}$) or a real analytic function (if $K = \mathbb{R}$). (A real analytic function is assumed to have domain an open subset of $\mathbb{R}^n$, for some n). Let $\mathbf{a} = \langle a_1, \ldots, a_n \rangle$ be a point of $dom(F)$. Then we say that a derivation $\delta : K \to K$ *respects F at the point $\mathbf{a}$* if $\delta(F(\mathbf{a})) = \sum_{i=1}^{n} \partial_i F(\mathbf{a}) \cdot \delta(a_i)$, where $\partial_i F$ denotes the partial derivative of F with respect to the i'th (real or complex) variable of F. If δ respects F at all points of its domain then we say that δ *respects F*. (Thus, a derivation in the usual sense is just a derivation in our sense that respects multiplication.)

Let $\mathcal{C}$ be a set of functions as above. We denote by $Der_K(\mathcal{C})$ the set of derivations that respect all $F \in \mathcal{C}$. It is clear that $Der_K(\mathcal{C})$ is a K-vector space (under pointwise operations).

Definition 3.1 Let $\mathcal{C}$ be as above. For a *finite* subset $X \subseteq K$ we define
$$DD_K^{\mathcal{C}}(X) := \{a \in K : \forall \delta \in Der_K(\mathcal{C}), \text{ if } \delta[X] = \{0\}, \text{ then } \delta(a) = 0\}.$$

For an arbitrary subset $X \subseteq K$ we define

$$DD_K^{\mathcal{C}}(X) := \{a \in K : a \in DD_K^{\mathcal{C}}(X') \text{ for some finite } X' \subseteq X\}.$$

Lemma 3.2 *For any* $\mathcal{C}$*, the operator* $DD_K^{\mathcal{C}}$ *is well defined (i.e., the two cases in 3.1 agree when* X *is finite) and is a pregeometry on* $\mathbb{C}$*.*

Proof It is trivial to check that $DD_K^{\mathcal{C}}$ is a well defined operator and that it satisfies all the axioms for a pregeometry apart from, possibly, the Exchange Principle. (The axiom of finite character is built into the definition.) To see that the Exchange Principle holds too, let $X \subseteq K$, $a, b \in K$ and suppose that $a \notin DD_K^{\mathcal{C}}(X)$ and that $b \notin DD_K^{\mathcal{C}}(X \cup \{a\})$. Let X' be an arbitrary finite subset of X and choose $\delta_1, \delta_2 \in Der_K(\mathcal{C})$ such that $\delta_1[X'] = \{0\}$, $\delta_1(a) \neq 0$, and $\delta_2[X' \cup \{a\}] = \{0\}$, $\delta_2(b) \neq 0$. Let $\delta := \delta_2(b) \cdot \delta_1 - \delta_1(b) \cdot \delta_2$. Then δ lies in the K-vector space $Der_K^{\mathcal{C}}$. Further, $\delta[X' \cup \{b\}] = \{0\}$ and $\delta(a) \neq 0$. So $a \notin DD_K^{\mathcal{C}}(X' \cup \{b\})$, and since X' was an arbitrary finite subset of X, $a \notin DD_K^{\mathcal{C}}(X \cup \{b\})$ as required. $\square$

We now concentrate on the case $K = \mathbb{C}$, and we write DD for $DD_{\mathbb{C}}^{\mathcal{F}}$. Our aim for the rest of this section is to show that DD and $\tilde{D}$ are the same pregeometry on $\mathbb{C}$. So we first prove the following

Lemma 3.3 $Der_{\mathbb{C}}(\tilde{\mathcal{F}}) = Der_{\mathbb{C}}(\mathcal{F})$.

Proof Obviously $Der_{\mathbb{C}}(\tilde{\mathcal{F}}) \subseteq Der_{\mathbb{C}}(\mathcal{F})$, so suppose that $\delta \in Der_{\mathbb{C}}(\mathcal{F})$. Then δ respects every function in $\mathcal{F}$ and we must show that this is preserved by implicit definability and by composition. So suppose that f, F are as in 1.4 and that δ respects F. Let $\mathbf{a} \in dom(f)$. Then it follows that

$$0 = \delta(F(\mathbf{a}, f(\mathbf{a}))) = \sum_{i=1}^n \frac{\partial F}{\partial z_i}(\mathbf{a}, f(\mathbf{a})) \cdot \delta(a_i) + \frac{\partial F}{\partial z_{n+1}}(\mathbf{a}, f(\mathbf{a})) \cdot \delta(f(\mathbf{a})).$$

However, by differentiating the identity in 1.4 and evaluating at the point $\mathbf{w} = \mathbf{a}$ we see that $\frac{\partial F}{\partial z_i}(\mathbf{a}, f(\mathbf{a})) = -\frac{\partial F}{\partial z_{n+1}}(\mathbf{a}, f(\mathbf{a})) \cdot \frac{\partial f}{\partial z_i}(\mathbf{a})$ for all $i = 1, \ldots, n$. By substituting these equations into the equation above, and cancelling the non-zero term $\frac{\partial F}{\partial z_{n+1}}(\mathbf{a}, f(\mathbf{a}))$, we obtain $\delta(f(\mathbf{a})) = \sum_{i=1}^n \frac{\partial f}{\partial z_i}(\mathbf{a}) \cdot \delta(a_i)$. Since $\mathbf{a} \in dom(f)$ was arbitrary, this shows that δ respects f, as required.

Now suppose that, in the notation of 1.6, δ respects the functions F and $G_1, \ldots, G_n$. Let $\mathbf{b} = \langle b_1, \ldots, b_m \rangle \in dom(F \circ \langle G_1, \ldots, G_n \rangle)$. Then

$\langle G_1(\mathbf{b}), \ldots, G_n(\mathbf{b}) \rangle \in dom(F)$ and

$$\delta(F(G_1(\mathbf{b}), \ldots, G_n(\mathbf{b}))) = \sum_{i=1}^{n} \frac{\partial F}{\partial z_i}(G_1(\mathbf{b}), \ldots, G_n(\mathbf{b})) \cdot \delta(G_i(\mathbf{b})).$$

Also, $\mathbf{b} \in dom(G_i)$ and $\delta(G_i(\mathbf{b})) = \sum_{j=1}^{m} \frac{\partial G_i}{\partial z_j} \cdot \delta(b_j)$ for each $i = 1, \ldots n$.

By combining these equations and using the chain rule, we see that δ also respects the function $F \circ \langle G_1, \ldots, G_n \rangle$ at an arbitrary point $\mathbf{b}$ of its domain and this completes the proof. $\qquad \square$

Theorem 3.4 *The operators DD and $\tilde{D}$ are identical.*

Proof Let $X \subseteq \mathbb{C}$.

Suppose first that $\mathbf{a} = \langle a_1, \ldots, a_n \rangle$ is a tuple from X and that F is a function in $\tilde{\mathcal{F}}$ with $\mathbf{a} \in dom(F)$. Let δ be any derivation in $Der_{\mathbb{C}}(\mathcal{F})$ satisfying $\delta[\{a_1, \ldots, a_n\}] = \{0\}$. By 3.3, δ respects F and it follows immediately that $\delta(F(\mathbf{a})) = 0$. This shows that $\tilde{D}(X) \subseteq DD(X)$.

Now suppose that w is any complex number such that $w \notin \tilde{D}(X)$. By 2.2 and the general theory of pregeometries, we may choose a $\tilde{D}$ -basis for $\mathbb{C}$ of the form $B \cup \{w\}$, where $w \notin B$ and $X \subseteq \tilde{D}(B)$. We construct a derivation $\delta \in Der_{\mathbb{C}}(\mathcal{F})$ such that $\delta(w) = 1$ and $\delta[B] = \{0\}$ (so that $\delta[X] = \{0\}$ by the first part of this proof). This shows that $w \notin DD(X)$ and completes the proof of the theorem.

To construct δ, let c be an arbitrary complex number and, using the fact that $c \in \tilde{D}(B \cup \{w\})$, pick a tuple $\mathbf{a} = \langle a_1, \ldots, a_n \rangle$ from B and a function $F \in \tilde{\mathcal{F}}$ such that $\langle \mathbf{a}, w \rangle \in dom(F)$ and $F(\mathbf{a}, w) = c$. I claim that $\frac{\partial F}{\partial z_{n+1}}(\mathbf{a}, w)$ depends only on c. For if also $G(\mathbf{a}, w) = c$ with $G \in \tilde{\mathcal{F}}$ and $\langle \mathbf{a}, w \rangle \in dom(G)$ (and we may suppose that the n and $\mathbf{a}$ are the same as before by adding vacuous variables to F and G) then the function $F - G$ lies in $\tilde{\mathcal{F}}$ and vanishes at the $\tilde{D}$ -generic $(n+1)$-tuple $\langle \mathbf{a}, w \rangle$. But then by 2.3, it vanishes on some open neighbourhood of $\langle \mathbf{a}, w \rangle$. Hence so do its partial derivatives, and the claim follows. Thus we may set $\delta(c) := \frac{\partial F}{\partial z_{n+1}}(\mathbf{a}, w)$. It is clear that δ is $\mathbb{Q}(\sqrt{-1})$-linear, and by taking F to be the first and second projection function on $\mathbb{C}^2$ we see that $\delta[B] = \{0\}$ and $\delta(w) = 1$ respectively. Finally, the fact that δ respects all functions in $\mathcal{F}$ follows from (indeed, it is an instance of) the chain rule, and I leave the easy details to the reader. $\qquad \square$

4 Real versus complex derivations

Let $\mathcal{C}$ be any collection of holomorphic functions. For each n-ary function $F \in \mathcal{C}$ there are two real valued, real analytic functions - the real and imaginary parts of F - with domain the open subset of $\mathbb{R}^{2n}$ corresponding to $dom(F)$ under the usual identification of $\mathbb{C}$ with $\mathbb{R}^2$. Let us denote by $\mathcal{C}_{real}$ the collection of all the real functions obtained in this way. Our aim in this section is to investigate the relationship between the $\mathbb{R}$-vector space $Der_{\mathbb{R}}(\mathcal{C}_{real})$ of derivations on $\mathbb{R}$ respecting all the functions in $\mathcal{C}_{real}$ and the $\mathbb{C}$-vector space $Der_{\mathbb{C}}(\mathcal{C})$ of derivations on $\mathbb{C}$ respecting all the functions in $\mathcal{C}$.

We shall use the following convention. If $F : U \to \mathbb{C}$ is a holomorphic function, where $U \subseteq \mathbb{C}^n$, then the real and imaginary parts, u, v say, of F are the real analytic functions with domain

$$U_{real} := \{\langle \mathbf{x}, \mathbf{y} \rangle = \langle \langle x_1, \ldots, x_n \rangle, \langle y_1, \ldots, y_n \rangle \rangle \in \mathbb{R}^{2n} : \mathbf{x} + \sqrt{-1}\mathbf{y} \in U\}$$

satisfying

$$F(\mathbf{x} + \sqrt{-1}\mathbf{y}) = u(\mathbf{x}, \mathbf{y}) + \sqrt{-1}v(\mathbf{x}, \mathbf{y})$$

for $\langle \mathbf{x}, \mathbf{y} \rangle \in U_{real}$.

Definition 4.1 For $\lambda, \mu : \mathbb{R} \to \mathbb{R}$ any functions, we define the function $[\lambda : \mu] : \mathbb{C} \to \mathbb{C}$ by $[\lambda : \mu](x + \sqrt{-1}y) := (\lambda(x) - \mu(y)) + \sqrt{-1}(\lambda(y) + \mu(x))$ (for $x, y \in \mathbb{R}$).

Lemma 4.2 *If λ and μ are derivations on $\mathbb{R}$ then $[\lambda : \mu]$ is a derivation on $\mathbb{C}$. Further, if F is a holomorphic function with real and imaginary parts u, v, and domain $U \subseteq \mathbb{C}^n$, and if λ, μ both respect u and v at a point $\langle \mathbf{x}, \mathbf{y} \rangle \in U_{real}$, then $[\lambda : \mu]$ respects F at the point $\mathbf{x} + \sqrt{-1}\mathbf{y}$.*

Proof Suppose that λ and μ are derivations on $\mathbb{R}$. Then $[\lambda : \mu]$ is clearly $\mathbb{Q}$-linear. Further, for $x, y \in \mathbb{R}$,

$$
\begin{aligned}
[\lambda : \mu](\sqrt{-1}(x + \sqrt{-1}y)) &= [\lambda : \mu](-y + \sqrt{-1}x) \\
&= (-\lambda(y) - \mu(x)) + \sqrt{-1}(\lambda(x) - \mu(y)) \\
&= \sqrt{-1}((\lambda(x) - \mu(y)) + \sqrt{-1}(\lambda(y) + \mu(x))) \\
&= \sqrt{-1}[\lambda : \mu](x + \sqrt{-1}y),
\end{aligned}
$$

so $[\lambda : \mu]$ is also $\mathbb{Q}(\sqrt{-1})$-linear.

As for the second part, let us write u_i for $\frac{\partial u}{\partial x_i}(\mathbf{x}, \mathbf{y})$ and u_{n+i} for $\frac{\partial u}{\partial y_i}(\mathbf{x}, \mathbf{y})$ (for $i = 1, \ldots, n$) and similarly for v. Then by 4.1, the fact that λ, μ respect u, v at $\langle \mathbf{x}, \mathbf{y} \rangle$, and the Cauchy-Riemann equations in the form $u_{n+i} = -v_i$, $v_{n+i} = u_i$ (for $i = 1, \ldots, n$) we obtain

$$
\begin{aligned}
[\lambda &: \mu](F(\mathbf{x} + \sqrt{-1}\mathbf{y})) \\
&= (\lambda(u(\mathbf{x}, \mathbf{y})) - \mu(v(\mathbf{x}, \mathbf{y}))) + \sqrt{-1}(\lambda(v(\mathbf{x}, \mathbf{y})) + \mu(u(\mathbf{x}, \mathbf{y}))) \\
&= \sum_{i=1}^{n} [u_i \lambda(x_i) + u_{n+i}\lambda(y_i) - v_i\mu(x_i) - v_{n+i}\mu(y_i) \\
&\qquad\quad + \sqrt{-1}(v_i\lambda(x_i) + v_{n+i}\lambda(y_i) + u_i\mu(x_i) + u_{n+i}\mu(y_i))] \\
&= \sum_{i=1}^{n} (u_i + \sqrt{-1}v_i)(\lambda(x_i) - \mu(y_i) + \sqrt{-1}(\lambda(y_i) + \mu(x_i))) \\
&= \sum_{i=1}^{n} \frac{\partial F}{\partial z_i}(\mathbf{x} + \sqrt{-1}\mathbf{y}) \cdot ([\lambda : \mu](x_i + \sqrt{-1}y_i)).
\end{aligned}
$$

Thus $[\lambda : \mu]$ respects F at the point $\mathbf{x} + \sqrt{-1}\mathbf{y}$, as required. $\square$

Now suppose that δ is any derivation on $\mathbb{C}$. Then, in particular, there exist functions $\lambda, \mu : \mathbb{R} \to \mathbb{R}$ such that $\delta(x) = \lambda(x) + \sqrt{-1}\mu(x)$ for $x \in \mathbb{R}$. Clearly λ and μ are $\mathbb{Q}$-linear, i.e., they are derivations on $\mathbb{R}$. Further, for all x, $y \in \mathbb{R}$,

$$
\begin{aligned}
[\lambda : \mu](x + \sqrt{-1}y) &= (\lambda(x) - \mu(y)) + \sqrt{-1}(\lambda(y) + \mu(x)) \\
&= (\lambda(x) + \sqrt{-1}\mu(x)) + \sqrt{-1}(\lambda(y) + \sqrt{-1}\mu(y)) \\
&= \delta(x) + \sqrt{-1}\delta(y)
\end{aligned}
$$

(since δ is $\mathbb{Q}(\sqrt{-1})$-linear).

Thus $\delta = [\lambda : \mu]$. We now go on to prove the main result of this section.

Theorem 4.3 *Let $\mathcal{C}$ be any collection of holomorphic functions closed under Schwarz reflection (see 1.3). Then the elements of $Der_{\mathbb{C}}(\mathcal{C})$ are precisely the maps of the form $[\lambda : \mu] : \mathbb{C} \to \mathbb{C}$ for $\lambda, \mu \in Der_{\mathbb{R}}(\mathcal{C}_{real})$.*

Proof It follows from 4.2 that if λ, $\mu \in Der_{\mathbb{R}}(\mathcal{C}_{real})$ then $[\lambda : \mu] \in Der_{\mathbb{C}}(\mathcal{C})$. So let $\delta \in Der_{\mathbb{R}}(\mathcal{C}_{real})$. We have observed above that $\delta = [\lambda : \mu]$ for some derivatives λ, μ on $\mathbb{R}$ and it remains to show that λ and μ both respect $\mathcal{C}_{real}$. To this end, let u, v be the real and imaginary parts of some function $F : U \to \mathbb{C}$ lying in $\mathcal{C}$. Let $\langle \mathbf{x}, \mathbf{y} \rangle \in U_{real}$ (see the convention immediately preceding 4.1) and, to ease the notation, temporarily write u, u_i and u_{n+i} for the complex numbers $u(\mathbf{x}, \mathbf{y})$, $\frac{\partial u}{\partial x_i}(\mathbf{x}, \mathbf{y})$, and $\frac{\partial u}{\partial y_i}(\mathbf{x}, \mathbf{y})$ respectively (for $i = 1, \ldots, n$) and similarly for v.

Then since δ respects F we have

$$(1) \qquad \delta(F(\mathbf{x} + \sqrt{-1}\mathbf{y})) = \sum_{i=1}^{n} \frac{\partial F}{\partial z_i}(\mathbf{x}, \mathbf{y}) \cdot \delta(x_i + \sqrt{-1}y_i).$$

But $\delta = [\lambda : \mu]$, $F(\mathbf{x} + \sqrt{-1}\mathbf{y}) = u + \sqrt{-1}v$ and $\frac{\partial F}{\partial z_i}(\mathbf{x}, \mathbf{y}) = u_i + \sqrt{-1}v_i$ (for $i = 1, \ldots n$), and hence it follows from 4.1 by equating real and imaginary parts in (1) that

$$(2) \qquad \lambda(u) - \mu(v) = \sum_{i=1}^{n} u_i(\lambda(x_i) - \mu(y_i)) - v_i(\lambda(y_i) + \mu(x_i))$$

and

$$(3) \qquad \lambda(v) + \mu(u) = \sum_{i=1}^{n} u_i(\lambda(y_i) - \mu(x_i)) + v_i(\lambda(x_i) - \mu(y_i)).$$

Now consider the Schwarz Reflection $F^{SR} : U' \to \mathbb{C}$ of F (see 1.3). We have $\mathbf{x} - \sqrt{-1}\mathbf{y} \in U'$ and $F^{SR}(\mathbf{x} - \sqrt{-1}\mathbf{y}) = \overline{F(\mathbf{x} + \sqrt{-1}\mathbf{y})} = u - \sqrt{-1}v$. Further, $\frac{\partial F^{SR}}{\partial z_i}(\mathbf{x} - \sqrt{-1}\mathbf{y}) = u_i - \sqrt{-1}v_i$ for $i = 1, \ldots, n$. Also, by hypothesis, $F^{SR} \in \mathcal{C}$ and so δ respects F^{SR} at the point $\mathbf{x} - \sqrt{-1}\mathbf{y}$. Hence, by applying the argument above with F^{SR} in place of F and $\mathbf{x} - \sqrt{-1}\mathbf{y}$ in place of $\mathbf{x} + \sqrt{-1}\mathbf{y}$ we obtain the equations

$$(4) \qquad \lambda(u) + \mu(v) = \sum_{i=1}^{n} u_i(\lambda(x_i) + \mu(y_i)) + v_i(-\lambda(y_i) + \mu(x_i))$$

and

$$(5) \qquad -\lambda(v) + \mu(u) = \sum_{i=1}^{n} u_i(-\lambda(y_i) - \mu(x_i)) - v_i(\lambda(x_i) + \mu(y_i)).$$

From (2), (4) and the Cauchy-Riemann equations in the form $v_i = -u_{n+i}$, $u_i = v_{n+i}$ (for $i = 1, \ldots, n$) we obtain the equations

$$\lambda(u) = \sum_{i=1}^{n} u_i\lambda(x_i) - v_i\lambda(y_i) = \sum_{i=1}^{n} u_i\lambda(x_i) + u_{n+i}\lambda(y_i),$$

and

$$\mu(v) = \sum_{i=1}^{n} u_i\mu(y_i) + v_i\mu(x_i) = \sum_{i=1}^{n} v_{n+i}\mu(y_i) + v_i\mu(x_i),$$

which show that λ respects the function u at the point $\langle \mathbf{x}, \mathbf{y} \rangle$ and that μ respects the function v at $\langle \mathbf{x}, \mathbf{y} \rangle$. The corresponding conclusions for λ, v and μ, u follow similarly from (3) and (5). $\qquad\qquad \square$

Remark 4.4 If δ is a derivation on the field $\mathbb{C}$ in the usual sense, and $\delta = [\lambda : \mu]$ (which determines λ and μ uniquely: just consider $\delta | \mathbb{R}$), then λ, μ are derivations on the field $\mathbb{R}$ in the usual sense. This follows either by direct calculation or from 4.5 by taking $\mathcal{C} = \{h\}$, where $h : \mathbb{C} \to \mathbb{C}$, $z \mapsto \frac{z^2}{2}$, and observing that real multiplication is the imaginary part of h.

5 The proof of the main theorem

We first observe that the results of the first three sections have versions for real analytic functions (defined on open subsets of $\mathbb{R}^n$, for various n). So let us fix a collection, $\mathcal{E}$ say, of such functions. We assume that $\mathcal{E}$ is closed under partial differentiation and that it contains all polynomials with rational coefficients. We let $PR(\mathcal{E})$ denote the collection of all functions $f | \Delta$, where $f \in \mathcal{E}$ and where Δ is *suitable* for f, i.e., it is a product of open intervals, with rational endpoints, such that $\bar{\Delta} \subseteq dom(f)$. Then $\mathbb{R}(PR(\mathcal{E}))$ denotes the expansion of the ordered field of real numbers by all functions in $PR(\mathcal{E})$. The definition of the notion of a function being *locally definable* from $\mathcal{E}$, and of the closure of $\mathcal{E}$ under composition and implicit definability, which we denote by $\tilde{\mathcal{E}}$, go through as before: just replace "holomorphic" everywhere by "real analytic". Similarly, one defines the operators $\tilde{E}(\cdot)$ and $LED(\cdot)$, of closure under functions in $\tilde{\mathcal{E}}$ and under functions locally definable from $\mathcal{E}$ respectively, and proves that they are both pregeometries on $\mathbb{R}$ with $\tilde{E}(X) \subseteq LED(X)$ for all $X \subseteq \mathbb{R}$. The analogue of 2.3 also holds.

We now establish the real version of 1.10.

Lemma 5.1 $\tilde{E}$ *and* LED *are identical pregeometries on* $\mathbb{R}$.

Proof Let $X \subseteq \mathbb{R}$. We must show that $LED(X) \subseteq \tilde{E}(X)$ and it is clearly sufficient to consider the case where $X = \{s_1, \ldots, s_n\}$ is finite and $\mathbf{s} := \langle s_1, \ldots, s_n \rangle$ is $\tilde{E}$-generic (i.e., $s_1, \ldots, s_n$ are $\tilde{E}$-independent real numbers).

We shall use Gabrielov's Theorem (see [2]) which tells us that any reduct of the structure $\mathbb{R}_{an}$ in which the collection of basic functions of the language is closed under differentiation, has a model complete theory. This clearly applies to our structure $\mathbb{R}(PR(\mathcal{E}))$.

We set $k = \tilde{E}(\{s_1, \ldots, s_n\})$ and observe that k is a subfield of $\mathbb{R}$ (by the real version of 2.1(i)) and is closed under all functions in $\tilde{\mathcal{E}}$. The proof of the lemma will be complete if we can show that the expansion of the field k by the restriction to k of (the graphs of) all functions in $PR(\mathcal{E})$ is existentially closed in $\mathbb{R}(PR(\mathcal{E}))$. For, by model completeness, this implies that this expansion is an *elementary* substructure of $\mathbb{R}(PR(\mathcal{E}))$ and hence closed under all (parameter-free) $\mathbb{R}(PR(\mathcal{E}))$-definable functions, whence $LED(\{s_1, \ldots, s_n\}) \subseteq k$, as required.

Now, by standard manipulations of existential formulas in languages expanding that of ordered fields, it is sufficient (in order to establish the required existential closedness) to prove the following

Claim Let $\mathbf{a} \in k^r$, $\mathbf{b} \in \mathbb{R}^m$ and $f \in \tilde{\mathcal{E}}$. Suppose further that Δ is suitable for f, $\langle \mathbf{a}, \mathbf{b} \rangle \in \Delta$ and $f(\mathbf{a}, \mathbf{b}) = 0$. Then there exists $\mathbf{b}' \in k^m$ such that $\langle \mathbf{a}, \mathbf{b}' \rangle \in \Delta$ and $f(\mathbf{a}, \mathbf{b}') = 0$.

In fact, it is sufficient for our purposes to prove the claim just for functions f lying in the compositional closure of $\mathcal{E}$, but, by stating the claim as we have, we may assume straight away that $r = n$ and that $\mathbf{a} = \mathbf{s}$. Now, to prove the claim, we pick a maximal $\tilde{E}$-independent subset of $\{b_1, \ldots, b_m\}$ over $\mathbf{s}$, where $\mathbf{b} = \langle b_1, \ldots, b_m \rangle$. Let us suppose, for notational convenience, that it is $\{b_1, \ldots, b_l\}$ (for some $l = 0, \ldots, m$), so that $b_{l+1}, \ldots, b_m \in \tilde{E}(\{s_1, \ldots, s_n, b_1, \ldots, b_l\})$. Say $\phi_i(\mathbf{s}, b_1, \ldots, b_l) = b_{l+i}$, where $\phi_i \in \tilde{\mathcal{E}}$, for $i = 1, \ldots, m-l$. Define $g \in \tilde{\mathcal{E}}$ by $g(x_1, \ldots, x_{n+l}) = g(\mathbf{x}) := f(\mathbf{x}, \phi_1(\mathbf{x}), \ldots, \phi_{m-l}(\mathbf{x}))$, so that $g(\mathbf{s}, b_1, \ldots, b_l) = 0$. However, since $\langle \mathbf{s}, b_1, \ldots, b_l \rangle$ is an $\tilde{E}$-generic point of $\mathbb{R}^{n+l}$, it follows from the real version of 2.3 that g vanishes on some open subset, V say, of $dom(g)$ with $\langle \mathbf{s}, b_1, \ldots, b_l \rangle \in V$. Thus we may pick rationals $q_1, \ldots, q_l$ sufficiently close to $b_1, \ldots, b_l$ (respectively) so that both $\langle \mathbf{s}, q_1, \ldots, q_l \rangle \in V$ and $\mathbf{b}' \in \Delta$, where $\mathbf{b}' := \langle q_1, \ldots, q_l, \phi_1(\mathbf{s}, q_1, \ldots, q_l), \ldots, \phi_{m-l}(\mathbf{s}, q_1, \ldots, q_l) \rangle$. This choice of $\mathbf{b}'$ clearly satisfies the conclusion of the claim, and hence the proof of the lemma is complete. $\qquad \square$

Corollary 5.2 *Let $s_1, \ldots, s_n$ be $\tilde{E}$-independent real numbers and suppose that g is a function locally definable from $\mathcal{E}$ with $\mathbf{s} := \langle s_1, \ldots, s_n \rangle \in dom(g)$. Then there exists a function $f \in \tilde{\mathcal{E}}$ with $\mathbf{s} \in dom(f)$ such that $f = g$ on some open neighbourhood of $\mathbf{s}$.*

Proof Since, by definition, $g(\mathbf{s}) \in LED(\{s_1, \ldots, s_n\})$, it follows from 5.1 that there exists a function $f \in \tilde{\mathcal{E}}$ with $\mathbf{s} \in dom(f)$ such that $f(\mathbf{s}) = g(\mathbf{s})$. Then the function $f - g$ is locally definable from $\mathcal{E}$ and vanishes at the point $\mathbf{s}$. But $\mathbf{s}$ is also LED-generic (by 5.1), so the result follows from the real version of 2.3. $\qquad \square$

Suppose now that λ is a derivation on $\mathbb{R}$ respecting every function in $\mathcal{E}$, i.e., $\lambda \in Der_{\mathbb{R}}(\mathcal{E})$. The proof of 3.3 goes through with complex variables replaced by real (and no other changes), and so $\lambda \in Der_{\mathbb{R}}(\tilde{\mathcal{E}})$. It now follows immediately from 5.2 that λ respects every function locally definable from $\mathcal{E}$ at *generic* points of their domains. One has to work a little harder at non-generic points:

Lemma 5.3 *Let* $\lambda \in Der_{\mathbb{R}}(\mathcal{E})$. *Then* λ *respects every function locally definable from* $\mathcal{E}$.

Proof Let f be a function locally definable from $\mathcal{E}$ and $\mathbf{s} = \langle s_1, \ldots, s_n \rangle$ a point of $dom(f)$. Choose a maximal $\tilde{E}$-independent subset of $\{s_1, \ldots, s_n\}$ and suppose, for notational convenience, that it is $\{s_1, \ldots, s_l\}$, where $0 \le l \le n$. Now choose functions $\phi_{l+1}, \ldots, \phi_n$ in $\tilde{\mathcal{E}}$ such that $\phi_i(\mathbf{s}') = s_i$ for $i = l+1, \ldots, n$, where $\mathbf{s}' := \langle s_1, \ldots, s_l \rangle$.

Define the function g by $g(\mathbf{x}') = f(\mathbf{x}', \phi_{l+1}(\mathbf{x}'), \ldots, \phi_n(\mathbf{x}'))$, so that g is locally definable from $\mathcal{E}$ and $g(\mathbf{s}') = f(\mathbf{s})$. Now by the discussion before the statement of the lemma, λ respects the ϕ_i's and g at the point $\mathbf{s}'$, so we have

$$(*) \quad \lambda(s_i) = \lambda(\phi_i(\mathbf{s}')) = \sum_{j=1}^{l} \frac{\partial \phi_i}{\partial x_j}(\mathbf{s}') \cdot \lambda(s_j), \text{ for } i = l+1, \ldots, n,$$

and

$$\begin{aligned} \lambda(f(\mathbf{s})) &= \lambda(g(\mathbf{s}')) \\ &= \sum_{j=1}^{l} \frac{\partial g}{\partial x_j}(\mathbf{s}') \cdot \lambda(s_j) \\ &= \sum_{j=1}^{l} \left(\frac{\partial f}{\partial x_j}(\mathbf{s}) + \sum_{i=l+1}^{n} \left(\frac{\partial f}{\partial x_i}(\mathbf{s}) \cdot \frac{\partial \phi_i}{\partial x_j}(\mathbf{s}') \right) \right) \cdot \lambda(s_j), \end{aligned}$$

(by the chain rule)

$$\begin{aligned} &= \sum_{j=1}^{l} \left(\frac{\partial f}{\partial x_j}(\mathbf{s}) \cdot \lambda(s_j) \right) + \sum_{i=l+1}^{n} \left(\frac{\partial f}{\partial x_i}(\mathbf{s}) \cdot \left(\sum_{j=1}^{l} \frac{\partial \phi_i}{\partial x_j}(\mathbf{s}') \cdot \lambda(s_j) \right) \right), \\ &= \sum_{j=1}^{n} \left(\frac{\partial f}{\partial x_j}(\mathbf{s}) \cdot \lambda(s_j) \right), \text{ by } (*). \end{aligned}$$

So λ respects f at $\mathbf{s}$, as required. $\qquad\square$

We can now prove that LD and $\tilde{D}$ are identical pregeometries on $\mathbb{C}$. Let $X \subseteq \mathbb{C}$. It remains to show that $LD(X) \subseteq \tilde{D}(X)$. So let $w \in$

$LD(X)$ and choose a function F, locally definable from $\mathcal{F}$, and elements $a_1, \ldots, a_n$ of X such that $\mathbf{a} := \langle a_1, \ldots, a_n \rangle \in dom(F)$ and $F(\mathbf{a}) = w$. I shall show that $w \in DD(X)$, which suffices by 3.4. Indeed, I shall show that $w \in DD(\{a_1, \ldots, a_n\})$.

So let δ be an element of $Der_{\mathbb{C}}(\mathcal{F})$ vanishing on the set $\{a_1, \ldots, a_n\}$. Then by 4.3 (with $\mathcal{C} = \mathcal{F}$ - recall our assumption on $\mathcal{F}$ stated just after 1.3) we may choose $\lambda, \mu \in Der_{\mathbb{R}}(\mathcal{F}_{real})$ such that $\delta = [\lambda : \mu]$. Now since $\mathcal{F}$ contains (complex) multiplication it follows that δ is a derivation on $\mathbb{C}$ in the usual sense and hence, by 4.4, both λ and μ are derivations on $\mathbb{R}$ in the usual sense. So if we define $\mathcal{E}$ to be the union of $\mathcal{F}_{real}$ with the set of all polynomials with rational coefficients, it follows that $\lambda, \mu \in Der_{\mathbb{R}}(\mathcal{E})$.

Now the function F is locally definable from $\mathcal{F}$. So its real and imaginary parts are certainly locally definable from $\mathcal{E}$, and hence, by 5.3, are respected by both λ and μ. But then, by 4.2, δ respects F at every point of its domain, in particular at the point $\mathbf{a}$. Since $\delta(a_i) = 0$ for each $i = 1, \ldots, n$ it follows that $\delta(F(\mathbf{a})) = 0$, i.e., $\delta(w) = 0$, and we are done.

The second part of 1.10 now follows by the same argument used to deduce 5.2 from 5.1.

References

[1] L. van den Dries, On the elementary theory of restricted elementary functions, *J. Symbolic Logic* 53 (1988), 796-808.

[2] A. Gabrielov, Complements of subanalytic sets and existential formulas for analytic functions, *Inv. Math.* 125 (1996), 1-12.

[3] Y. Peterzil and S. Starchenko, "Complex-like" analysis in o-minimal structures, in *Proceedings of the RAAG Summer School Lisbon 2003: O-minimal Structures*, Eds M. Edmundo, D. Richardson, A.J. Wilkie (2005), 77-103.

[4] A.J. Wilkie, Model completeness results for expansions of the ordered field of real numbers by restricted Pfaffian functions and the exponential function, *J. Amer. Math. Soc.* 9, No. 4, (1996), 1051-1094.

[5] B. Zilber, Generalized analytic sets, *Translation from Algebra i Logika*, Novosibirsk, 36(4) (1997), 361-380.

Some observations about the real and imaginary parts of complex Pfaffian functions

Angus Macintyre

Queen Mary University of London

1 Introduction

The main result of this paper has a simple proof, but is a central component in a large-scale project I have recently completed on the model theory of elliptic functions [7, 8]. In that project I take up important work of Bianconi [2] from around 1990, on the Weierstrass $\wp$ functions on an appropriate domain, and carry it to a decidability result modulo André's conjecture on 1-motives [1]. Bianconi proved model-completeness results, nonconstructively, for the basic situation, and I can see no way to constructivize the method he uses. Instead, I use ideas from two major developments subsequent to Bianconi's work, namely the work of Wilkie [10] and Macintyre-Wilkie [9], and the work of Gabrielov [4] and Gabrielov-Vorobjov [5]. To link with these papers, I interpret Bianconi's formulations in one based on taking the compositional inverse $\wp^{-1}$, on an appropriate compact, as primitive. The latter function, in contrast to $\wp$, is <u>complex Pfaffian</u>, and this alone yields, by a result of Gabrielov [5] an important constructive multiplicity bound. But I need more, namely that the real and imaginary parts of $\wp^{-1}$ are <u>real Pfaffian</u> and this is what I prove below. I doubt that there is any nontrivial general result allowing one to deduce that the real and imaginary parts of a complex Pfaffian function are real Pfaffian, and it is for this reason that I choose to publish the simple, useful result below.

2 Pfaffian functions

Let U be an open set in $\mathbb{R}^n$ (resp. $\mathbb{C}^n$), and let $\langle f_1, \ldots, f_m \rangle$ be a sequence of real (resp. complex) analytic functions on U. We say that $\langle f_1, \ldots, f_m \rangle$ is a *Pfaffian chain* if for each $j \leq m$ and $k \leq n$ there is a polynomial

$P_{jk}(s_1, \ldots, s_n, t_1, \ldots, t_j)$ over $\mathbb{R}$ (resp. $\mathbb{C}$) so that

$$\frac{\partial f_j}{\partial s_k}(s_1, \ldots, s_n) = P_{jk}(s_1, \ldots, s_n, f_1(s_1, \ldots, s_n), \ldots, f_j(s_1, \ldots, s_n)).$$

A function on U is called *Pfaffian* if it occurs in a Pfaffian chain.

For the basic Finiteness Theorems that make the real case so important for model theory, one should consult Khovanskii's [6], and for an important multiplicity estimate in the complex case one should consult Gabrielov-Vorobjov [5].

Polynomials are obviously Pfaffian in either sense, as is the exponential function. However $f(x,y) = exp(x)cos(y)$, the real part of the complex exponential, is not Pfaffian on $\mathbb{R}^2$. For if it were, so would be

$$g(x,y) = f(x, y + \frac{\pi}{2}) = \exp(x)\sin(y).$$

Then, by Khovanskii [6] the system

$$f(x,y) = 1$$
$$g(x,y) = 0$$

would have only finitely many solutions in $\mathbb{R}^2$ with nonsingular Jacobian matrix. The Jacobian matrix is

$$\begin{vmatrix} e^x \cos y & -e^x \sin y \\ e^x \sin y & -e^x \cos y \end{vmatrix}$$

with determinant $\exp(2x)$, and so is nonsingular. Then $(0, 2n\pi)$ is a solution, for $n \in \mathbb{Z}$

Because of such an immediate example, I see no prospect of finding any really interesting hypotheses guaranteeing that the real and imaginary parts of a complex Pfaffian function are real Pfaffian.

2.1

The very special case that led me to the simple result below concerns the Weierstrass $\wp$ function. Such a function satisfies a differential equation

$$(\wp')^2 = 4(\wp(z) - e_1)(\wp(z) - e_2)(\wp(z) - e_3)$$
$$= 4\wp(z)^3 - g_2\wp(z) - g_3 = g(\wp(z))$$

and is meromorphic, with poles (of multiplicity 2) exactly on its period lattice Λ. On the torus $\mathbb{C}/\Lambda$, $\wp$, qua function to the projective line, takes each value twice, counting multiplicities. At a point w of $\mathbb{C}$ there are generally two branches of an inverse f for $\wp$, satisfying the differential equation

$$f'(w) = \frac{1}{\sqrt{g(w)}}.$$

(For some of the copious information about $\wp^{-1}$ and integrals of the first kind, see [3]).

Lemma 2.1 *Any such f is complex Pfaffian on any open set excluding e_1, e_2 and e_3.*

Proof. Obviously it is enough to show that $h(w) = \frac{1}{\sqrt{g(w)}}$ is Pfaffian. But

$$h' = -\frac{1}{2}g'(w)((h(w))^3,$$

and g' is a polynomial. $\qquad\square$

This result, though trivial, is enough to yield a uniform multiplicity estimate for polynomials in $\wp^{-1}$ (see 4.2 of [5]).

2.2

For my work in [8] one has to consider the real and imaginary parts of $\wp^{-1}$ on appropriate open U. In view of 2.1 one certainly needs an argument to show that they are real Pfaffian.

The argument is purely algebraic, so let

$$\wp^{-1} = u(x, y) + iv(x, y)$$

with the usual convention that $z = x + iy$, with x, y, u, v real.

Now

$$\frac{d}{dz}\wp^{-1} = \frac{\partial u}{\partial x} + i\frac{\partial v}{\partial x} = \frac{\partial v}{\partial y} - i\frac{\partial u}{\partial y}$$

by Cauchy-Riemann. So

$$\frac{\partial u}{\partial x} = \frac{1}{2}\left(\frac{d}{dz}\wp^{-1}(z) + \overline{\frac{d}{dz}\wp^{-1}(z)}\right)$$

$$\frac{\partial v}{\partial x} = \frac{1}{2i}\left(\frac{d}{dz}\wp^{-1}(z) - \overline{\frac{d}{dz}\wp^{-1}(z)}\right)$$

with related expressions for the other two partial derivatives.

So

$$\frac{\partial u}{\partial x} = \frac{1}{2}\left(\frac{1}{\sqrt{g(z)}} + \frac{1}{\overline{\sqrt{g(z)}}}\right) = \frac{\mathrm{Re}(\sqrt{(g(z)})}{|g(z)|}$$

(here we make a definite choice of $\sqrt{(g(z))}$).

Now evidently

$$g(z) = A(x, y) + iB(x, y),$$

where $A(x, y)$ and $B(x, y)$ are polynomials with their coefficients in $\mathbb{Q}(\mathrm{Re}(g_2), \mathrm{Im}(g_2), \mathrm{Re}(g_3), \mathrm{Im}(g_3))$. Now

$$g(z) = \sqrt{A^2 + B^2}\left(\frac{A}{\sqrt{A^2 + B^2}} + i\frac{B}{\sqrt{A^2 + B^2}}\right) = re^{i\theta},$$

$$r = \sqrt{A^2 + B^2}, \quad \cos\theta = \frac{A}{\sqrt{A^2 + B^2}}, \quad \sin\theta = \frac{B}{\sqrt{A^2 + B^2}},$$

so

$$\sqrt{g(z)} = \pm r^{1/2}e^{i\theta/2}.$$

Without loss of generality fix θ, and assume we take the positive square root of r. Then

$$
\begin{aligned}
\mathrm{Re}(\sqrt{g(z)}) &= r^{1/2}\cos(\theta/2) \\
&= (A^2 + B^2)^{1/4}\sqrt{\frac{1}{2}\left(1 + \frac{A}{\sqrt{A^2 + B^2}}\right)} \\
&= \frac{1}{\sqrt{2}}\sqrt{A + \sqrt{A^2 + B^2}}.
\end{aligned}
$$

Notice that there is again a $\pm$ issue, with the second square root, a point to which we return below. For now we work formally. We have:

$$\frac{\partial u}{\partial x} = \frac{1}{\sqrt{2}}\frac{1}{(A^2 + B^2)^{1/2}}\sqrt{A + \sqrt{A^2 + B^2}}.$$

Similarly, up to a choice of sign,

$$
\begin{aligned}
\frac{\partial u}{\partial y} &= \frac{\mathrm{Im}(\sqrt{g(z)})}{|g(z)|} \\
&= \frac{1}{\sqrt{2}}\frac{1}{(A^2 + B^2)^{1/2}}\sqrt{\sqrt{A^2 + B^2} - A}.
\end{aligned}
$$

again with a sign choice.

To get u and v Pfaffian when $g \neq 0$ (i.e when $A^2 + B^2 \neq 0$) we need only show that the algebraic functions

$$\frac{1}{(A^2 + B^2)^{1/2}}, \quad \sqrt{A + \sqrt{A^2 + B^2}}$$

are Pfaffian, for polynomial A and B.

Pfaffian functions are closed under $+$, $-$, $\cdot$ [5]. Less obviously, one has also

Lemma 2.2 *If f is Pfaffian on U, and nonzero there, $\frac{1}{f}$ is Pfaffian on U, and if $f > 0$ on U then $\pm\sqrt{f}$ is Pfaffian on U.*

Proof. In both cases, the idea is to extend a Pfaffian chain for f to one for $\frac{1}{f}$ or $\pm\sqrt{f}$. All one needs is

$$\frac{\partial}{\partial v}\left(\frac{1}{f}\right) = -\frac{\partial f}{\partial v}\left(\frac{1}{f}\right)^2$$

$$\frac{\partial}{\partial v}\left(f^{-1/2}\right) = -\frac{1}{2}\frac{\partial f}{\partial v}\left(f^{-1/2}\right)^3$$

and then put the chain for $\left(f^{-1/2}\right)^{-1}$ on the end of the latter.   □

Corollary 2.3 *The algebraic functions*

$$\frac{1}{(A^2 + B^2)^{3/4}}, \quad \sqrt{A + \sqrt{A^2 + B^2}}$$

are Pfaffian, on any open set where $A^2 + B^2 \neq 0$, once definite choices of square roots are made.

Before stating the main result for $\wp^{-1}$, I clear up the point about the sign of the square root. Consider a point w_0 where $g(w_0) \neq 0$, and a neighbourhood U of w_0 on which g does not vanish. On an open U_0 contained in U, with $w_0 \in U_0$, there will be two distinct analytic branches $\pm\sqrt{g(z)}$, giving a $\pm$ ambiguity in the integral of the first kind for $\wp^{-1}$, corresponding to the fact that $\wp$ is an even function. For my purposes in [8], one needs only make a definite choice of branch for a suitably chosen w_0. The signs of $\sqrt{\sqrt{A^2 + B^2} - A}$ will then be determined.

Now:

Theorem 2.4 *For each w_0 with $g(w_0) \neq 0$ and each analytic branch of $\sqrt{g(w)}$ on an open neighbourhood of $g(w_0$, the corresponding branch* of $\wp^{-1}$ has its real and imaginary parts Pfaffian. Moreover, there is a bound, uniform in $\wp$, for the complexity of Pfaffian chains for the real and imaginary parts.

Proof. I have shown that $Re(\wp^{-1})$ is Pfaffian, and a similar argument (or Cauchy-Riemann) then gives the result for $Im(\wp^{-1})$.   □

3 Integrals of the second kind

In [8] I will need to consider also the Weierstrass ζ function, which satisfies

$$\frac{d}{dz}\zeta = -\wp.$$

This is associated with the integral

$$\int \frac{zdz}{\sqrt{g(z)}}$$

(integral of second kind), via

$$G(\wp(z)) = -\zeta(z)$$

where

$$G(z) = \int^z \frac{zdz}{\sqrt{g(z)}}.$$

For all this, see [3].

Now G is obviously complex Pfaffian, and a minor variant of my previous argument gives

Theorem 3.1 *The real and imaginary parts of G are Pfaffian.*

A precise statement along the lines of Theorem 2.4 is easily given.

4 The general Theorem

I isolated above the special cases of the Weierstrass functions, since one can be very explicit about the Pfaffian chain. Now I sketch a proof of the most general result I know on the topic.

I suppose, generalizing our previous assumptions, that f is analytic on a neighbourhood U of z_0, and that f' is algebraic in that neighbourhood. Suppose $F(w, z)$ is a polynomial over $\mathbb{C}$ in w, z so that

$$F(f'(z), z) = 0$$

in U, and suppose moreover that

$$\frac{\partial F}{\partial w}(f'(z_0), z_0) \neq 0 \quad \text{on } U.$$

Now let $f(z) = u(x, y) + iv(x, y)$ as usual. So

$$f'(z) = u_x(x, y) + iv_x(x, y) = v_y(x, y) - iu_y(x, y).$$

Now if $z_0 = x_0 + iy_0$

$$F(u_x(x_0, y_0) + iv_x(x_0, y_0), x_0 + iy_0) = 0$$

and

$$\bar{F}(u_x(x_0, y_0) - iv_x(x_0, y_0), x_0 - iy_0) = 0,$$

where $\bar{F}$ is the polynomial complex conjugate to F.

Thus we have two polynomial equations for the unknowns u_x, v_x in terms of x and y. The Jacobian matrix of the above system, with respect to the first two variables, is

$$\begin{pmatrix} \frac{\partial F}{\partial w}(u_x + iv_x, x + iy) & i\frac{\partial F}{\partial w}(u_x + iv_x, x + iy) \\ \frac{\partial \bar{F}}{\partial w}(u_x - iv_x, x - iy) & -i\frac{\partial \bar{F}}{\partial w}(u_x - iv_x, x - iy) \end{pmatrix}$$

with determinant

$$-2i\left|\frac{\partial F}{\partial w}(u_x + iv_x, x + iy)\right| \neq 0 \text{ on } U.$$

It follows easily that u_x and v_x are algebraic functions of x and y on a neighbourhood of (x_0, y_0). Indeed, they are real algebraic functions of x and y.

As in the special cases, one can deduce that u and v are Pfaffian on a neighbourhood of (x_0, y_0), once one shows that real algebraic functions of x, y are Pfaffian.

Lemma 4.1 *Let $g(x, y)$ be a real algebraic function on a neighbourhood U of (x_0, y_0). Then g is Pfaffian on a dense open subset of a neighbourhood of (x_0, y_0).*

Proof. Suppose $H(s, x, y) \in \mathbb{R}[s, x, y]$ is of minimal s-degree so that $H(g(x, y), x, y) = 0$ on a neighbourhood of (x_0, y_0). Then H is clearly irreducible over $\mathbb{R}(x, y)$. Now we have

$$\frac{\partial}{\partial x}H(g(x, y), x, y) = 0 \text{ and } \frac{\partial}{\partial y}H(g(x, y), x, y) = 0$$

on the appropriate neighbourhood of (x_0, y_0). So

$$\frac{\partial H}{\partial s}(g(x,y),x,y)\frac{\partial g}{\partial x} + \frac{\partial H}{\partial x}(g(x,y),x,y) = 0$$

and

$$\frac{\partial H}{\partial s}(g(x,y),x,y)\frac{\partial g}{\partial y} + \frac{\partial H}{\partial y}(g(x,y),x,y) = 0.$$

Now there is a polynomial $B(s,x,y)$ in s over $\mathbb{R}(x,y)$ so that

$$B \cdot \frac{\partial H}{\partial s} - 1$$

is in the ideal generated by $H(s,x,y)$ in $\mathbb{R}(x,y)[s]$. B will involve inverting finitely many polynomials in $\mathbb{R}[x,y]$. There is no guarantee that none of those vanishes at (x_0, y_0), but on a dense open subset of any open neighbourhood of (x_0, y_0) none of them vanishes. So I switch to considering any (x_1, y_1) in this dense open set. Then, near (x_1, y_1)

$$\frac{\partial g}{\partial x} = -B(g(x,y),x,y)\frac{\partial H}{\partial x}(g(x,y),x,y)$$
$$\frac{\partial g}{\partial y} = -B(g(x,y),x,y)\frac{\partial H}{\partial y}(g(x,y),x,y).$$

The only obstruction to this giving a Pfaffian is the presence in B of rational functions of x, y (due to the above-mentioned divisions). But on open sets where they are defined, rational functions are Pfaffian by Lemma 2.2. This proves the lemma. $\qquad\square$

From this lemma, one immediately deduces:

Theorem 4.2 *Suppose $z_0 = x_0 + iy_0$, f analytic on a neighbourhood U of z_0 and f' algebraic in that neighbourhood (with a fixed relation $F(f'(z), z) = 0$ as above). Then $Re(f)$ and $Im(f)$ are Pfaffian on a dense open subset of U.*

5 Concluding remarks

Until the need for an improvement of Theorem 4.2 arises, I am content to leave the matter here.

References

[1] C. Bertolin, Périodes de 1-motifs et transcendance (French), *J. Number Theory* 97 (2002), no. 2, 204–221.

[2] R. Bianconi, *Some Results in the Model Theory of Analytic Functions*, Thesis, Oxford, 1990.

[3] P. Du Val, *Elliptic Functions and Elliptic Curves*, Cambridge University Press, 1973

[4] A. Gabrielov, Complements of subanalytic sets and existential formulas for analytic functions, *Invent. Math.* 125 (1996), no. 1, 1–12.

[5] A. Gabrielov and N. Vorobjov, Complexity of computations with Pfaffian and Noetherian functions, in *Normal forms, bifurcations and finiteness problems in differential equations*, 211–250, NATO Sci. Ser. II Math. Phys. Chem., 137, Kluwer Acad. Publ., Dordrecht, 2004.

[6] A.G. Khovanskii, *Fewnomials*, Translations of Mathematical Monographs, vol 88, AMS, 1991.

[7] A. Macintyre, Elementary theory of elliptic functions 1; the formalism and a special case, 104-131 in *O-minimal Structures*, Proceedings of the RAAG Summer School Lisbon 2003, Lecture Notes in Real Algebraic and Analytic Geometry (M. Edmundo, D. Richardson and A. Wilkie eds.,) Cuvillier Verlag 2005.

[8] A. Macintyre, The elementary theory of elliptic functions 2, in preparation.

[9] A. Macintyre and A.J. Wilkie, On the decidability of the real exponential field, in *Kreiseliana*, 441–467, A.K. Peters, Wellesley, MA, 1996.

[10] A. J. Wilkie, Model completeness results for expansions of the ordered field of real numbers by restricted Pfaffian functions and the exponential function, *J. Amer. Math. Soc.* 9 (1996), no. 4, 1051–1094.

Fusion of structures of finite Morley rank

Martin Ziegler
Universität Freiburg

Summary

Let T_1 and T_2 be two countable complete theories in disjoint languages, of finite Morley rank, the same Morley degree, with definable Morley rank and degree. Let N be a common multiple of the ranks of T_1 and T_2. We show that $T_1 \cup T_2$ has a nice complete expansion of Morley rank N.

1 Introduction

We call a countable complete L–theory T *good* if it has finite definable rank[1] $n > 0$ and definable degree[2]. A *conservative* expansion T' of T is a complete expansion of T, whose rank n' is a multiple of n, such that for all L–formulas $\phi(x, b)$,

$$
\begin{aligned}
\mathrm{MR}_{T'}\, \phi(x, b) &= \frac{n'}{n}\, \mathrm{MR}_T\, \phi(x, b) \\
\mathrm{MD}_{T'}\, \phi(x, b) &= \mathrm{MD}_T\, \phi(x, b).
\end{aligned}
$$

We call the quotient $\frac{n'}{n}$ the *index* of the expansion.

In this note we will prove the following theorem.

Theorem 1.1 *Let T_1 and T_2 be two good theories in disjoint languages of the same degree e and let N be a common multiple of their ranks. Then T_1 and T_2 have a common good conservative expansion T of rank N.*

Furthermore, if in T_i the predicates $P_i^1, \ldots, P_i^e$ define a partition of

---

1 By "rank" we always mean "Morley rank", "degree" is "Morley degree".
2 I.e. the DMP, the definable multiplicity property.

the universe into sets of degree 1, T *can be chosen to imply* $P_1^j = P_2^j$ *for* $j = 1, \dots, e$.

If both, T_1 and T_2, have rank and degree 1, this is Hrushovski's fusion [5], except that we allow the language of T to be larger than the union of the languages of T_1 and T_2. The core of our proof is an adaption of the exposition of Hrushovski's fusion given in [3] and (in Section 2.2) of ideas from Poizat's [6].

As an immediate application we get an explanation of the title of Poizat's [6]:

Corollary 1.2 ([6], [1]) *In any characteristic there is an algebraically closed field K with a subset N such that (K, N) has rank 2.*

Proof Apply 1.1 for T_1 the theory of algebraically closed fields of some fixed characteristic and for T_2 any good theory of rank 2 and degree 1, e.g. the "square of the identity". $\square$

For another account of 1.2 see [2].

Theorem 1.1 was motivated by the following surprising result of A. Hasson:

Corollary 1.3 ([4]) *Every good theory can be interpreted in a good strongly minimal set.*

Proof Let T_1 be a good theory of rank n and degree e. Consider any good theory T_2 of rank n and degree e which can be interpreted in a strongly minimal set X defined in T_2. Use 1.1 to obtain a good theory T of rank n which conservatively expands T_1 and T_2. T_2 is then interpreted in X, which is still strongly minimal in T. $\square$

The simplest example of a theory T_2 as used in the above proof is the "disjoint union of e–copies of the n–th power of the identity": Let X be an infinite set, $Y_1, \dots, Y_e$ be disjoint of copies of X^n and Δ the diagonal of Y_1. Then consider the structure

$$(M, Y_1, \dots, Y_e, \Delta, f_1, \dots, f_e)$$

where M is the disjoint union of the Y_j and f_j is the canonical bijection between Δ^n and Y_j.

The above proof shows that *every good theory of rank n and degree e*

with a partition $P_1 \cup \cdots \cup P_e$ into definable sets of degree 1 has a good conservative expansion of index 1 which contains a strongly minimal set X such that each P_j is in definable bijection with X^n. This yields

Corollary 1.4 *Let T be a good theory and X and Y be two sets of maximal rank and the same degree. Then T has a good conservative expansion of index 1 with a definable bijection between X and Y.* □

Let T be a good theory of rank N with a definable bijection between the universe and the N–th power of a strongly minimal set X. Then the rank of every good expansion of T is a multiple of N. This shows that in Theorem 1.1 one has to assume that N is a common multiple of the ranks of T_1 and T_2, even if one is not interested in the conservativeness of the expansions. A contrasting example is the case where the languages of the T_i have only unary predicates. Then the rank of a completion of $T_1 \cup T_2$ is bounded by $\mathrm{MR}(T_1) + \mathrm{MR}(T_2) - 1$. So, in 1.1, one has in general to increase the language to find an expansion whose rank is a common multiple of the ranks of T_1 and T_2.

I don't know if the last corollary remains true, if one assumes only that X and Y have the same rank (and degree). The following theorem can be used to prove a weaker result.

Theorem 1.5 *Let T be a two-sorted theory with sorts Σ_1 and Σ_2. Let $T_1 = T \restriction \Sigma_1$ be the theory of the full induced structure on Σ_1 and T_1^* a conservative expansion of T_1 of index 1. Assume that T and T_1^* have definable finite rank. Then $T^* = T_1^* \cup T$ is a conservative expansion of T of index 1 which has again definable rank.*

There are examples where T and T_1^* have the DMP, but T^* has not.

Corollary 1.6 *Let T be a good theory and X and Y be two sets of the same rank and the same degree. Then T has a conservative expansion of T^* of index 1 with a definable bijection between X and Y. T^* has definable rank.*

Proof Let T' be the following (good) theory with sorts Σ_1 and Σ_2: Σ_2 is a model of T; Σ_1 is a disjoint union of two predicates X' and Y'; there are bijections between X and X' and between Y and Y'. In $T_1' = T' \restriction \Sigma_1$, X' and Y' have maximal rank and same degree. By 1.4 T_1' has a good conservative expansion $T_1'^*$ of index 1 with a definable bijection between X' and Y'. $T^* = (T' \cup T_1'^*) \upharpoonright \Sigma_2$ has the required properties. $\quad\square$

In [4, Theorem 18] it is proved that for any m and n, any two good strongly minimal sets can be glued together to form a two–sorted structure, where both sets have rank one and there is a definable m-to-n function between them. By Remark 3 of [4] the proof "generalizes to finite-rank". A. Hasson has told me that the generalized proof shows that the union of two good theories of finite rank has a completion of finite rank. Since here the theories may have different degree, the expansions are in general not conservative.

2 Proof of Theorem 1.1

Theorem 1.1 follows from the next theorem, which we will prove in this section.

Theorem 2.1 *Let T_1 and T_2 be to good theories in disjoint languages L_1 and L_2 with ranks $N_1 \le N_2$ and of degree e, and N be the least common multiple of N_1 and N_2. In T_i let the predicates $P_i^1, \ldots, P_i^e$ define a partition of the universe into sets of degree 1. Assume also that T_1 satisfies*

(∗)   *If N_1 divides $N_2 = N$, then each non-algebraic element is interalgebraic with infinitely many other elements. Otherwise, the universe is a union of infinite $\emptyset$–definable $\mathbb{Q}$–vector spaces $V_0, \ldots, V_l$.*

Then $T_1 \cup T_2$ has a completion T of rank N which implies $P_1^j = P_2^j$ and is a good conservative expansion of T_1 and T_2.

Proof of 1.1. Denote the construction in 2.1 by $T_1 + T_2$. Let now T_1 and T_2 be as in 1.1. By adding constants we may assume that the predicates P_i^j are present. Let T_0 be the theory of the disjoint union of e infinite $\mathbb{Q}$–vector spaces. T_0 has rank 1 and degree e. Let N' be the least common multiple of the ranks of T_1 and T_2. Then

$$T' = (T_0 + T_1) + T_2$$

is a good conservative expansion of $T_1 \cup T_2$ of rank N'. Finally set $T = T' + T_3$ for any good theory T_3 of rank N and degree e. $\quad\square$

Actually we need the proposition only in the case that N_1 divides N_2. We have stated it in stronger form, since the proof can be given by a direct application of Hrushovski's fusion machinery to T_1 and T_2.

It is easy to see that, by naming parameters[3], we may assume the following.

$(**)$ If $N_1 = N_2$, for each j, the theory T_2 has infinitely many 1–types over $\emptyset$ of rank $N_2 - 1$ which contain $P_2^j(x)$.

2.1 Hrushovki's machinery

In this section we will develop the theory without using the assumptions $(*)$ and $(**)$. This is a straightforward[4] generalization of sections 2–6 of [3]. We will omit most of the proofs.

2.1.1 Codes (see [3], Section 2)

Let T be a good theory of degree e with predicates $P^1, \ldots, P^e$ which define a partition of the universe in sets of degree 1. We call a formula $\chi(x, b)$ *simple*, if

- it has degree 1,
- the components of a generic realization are pairwise different and not algebraic over b.

A *code* c is a parameter-free formula

$$\phi_c(x, y),$$

where $|x| = n_c$ and y lies in some sort of T^{eq}, with the following properties.

(i) $\phi_c(x, b)$ is either empty[5] or simple. Furthermore there are indices $e_{c,i}$ such that $\phi_c(x, y)$ implies that the x_i are pairwise different and $P^{e_{c,1}}(x_1) \wedge \cdots \wedge P^{e_{c,n_c}}(x_{n_c})$.

(ii) All non-empty $\phi_c(x, b)$ have Morley rank k_c and Morley degree 1.

(iii) For each subset s of $\{1, \ldots, n_c\}$ there exists an integer $k_{c,s}$ such that for every realization a of $\phi_c(x, b)$

$$\mathrm{MR}(a/ba_s) \leq k_{c,s},$$

---

3 We can forget the new constants after the construction of T. So, the language is not increased.

4 For the convenience of the reader many definition and statements are copied verbatim from [3].

5 We assume that $\phi_c(x, b)$ is non-empty for some b.

and equality holds for generic a.[6]

(iv) If both $\phi_c(x,b)$ and $\phi_c(x,b')$ are non-empty and $\phi_c(x,b) \sim^{k_c}$ $\phi_c(x,b')$[7], then $b = b'$.

Lemma 2.2 *Let* $\chi(x,d)$ *be a simple formula. Then there is some code* c *and some* $b_0 \in \mathrm{dcl}^{\mathrm{eq}}(d)$ *such that* $\chi(x,d) \sim^{k_c} \phi_c(x,b_0)$.

We say that c *encodes* $\chi(x,d)$.

Proof As the proof of [3, 2.2]. Note that, by definability of rank, the rank is *additive*

$$\mathrm{MR}(ab/B) = \mathrm{MR}(a/Bb) + \mathrm{MR}(b/B).$$

(see e.g. [7, 4.4]). □

Let c be a code, $\phi_c(x,b)$ non-empty and $p \in S(b)$ the (stationary) type of rank k_c determined by $\phi_c(x,b)$. (iv) implies that b is the canonical base of p. Hence, b lies in the definable closure of a sufficiently large segment of a Morley sequence of p (which we call a *Morley sequence of* $\phi_c(x,b)$.) Let m_c be some upper bound for the length of such a segment. Note that one can always bound m_c by the rank of the sort of y in $\phi_c(x,y)$.

Lemma 2.3 *For every code* c *and every integer* $\mu \geq m_c - 1$ *there exists some formula* $\Psi_c(x_0, \ldots, x_\mu, y)$ *without parameters satisfying the following:*

(v) *Given a Morley sequence* $e_0, \ldots, e_\mu$ *of* $\phi_c(x,b)$, *then*

$$\models \Psi_c(e_0, \ldots, e_\mu, b).$$

(vi) *For all* $e_0, \ldots, e_\mu, b$ *realizing* Ψ_c *the* e_i's *are pairwise disjoint realizations of* $\phi_c(x,b)$.

(vii) *Let* $e_0, \ldots, e_\mu, b$ *realize* Ψ_c. *Then* b *lies in the definable closure of any* m_c *many of the* e_i's.

We say for $\Psi_c(x_0, \ldots, x_\mu, y)$ that "$x_0, \ldots, x_\mu$ is a *pseudo Morley sequence of* c *over* y".

Proof As the proof of [3, 2.3]. □

---

6 $a_s = \{a_i \mid i \in s\}$

7 This means that the Morley rank of the symmetric difference of $\phi_c(x,b)$ and $\phi_c(x,b')$ is smaller than k_c.

We choose for every code (and every μ) a formula Ψ_c as above.

Let c be a code and σ some permutation of $\{1, \ldots, n_c\}$. Then c^σ defined by

$$\phi_{c^\sigma}(x^\sigma, y) = \phi_c(x, y)$$

is also a code. Similarly,

$$\Psi_{c^\sigma}(\bar{x}^\sigma, y) = \Psi_c(\bar{x}, y)$$

defines a pseudo Morley sequence of c^σ.

We call two codes c and c' *equivalent* if $n_c = n_{c'}$, $m_c = m_{c'}$ and

- for every b there is some b' such that $\phi_c(x, b) \equiv \phi_{c'}(x, b')$ and $\Psi_c(\bar{x}, b) \equiv \Psi_{c'}(\bar{x}, b')$ in T,
- similarly permuting c and c'.

Theorem 2.4 *There is a collection of codes C such that:*

(viii) Every simple formula can be encoded by exactly one $c \in C$.

(ix) For every $c \in C$ and every permutation σ, c^σ is equivalent to a code in C.

Proof As the proof of [3, 2.4]. Note that we may have to change the Ψ_c. □

2.1.2 The δ-function (see [3], Section 3)

Let T_1 and T_2 be two good theories as in Theorem 1.1. We assume that the T_i has quantifier elimination in the relational language L_i. To deal with the predicates P_i^j in an effective way we replace both P_1^j and P_2^j by P^j. Then L_1 and L_2 intersect in $L_0 = \{P_1, \ldots, P_e\}$ and T_1 and T_2 intersect in the theory of a partition of the universe into e infinite sets.

Define $\mathcal{K}$ to be the class of all models of $T_{1,\forall} \cup T_{2,\forall}$. We allow also $\emptyset$ to be in $\mathcal{K}$.

Let N_i be rank of T_i, $N = \mathrm{lcm}(N_1, N_2)$ and $N = \nu_1 N_1 = \nu_2 N_2$. We define for finite $A \in \mathcal{K}$

$$(2.1) \qquad \delta(A) = \nu_1 \, \mathrm{MR}_1(A) + \nu_2 \, \mathrm{MR}_2(A) - N \cdot |A|.$$

By additivity of rank δ has the following properties.

(2.2) $\delta(\emptyset) = 0$

(2.3) $\delta(\{a\}) \leq N$ for single elements a

(2.4) $\delta(A \cup B) + \delta(A \cap B) \leq \delta(A) + \delta(B)$

(2.3) is a special case of

(2.5) $\delta(a/B) \leq \nu_i \operatorname{MR}_i(a/B), \quad (i = 1, 2),$

which holds for arbitrary tuples a.

If $A \setminus B$ is finite, we set

$$\delta(A/B) = \nu_1 \operatorname{MR}_1(A/B) + \nu_2 \operatorname{MR}_2(A/B) - N|A \setminus B|.$$

For finite B, it follows that $\delta(A/B) = \delta(A \cup B) - \delta(B)$.

B is *strong* in A if $B \subset A$ and $\delta(A'/B) \geq 0$ for all finite $A' \subset A$. We denote this by

$$B \leq A.$$

$B \nleq A$ is *minimal* if $B \leq A' \leq A$ for no A' properly contained between B and A. a is *algebraic* over B, if a/B is algebraic in the sense of T_1 or T_2. A/B is non-algebraic if no $a \in A \setminus B$ is algebraic over B.

Lemma 2.5 $B \leq A$ *is minimal iff* $\delta(A/A') < 0$ *for all* A' *which lie properly between* B *and* A.

Proof As the proof of [3, 3.1].   □

Lemma [3, 3.2] is not longer true, instead we have

Lemma 2.6 *Let* $B \leq A$ *be a minimal extension. There are three cases*

   (i) $\delta(A/B) = 0$, $A = B \cup \{a\}$ *for an element* $a \in A \setminus B$, *which is algebraic over* B. *(*algebraic simple *extension)*
   (ii) $\delta(A/B) = 0$, A/B *is non-algebraic. (*prealgebraic *extension)*
   (iii) A/B *is non-algebraic and* $1 \leq \delta(A/B) \leq N$, *(*transcendental extension*). If* $\delta(A/B) = N$, *we have* $A = B \cup \{a\}$ *for an element* a *with* $\operatorname{MR}_i(a/B) = N_i$ *for* $i = 1, 2$. *(*transcendental simple extension[8]*)*

---

8 A transcendental simple extension is a transcendental extension by a single element. Note that simple extensions are *not* related to simple formulas.

Proof Assume first that A/B is algebraic. That means that some element $a \in A \setminus B$ is algebraic over B. This implies $\delta(a/B) = 0$ and $B \cup \{a\} \le A$. So we are in case (i).

Now assume that A/B is transcendental and $\delta(A/B) \ge N$. Since $\delta(a/B) \le N$ for all elements $a \in A \setminus B$, Lemma 2.5 implies $B \cup \{a\} = A$. $\square$

Note that, unlike the situation in [3], there may be prealgebraic extensions A/B by single elements if N_1 and N_2 are not relatively prime. We do not call these extensions "simple".

Remark. *If N_1 and N_2 are relatively prime, each strong extension by a single element is simple.*

Proof Let $A = B \cup \{a\}$ be a strong extension of B. If $\delta(A/B) > 0$, the extension is transcendental simple. Otherwise

$$\nu_1 \mathrm{MR}_1(a/A) + \nu_2 \mathrm{MR}_2(a/A) = N_2 \mathrm{MR}_1(a/A) + N_1 \mathrm{MR}_2(a/A) = N.$$

It follows that $\mathrm{MR}_1(a/A)$ is divisible by N_1 and $\mathrm{MR}_2(a/A)$ is divisible by N_2. Whence either $\mathrm{MR}_1(a/A)$ or $\mathrm{MR}_2(a/A)$ must be zero. So A/B is algebraic simple. $\square$

We will work in the class

$$\mathcal{K}^0 = \{M \in \mathcal{K} \mid \emptyset \le M\}.$$

Fix an element M of $\mathcal{K}^0$. We define for finite subsets of M.

$$\mathrm{d}(A) = \min_{A \subset A' \subset M} \delta(A').$$

d satisfies (2.2), (2.3), (2.4) and

(2.6) $$\mathrm{d}(A) \ge 0$$

(2.7) $$A \subset B \Rightarrow \mathrm{d}(A) \le \mathrm{d}(B)$$

$\square$

We define

$$\mathrm{d}(A/B) = \mathrm{d}(AB) - \mathrm{d}(B) = \delta(\mathrm{cl}(AB)/\mathrm{cl}(B)),$$

where $\mathrm{cl}(A)$, the *closure* of A, is the smallest strong subset of M which extends A. Note that the closure of a finite set is again finite (cf. [3, 3.4]).

2.1.3 Prealgebraic codes (see [3], Section 4)

For each T_i fix a set C_i of codes as in 2.4. We may assume that all ϕ_c and Ψ_c are quantifier free.

A *prealgebraic code* is a pair $c \in C_1 \times C_2$ such that

- $n_c = n_{c_1} = n_{c_2}$
- $e_{c_1,j} = e_{c_2,j}$ for all $j \in \{1, \ldots, n_c\}$.
- $\nu_1 k_{c_1} + \nu_2 k_{c_2} - N \cdot n_c = 0$
- $\nu_1 k_{c_1,s} + \nu_2 k_{c_2,s} - N(n_c - |s|) < 0$ for all non-empty proper subsets s of $\{1, \ldots, n_c\}$.

Set $m_c = \max(m_{c_1}, m_{c_2})$ and for each permutation σ, $c^\sigma = (c_1^\sigma, c_2^\sigma)$. c^σ is again prealgebraic.

Some explanatory remarks: T_1^{eq} and T_2^{eq} share only their home sort. An element $b \in \mathrm{dcl}^{\mathrm{eq}}(B)$ is a pair $b = (b_1, b_2)$ with $b_i \in \mathrm{dcl}^{\mathrm{eq}}{}_i(B)$ for $i = 1, 2$. Likewise for $\mathrm{acl}^{\mathrm{eq}}(B)$. A *generic realization* of $\phi_c(x, b)$ (over B) is a generic realization of $\phi_{c_i}(x, b_i)$ (over B) in T_i for $i = 1, 2$. A *Morley sequence* of $\phi_c(x, b)$ is a Morley sequence both of $\phi_{c_1}(x, b_1)$ and $\phi_{c_2}(x, b_2)$. A *pseudo Morley sequence* of c over b is a realization of both $\Psi_{c_1}(\bar{x}, b_1)$ and $\Psi_{c_2}(\bar{x}, b_2)$. We say that M is *independent* from A over B if M is independent from A over B both in T_1 and T_2.

The following three lemmas are proved as Lemmas 4.1, 4.2 and 4.3 in [3].

Lemma 2.7 *Let $B \leq B \cup \{a_1, \ldots, a_n\}$ be a prealgebraic minimal extension and $a = (a_1, \ldots, a_n)$. Then there is some prealgebraic code c and $b \in \mathrm{acl}^{\mathrm{eq}}(B)$ such that a is a generic realization of $\phi_c(a, b)$.* $\square$

Lemma 2.8 *Let $B \in \mathcal{K}$, c a prealgebraic code and $b \in \mathrm{acl}^{\mathrm{eq}}(B)$. Take a generic realization $a = (a_1, \ldots, a_{n_c})$ of $\phi_c(x, b)$ over B. Then $B \cup \{a_1, \ldots, a_{n_c}\}$ is a prealgebraic minimal extension of B.* $\square$

Note that the isomorphism type of a over B is uniquely determined.

Lemma 2.9 *Let $B \subset A$ in $\mathcal{K}$, c a prealgebraic code, b in $\mathrm{acl}^{\mathrm{eq}}(B)$ and $a \in A$ a realization of $\phi_c(x, b)$ which does not lie completely in B. Then*

1. *$\delta(a/B) \leq 0$.*

2. If $\delta(a/B) = 0$, then a is a generic realization of $\phi_c(x, b)$ over B.

□

The next Lemma is the analogue of [3, 4.4]

Lemma 2.10 *Let $M \leq N$ an extension in $\mathcal{K}$ and $e_0, \ldots, e_\mu \in N$ a pseudo Morley sequence of c over b. Then one of the following holds:*

- $b \in \mathrm{dcl}^{\mathrm{eq}}(M)$
- *more than $\mu - m_c \cdot (N(n_c - 1) + 1)$ many of the e_i lie in $N \setminus M$.*

Proof If b is not in $\mathrm{dcl}^{\mathrm{eq}}(M)$, less than m_c many of the e_i lie in M. Let r be the number of elements not in $N \setminus M$. We change the indexing so that $e_i \in N \setminus M$ implies $i \geq r$ and $e_i \in M$ implies $i < (m_c - 1)$. By Lemma 2.9 we have $\delta(e_i/Me_0, \ldots, e_{i-1}) < 0$ for all $i \in [m_c, r)$. This implies, for $m = \min(m_c, r)$,

$$0 \leq \delta(e_0, \ldots, e_{r-1}/M) \leq \delta(e_0, \ldots, e_{m-1}/M) - (r - m_c).$$

On the other hand we have $\delta(e_0, \ldots, e_{m-1}/M) \leq N \cdot m \cdot (n_c - 1)$, which implies

$$r \leq N \cdot m \cdot (n_c - 1) + m_c \leq N \cdot m_c \cdot (n_c - 1) + m_c.$$

□

2.1.4 The class $\mathcal{K}^\mu$ (see [3], Section 5)

Choose a function μ^* from prealgebraic codes to natural numbers similar to section 5 of [3]. μ^* must satisfy $\mu^*(c) \geq m_c - 1$ and be finite-to-one for every fixed n_c. Also we must have $\mu^*(c) = \mu^*(d)$, if c is equivalent to a permutation of d. Then set

$$\mu(c) = m_c \cdot (N(n_c - 1) + 1) + \mu^*(c).$$

From now on, a *pseudo Morley sequence* denotes a pseudo Morley sequence of length $\mu(c) + 1$ for a prealgebraic code c.

The class $\mathcal{K}^\mu$ consists of the all structures in $\mathcal{K}^0$ which do not contain any pseudo Morley sequence.

The following lemma and its corollary have the same proofs as their analogues [3, 5.1] and [3, 5.2].

Lemma 2.11 *Let B be a finite strong subset of $M \in \mathcal{K}^\mu$ and A/B a prealgebraic minimal extension. Then there are only finitely many B-isomorphic copies of A in M.*

□

Corollary 2.12 *Let $B \leq M \in \mathcal{K}^\mu$, $B \subset A$ finite with $\delta(A/B) = 0$. Then there are only finitely many A' such that: $B \leq A' \subset M$ and A' is B-isomorphic to A.* $\square$

Lemma [3, 5.4] may be wrong here. We have instead:

Lemma 2.13 *Let $M \in \mathcal{K}^\mu$ and N a simple extension of M. Then $N \in \mathcal{K}^\mu$.*

Proof Let $(e_i) \in N$ a pseudo Morley sequence of c over b. At least $\mu(c)$ of the e_i lie in M. Since $\mu(c) \geq m_c$, we have $b \in \mathrm{dcl}^{\mathrm{eq}}(M)$. Since M belongs to $\mathcal{K}^\mu$, one e_i does not lie in M. By 2.9 we conclude that e_i is disjoint from M and a generic realization of $\phi_c(x, b)$. So $n_c = 1$ and N/M is prealgebraic, i.e. not simple. $\square$

Proposition 2.14 $\mathcal{K}^\mu$ *has the amalgamation property with respect to strong embeddings.*

Proof The proof is the same as the proof of [3, 5.5], the main ingredient being Lemma 2.10. Only one point has to be checked: If A/B is strong and $a \in A$ is algebraic over b, say in the sense of T_1, then $\mathrm{tp}_2(a/B)$ is uniquely determined. This is the case, since

$$0 \leq \delta(a/B) = \nu_2 \, \mathrm{MR}_2(a/B) - N \leq \nu_2 N_2 - N = 0$$

implies that $\mathrm{MR}_2(a/B) = N_2$. On the other hand, $tp_1(a/B)$ implies $a \in P^j$ for some j. So the T_2–type of a/B is uniquely determined since P^j has degree 1 in T_2. $\square$

The proof has the following corollary.

Corollary 2.15 *Two strong extensions $B \leq M$ and $B \leq A$ in $\mathcal{K}^\mu$ can be amalgamated in $M, A \leq M' \in \mathcal{K}^\mu$ such that $\delta(M'/M) = \delta(A/B)$ and $\delta(M'/A) = \delta(M/B)$.* $\square$

A structure $M \in \mathcal{K}^\mu$ is *rich* if for every finite $B \leq M$ and every finite $B \leq A \in \mathcal{K}^\mu$ there is some B-isomorphic copy of A in M. We will show in the next section that rich structures are models of $T_1 \cup T_2$.

Corollary 2.16 *There is a unique (up to isomorphism) countable rich structure K^μ. Any two rich structures are $(L_1 \cup L_2)_{\infty,\omega}$–equivalent.* $\square$

2.1.5 The theory T^μ (see [3], Section 6)

Lemma 2.17 *Let $M \in \mathcal{K}^\mu$, $b \in \mathrm{acl}^{\mathrm{eq}}(M)$, $a \models \phi_c(x,b)$ generic over M and M' the prealgebraic minimal extension $M \cup \{a_1, \cdots a_{n_c}\}$. If M' is not in $\mathcal{K}^\mu$, then one of the following holds.*

(a) *M' contains a pseudo Morley sequence of c over b, all whose elements but possibly one are contained in M.*

(b) *M' contains a pseudo Morley sequence for some code c' with more than $\mu^*(c')$ many elements in $M' \setminus M$.*

Proof As in the proof of [3, 6.1], this follows from 2.9 and 2.10. □

As in [3], Lemmas 2.7, 2.8 and 2.17 imply that we can describe all M with the following properties by an elementary theory T^μ.

Axioms of T^μ

(a) $M \in \mathcal{K}^\mu$

(b) $T_1 \cup T_2$

(c) M has no prealgebraic minimal extension in $\mathcal{K}^\mu$.

To prove the analogue of Theorem [3, 6.3], which says that the rich structures are the ω–saturated models of T^μ we need the assumptions $(*)$ and $(**)$. Whithout this we can only show[9]

Lemma 2.18 *Rich structures are models of T^μ.*

Proof Let K be rich. Consider a quantifier free L_1–formula $\chi(x)$ with parameters in K which is T_1–consistent. Let B be a finite strong subset of K which contains the parameters. If $\chi(x)$ is not realized in B, realize $\chi(x)$ by a new element a and define the structure $A = B \cup \{a\}$ in such a way that $\mathrm{MR}_2(a/B) = N_2$. Then $\delta(a/B) = \nu_1 \mathrm{MR}_1(a/B)$, so $B \leq A$ and A/B is simple. So by 2.13 B belong to $\mathcal{K}^\mu$. Since K is rich, it contains a copy of A/B. This proves that $\chi(x)$ is realized in K. This shows that K is model of T_1. The same proof shows that K is also a model of T_2.

Axiom (c) is proved like in the proof of [3, 6.3]. □

[9] It is conceivable that T^μ might be incomplete. We even do not know wether T^μ has an ω–stable completion. (This question was raised by the referee.)

2.2 Poizat's argument

We assume now conditions $(*)$ and $(**)$ of Theorem 2.1. We want to show that ω–saturated models of T^μ are rich. We start with two lemmas.

Lemma 2.19 T_1 *has the following property. Let* $M_1 > 0$ *and* M_2 *be two natural numbers, a an element of an* $\emptyset$–*definable* $\mathbb{Q}$–*vector space* V_α. *Let B be a set of parameters such that* V_α *contains elements which are of rank 1 over B. Then there are elements* $c_1, \ldots, c_{M_2}$ *of* V_α *such that for all* $s \subset \{1, \ldots, M_2\}$

$$(2.8) \qquad \min(M_1, |s|) \leq \mathrm{MR}_1(c_s/Ba) \leq M_1$$

and, if $|s| > M_1$

$$(2.9) \qquad \mathrm{MR}_1(c_s/B) = \mathrm{MR}_1(c_s/Ba) + \mathrm{MR}_1(a/B).$$

Proof We start with a sequence $v_1, \ldots, v_{M_2}$ of elements of $\mathbb{Q}^{M_1}$ such that

- any M_1 elements of the sequence are $\mathbb{Q}$–linearly independent,
- any $M_1 + 1$ elements of the sequence are linearly dependent, but affinely independent.

Then we choose any B–independent sequence $\bar{e} = (e_1, \ldots, e_{M_1})$ of elements of V_α which have rank 1 over B, such that $\bar{e}$ is independent from a over B. We consider $\bar{e}$ as a column vector and the v_i as a row vectors and define

$$c_i = v_i \cdot \bar{e} + a.$$

Since all c_i are algebraic over $Ba\bar{e}$, it is clear that

$$\mathrm{MR}_1(c_s/Ba) \leq \mathrm{MR}_1(\bar{e}/Ba) = M_1.$$

To show $\min(M_1, |s|) \leq \mathrm{MR}_1(c_s/Ba)$, we may assume that $|s| \leq M_1$. Since the v_i, $i \in s$ are linearly independent there is a subsequence $\bar{e}'$ of $\bar{e}$ of length $M_1 - |s|$ such that the elements of $\bar{e}'$ and $v_s \cdot \bar{e}$ span the same $\mathbb{Q}$–vector space as the elements of $\bar{e}$. So we have

$$M_1 = \mathrm{MR}_1(\bar{e}/Ba) = \mathrm{MR}_1(\bar{e}', v_s \cdot \bar{e}/Ba) \leq (M_1 - |s|) + \mathrm{MR}_1(v_s \cdot \bar{e}/Ba)$$

and hence

$$|s| \leq \mathrm{MR}_1(v_s \cdot \bar{e}/Ba) = \mathrm{MR}_1(c_s/Ba).$$

The last equation follows from the fact that each $M_1 + 1$ many of the

e_i span an affine subspace which contains a. The reason for this is that the according v_i are linearly dependent, but affinely independent, and therefore span an affine space which contains 0. □

Lemma 2.20 *If $N_1 = N_2$, T_2 has the following property. Let B be any set of parameters, and p be the type over B of an M_2–tuple of independent elements of rank N_2 over B. Then p is the limit of types of tuples of independent elements of rank $N_2 - 1$ over B.*

Proof We indicate the proof for $M_2 = 2$. Let $p = \text{tp}(a_1 a_2 / B)$ and $\phi(x_1, x_2) \in p$. The formula $\phi_1(x_1) =$ "$\text{MR}_{x_2}\, \phi(x_1, x_2) \geq N_2$" has rank N_2. Therefore, by $(**)$, there is a type q_1 over B which has rank $N_2 - 1$ and contains $\phi_1(x_1)$. Let b_1 be a realization of q_1. By the open mapping theorem, and $(**)$ again, $\phi(b_1, x_2)$ contains a type q_2 over Bb_1, of rank $N_2 - 1$ which does not fork over B. Realize q_2 by b_2. The type of $b_1 b_2$ over B contains ϕ, b_1 and b_2 are independent and of rank $N_2 - 1$ over B. □

Proposition 2.21 *The rich structures are exactly the ω–saturated models of T^μ.*

Proof That rich structures are models of T^μ was proved in 2.18. As in the proof of [3, 6.3] one sees that it suffices to prove that ω–saturated models of T^μ are rich. So let K be an ω–saturated model, $B \leq K$ finite and $B \leq A$ a minimal extension which belongs to $\mathcal{K}^\mu$. We show that A/B can be strongly embedded in K by induction over $d = \delta(A/B)$.

If $d = 0$ the extension is algebraic or prealgebraic and the claim follows from 2.14, since K has no algebraic or prealgebraic extensions. So we assume $d > 0$. All we use from the minimality of A/B in this case is that $A \neq B$ and $\delta(X/B) > 0$ for all subsets of A, which are not contained in B.

We may assume that B is large enough to have, for each j, parameters for an L_2–formula in P^j which has rank $N_2 - 1$ in T_2. Choose two numbers M_1 and M_2 such that

$$\nu_1 M_1 - \nu_2 M_2 = -1.$$

The M_i are uniquely determined if we impose the condition $0 \leq M_1 < \nu_2$. We have then

$$M_1 = \frac{\nu_2 M_2 - 1}{\nu_1} < M_2,$$

since $\nu_2 \leq \nu_1$.

Let a be an arbitrary element of $A \setminus B$. Since $\delta(a/B) > 0$, a is not algebraic over B.

If N_1 divides N_2, i.e. if $\nu_2 = M_2 = 1$ and $M_1 = 0$, we choose an element $c_1 \notin A$, which is in the sense of T_1 interalgebraic with a and has rank N_2 over A in the sense of T_2. We set $C = A \cup \{c_1\}$. If N_1 does not divide N_2, we have $M_1 > 0$. We define then $C = A \cup \{c_1, \ldots, c_{M_2}\}$ where the c_i are given by Lemma 2.19 and are – in the sense of T_1 – independent from A over Ba. In the sense of T_2 they are chosen to be A–independent and of rank $N_2 - 1$ over A.

We compute

$$\delta(C/A) = \nu_1 M_1 + \nu_2 M_2 (N_2 - 1) - N M_2 = \nu_1 M_1 - \nu_2 M_2 = -1.$$

Claim 1 $B \leq C$.

Proof. Let X be a set between B and A and Y be a subset of $\{c_1, \ldots, c_{M_2}\}$ of size y. Note that $\delta(XY/B) \geq \delta(Y/A) + \delta(X/B)$ and by equation (2.8) we have

$$\delta(Y/A) \geq \nu_1 \min(M_1, y) + \nu_2 y (N_2 - 1) - N y = \nu_1 \min(M_1, y) - \nu_2 y.$$

Case 1: $y \leq M_1$. Then $\delta(XY/B) \geq \delta(Y/A) \geq (\nu_1 - \nu_2)y \geq 0$.

Case 2: $M_1 < y$. Then we have

$$\delta(Y/A) = \nu_1 M_1 - \nu_2 y \geq \nu_1 M_1 - \nu_2 M_2 = -1$$

and distinguish two cases: If $X = B$, then, by (2.9), $\mathrm{MR}_1(Y/B) > \mathrm{MR}_1(Y/A)$ and therefore $\delta(XY/B) = \delta(Y/B) > \delta(Y/A) \geq -1$. If X is different from B we have $\delta(XY/B) \geq -1 + \delta(X/B) \geq 0$. This proves the claim. $\qquad \square$

Claim 2 The closure of A in C equals C.

Proof. Let Y be a proper subset of $\{c_1, \ldots, c_{M_2}\}$ of size y. We have to show that $\delta(Y/A) > -1$. By the above this is clear if $y \leq M_1$. Otherwise we have

$$\delta(Y/A) = \nu_1 M_1 - \nu_2 y > \nu_1 M_1 - \nu_2 M_2 = -1.$$

This proves the claim. $\qquad \square$

It follows (if N_1 does not divide N_2, from the proof of Lemma 2.19) that one can produce a sequence of extensions $A \subset C_i$ like above such that the types $\mathrm{tp}_1(C_i/A)$ converge against a type $\mathrm{tp}_1(D/A)$ where the elements $d_0, \ldots, d_{M_2}$ are of rank ≥ 1 and algebraically independent[10] over A in the sense of T_1. If $N_1 < N_2$ we simply choose the types $\text{tp}_2(C_i/A)$ and $\text{tp}_2(D/A)$ to be all the same and with components of rank $N_2 - 1$ independent over A in the sense of T_2. If $N_1 = N_2$, it follows from Lemma 2.20 that we may assume that the types $\text{tp}_2(C_i/A)$ converge to $\text{tp}_2(D/A)$ and that the d_i have rank N_2 over A and are independent over A in the sense of T_2.

If $N_1 < N_2$, we have

$$\delta(d_i/Ad_0 \ldots d_{i-1}) \geq \nu_1 \cdot 1 + \nu_2(N_2 - 1) - N = \nu_1 - \nu_2 > 0.$$

If $N_1 = N_2$ we have for every i

$$\delta(d_i/Ad_0 \ldots d_{i-1}) \geq \nu_1 \cdot 1 + \nu_2 N_2 - N = \nu_1 > 0.$$

So D is a strong extension of A which splits into a sequence of transcendental simple extensions. So, by Lemma 2.13, D belongs to $\mathcal{K}^\mu$.

Claim 3 For large enough i we have $C_i \in \mathcal{K}^\mu$.

Proof. Proof: Since the C_i have all the same size, if C_i does not belong to $\mathcal{K}^\mu$ and μ is finite-to-1 for fixed n_c, there is a certain finite set of prealgebraic codes which can be responsible for this. Since $D \in \mathcal{K}^\mu$, almost all C_i belong to $\mathcal{K}^\mu$. $\qquad\square$

Now by induction for large enough i, C_i can be strongly embedded over B into K. Since K is ω–saturated this implies that D can be strongly embedded into K. Such an embedding also strongly embeds A, since $A \leq D$. $\qquad\square$

Corollary 2.22 T^μ *is complete. In models of T^μ two tuples have the same type iff they have isomorphic closures.*

Proof Same as the proof of [3, 7.1]. $\qquad\square$

2.3 Rank computation

Proposition 2.23 *In T^μ we have for tuples a*

$$\text{MR}(a/B) = \text{d}(a/B).$$

---

10 It suffices that d_i is not in $\text{acl}_1(Ad_0 \ldots d_{i-1})$.

Proof We prove first $\mathrm{MR}(a/B) \le \mathrm{d}(a/B)$. Since the closure is algebraic we may assume that B and $A = B \cup \{a\}$ are closed. Then $\mathrm{d}(a/B) = \delta(a/B)$, so it suffices to show that $\mathrm{MR}(a/B) \le \delta(a/B)$ for all closed B and arbitrary a. We do this by induction on $d = \delta(a/B)$.

Let M be an ω–saturated model, which contains B such that the (a priori infinite) rank of a over M is the same as the rank of a over B. Then $\delta(a/M) \le \delta(a/B)$ and by induction we may assume that $\delta(a/M) = d$. Also we may assume that a is disjoint from M. Write $a = (a_1, \ldots, a_n)$.

Choose for $i = 1, 2$ an $L_i(M)$-formula $\phi_i(x) \in \mathrm{tp}_i(a/M)$ with the following properties.

(i) ϕ_i has degree 1

If a' is any realization of $\phi(x)$, then

(ii) the components of a' are pairwise different
(iii) $\mathrm{MR}_i(a'/Ma'_s) \le k_{i,s}$, where s is any subset of $\{1, \ldots, n\}$ and $k_{i,s} = \mathrm{MR}_i(a/Ma_s)$.

It follows that $\mathrm{MR}_i \phi_i = k_{i,\emptyset} = \mathrm{MR}_i(a/M)$.

Let a' be any realization of $\phi(x, b) = \phi_1(x, b) \wedge \phi_2(x, b)$. The inequality $\mathrm{MR}(a/M) \le d$ follows the from ω–saturation of M and the next claim.

Claim Either $\mathrm{MR}(a'/M) < d$ or $\mathrm{tp}(a'/M) = \mathrm{tp}(a/M)$.

Proof.
Case 1. $\delta(a'/M) < d$. Then $\mathrm{MR}(a'/M) < d$ by induction.

Case 2. $\delta(a'/M) \ge d$. Set $s = \{i \mid a'_i \in M\}$ consider the inequality

$$\delta(a'/M) = \nu_1 \cdot \mathrm{MR}_1(a'/Ma'_s) + \nu_2 \, \mathrm{MR}_2(a'/Ma'_s) - N \cdot (n - |s|)$$
$$\le \nu_1 \cdot k_{1,s} + \nu_2 k_{2,s} - N \cdot (n - |s|)$$
$$= \delta(a/Ma_s) \le \delta(a/M).$$

Our assumption implies $\mathrm{MR}_i(a'/Ma'_s) = k_{i,s}$ and $\delta(a/Ma_s) = \delta(a/M)$. The latter implies $\delta(a_s/M) = 0$, so a_s/M is algebraic in the sense of T^μ (2.12), which is only possible if s is empty. So we have $\mathrm{MR}_i(a'/M) = \mathrm{MR}_i(a/M)$, which implies that a' and a are isomorphic over M, and $\delta(a'/M) = d$.

Case 2.1. $M \cup \{a'\}$ is not closed. Then a' has an extension a'' with $\delta(a''/M) < d$. It follows $\mathrm{MR}(a'/M) \le \mathrm{MR}(a''/M) < d$ by induction.

Case 2.2. $M \cup \{a'\}$ is closed. Then $\mathrm{tp}(a'/M) = \mathrm{tp}(a/M)$.

Now we prove $\mathrm{d}(a/B) \leq \mathrm{MR}(a/B)$ by induction on $d = \mathrm{d}(a/B)$. We may assume that B is finite, that B and $B \cup \{a\}$ are closed and (using 2.15) that B has, for each j, parameters for an L_2–formula in P^j which has rank $N_2 - 1$ in T_2. If $d = 0$, there is nothing to show. If $d > 0$, we decompose A/B into $B \leq B' \leq A$, where B' is maximal with $\delta(B'/B) = 0$.

Now we can use the construction in proof of 2.21 to obtain a sequence of extensions $A \subset C_i$ and $A \leq D$, such that $B' \leq C_i$, $\delta(C_i/A) = d - 1$, all in $\mathcal{K}^\mu$, such that C_i is the closure of A and the qf-types of the C_i over A converge against the qf-type of D over A. We may assume that D is closed (in the monster model). We also choose copies C'_i of C_i over B' which are closed. Let A'_i be the corresponding copy of A in C'_i. Since the types $\mathrm{tp}(C'_i/B)$ converge against $\mathrm{tp}(D/B)$, the types $\mathrm{tp}(A'_i/B)$ converge against $\mathrm{tp}(A/B)$. Now $\mathrm{d}(A'_i/B) = \delta(C'_i/B) = d - 1$, so by induction $d - 1 \leq \mathrm{MR}(C'_i/B)$, which implies $d \leq \mathrm{MR}(A/B)$. $\qquad\square$

The referee has pointed out that our proof of $\mathrm{MR}(a/B) \leq \mathrm{d}(a/B)$ can be rephrased as follows: It it easy to see that d–independence defines a notion of independence. The claim in the proof of 2.23 shows that types over ω–saturated models are isolated among the types of at least the same rank. This implies the above inequality.

Lemma 2.24 *Let $\phi(x)$ be an L_i–formula (with parameters). Then*

$$\mathrm{MR}\,\phi = \nu_i\,\mathrm{MR}_i\,\phi.$$

Proof Consider $i = 1$, the case $i = 2$ works the same. Let $\phi(x)$ be defined over the closed set B. If a is any realization of ϕ, we have by (2.5)

$$\mathrm{MR}(a/B) \leq \delta(a/B) \leq \nu_1\,\mathrm{MR}_1(a/B) \leq \nu_1\,\mathrm{MR}_1\,\phi.$$

So $\mathrm{MR}\,\phi \leq \nu_1\,\mathrm{MR}_1\,\phi$. For the converse choose a generic realization $a = (a_1, \ldots, a_n)$ of ϕ. Choose $\mathrm{tp}_2(a/B)$ of maximal possible rank[11]. Then clearly $\delta(a/B) = \nu_1\,\mathrm{MR}_1(a/B) = \nu_1\,\mathrm{MR}_1\,\phi$. Also, for every i, $B \cup \{a_1, \ldots, a_i\}$ is equal to, or a simple extension of, $B \cup \{a_1, \ldots, a_{i-1}\}$. So, by 2.13, $B \cup \{a\}$ belongs to $\mathcal{K}^\mu$. We can therefore find $B \cup \{a\}$ as a closed subset of a model of T^μ. This implies $\mathrm{MR}(a/B) = \delta(a/B) = \nu_1\,\mathrm{MR}_1\,\phi$. $\qquad\square$

11 This is N_2 times the number of different a_i's

Lemma 2.25 *Let $\phi(x)$ be an L_i-formula (with parameters). Then*

$$\mathrm{MD}\,\phi = \mathrm{MD}_i\,\phi.$$

Proof Consider $i = 1$. Let $\phi(x)$ be defined over the closed set B. We may assume that ϕ is simple in the sense of T_1. Let a be a realization of $\phi(x)$ with $\mathrm{MR}(a/B) = \mathrm{MR}\,\phi$. Then $\mathrm{MR}_1(a/B) = \mathrm{MR}_1\,\phi$, which determines $\mathrm{tp}_1(a/B)$ uniquely, since $\mathrm{MD}_1\,\phi = 1$. In the sense of T_2 the a_i are B–independent generic elements of certain P^j's, so the type $\mathrm{tp}_2(a/B)$ is uniquely determined. Finally $B \cup \{a\}$ must be closed. This implies that $\mathrm{tp}(a/B)$ is uniquely determined and $\mathrm{MD}\,\phi = 1$. $\square$

2.4 Definable rank and degree

It remains to show that T^μ has definable rank and degree. If N_1 does not divide N_2 the definability of rank follows from the fact that the universe of T^μ is covered by a finite set of definable groups. We give a proof which works also for the case $N_1 | N_2$.

We use the following observation, due to M. Hils. Call a formula $\phi(x, b)$ of rank n and degree 1 *normal* if b satisfies a formula $\theta(y)$ such that $\phi(x, b')$ has rank n and degree 1 for all realizations b' of θ. A type is *normal* if it contains a normal formula of the same rank. We have then

Lemma 2.26 *Let T be a complete theory of finite rank. Then*

1. *T has definable rank and degree iff every type over a model M is normal.*
2. *If $\mathrm{tp}(a, a'/M)$ is normal, and a' is algebraic over Ma, then also $\mathrm{tp}(a/M)$ is normal.* $\square$

In 1. it suffices to consider ω–saturated models M. Also, if M is ω–saturated and $b \in M$, then $\phi(x, b)$ is normal iff there is a $\theta(y)$ defined over M such that $\phi(x, b')$ has rank n and degree 1 for all b' in $\theta(M)$.

Consider an ω–saturated model M of T^μ and a type $p = \mathrm{tp}(a/M)$ of rank $d = \mathrm{d}(a/M)$. We want to show that p is normal. By 2.26.2. we may assume that $M \cup \{a\}$ is closed, i.e. $d = \delta(a/M)$. We may also assume that a is disjoint from M and that all components of a are different. Choose for each $i = 1, 2$ formulas $\phi_i(x, m) \in \mathrm{tp}_i(a/M)$ with properties (i), (ii), (iii) as in the first part of the proof of proposition 2.23. Choose a formula $\theta(x)$ over M, which is satisfied by m, such that for all $m' \in \theta(M)$ the formulas $\phi(x, m')$ satisfy (i), (ii), and (iii) and $\mathrm{MR}_i\, \phi_i(x, m') = k_{i,\emptyset}$ for $i = 1, 2$. Let a' be a generic realization of $\phi(x, m')$, which has a unique qf-type over M. Then $\delta(a'_s/M) = \delta(a_s/M)$ for all $s \subset \{1, \ldots, n\}$, especially $\delta(a'/M) = d$. This implies that $M_{m'} = M \cup \{a'\}$ is a strong extension of M. One sees easily, like in [3, 6.2], that we can strengthen θ to ensure that $M_{m'} \in \mathcal{K}^{\mu 12}$. So we can find a' with $M_{m'}$ closed in the universe. This implies $\mathrm{MR}(a'/M) = d$.

The proof of 2.23 shows that for all realizations a'' of $\phi(x, m')$ either $\mathrm{MR}(a''/M) < d$ or $\mathrm{tp}(a''/M) = \mathrm{tp}(a'/M)$. This shows that $\phi(x, m')$ has rank d degree 1 and that $\phi(x, m)$ is normal.

This completes the proof of Theorem 2.1.

3 Proof of Theorem 1.5

We start with an easy lemma.

Lemma 3.1 *Let T be a complete two–sorted theory with sorts Σ_1 and Σ_2. Then the following are equivalent.*

  a) Σ_1 *is stably embedded.*
  b) *Let T_1^* be a one–sorted complete expansion of $T_1 = T \upharpoonright \Sigma_1$. Then $T^* = T_1^* \cup T$ is complete.*

Proof a)→b): Consider $S = (S_1^*, S_2)$ and $S' = (S_1'^*, S_2')$ two saturated models of T^* of the same cardinality. Since T and T_1^* are complete, there are isomorphisms $f : (S_1, S_2) \to (S_1', S_2')$ and $g : S_1^* \to S_1'^*$. $f^{-1}g \upharpoonright S_1$ is an automorphism of the structure induced on S_1. Since S_1 is stably embedded, there is an extension of $f^{-1}g \upharpoonright S_1$ to an automorphism h of (S_1, S_2). Then fh is an isomorphism $S \to S'$.

b)→a): This is not used in this article and left to the reader. □

---

12 The argument is as follows. Decompose the extension $M \leq M \cup \{a\}$ into a sequence of minimal extensions, where the prealgebraic extensions are given by codes $c_1, \ldots, c_k$. Strengthen θ so that the extensions $M \leq M \cup \{a'\}$ are also composed of prealgebraic extension coming from $c_1, \cdots, c_k$. The argument of [3, 6.2] shows now that "$M \cup \{a'\} \in \mathcal{K}^\mu$" is an elementary property of m'.

We fix for the rest of the section T, T_1, T_1^* and T^* as in 1.5. Let L, L_1, L_1^* and $L^* = L_1^* \cup L$ be the respective languages. We may assume that T_1 has elimination of imaginaries.[13]

The following lemma is due to Anand Pillay. We need only that Σ_1 is stably embedded.

Corollary 3.2 *In T^* every L^*–formula $\Phi(x)$ is equivalent to a formula of the form*

$$\psi^*(t(x)),$$

where $\psi^*(y)$ is an L_1^*–formula and t is a T–definable function with values in some power of Σ_1.

Proof Let $S = (S_1, S_2)$ be a model of T, where S_1 is a model of T_1 and S^* be an expansion to a model of T^*. Let a be a tuple from S. Since S_1 is stably embedded and has elimination of imaginaries, every a–definable relation on S_1 has a canonical parameter in S_1. $B = \mathrm{dcl}(a) \cap S_1$ is the set of these parameters and $(S_1, b)_{b \in B}$ is the structure induced by (S, a) on S_1.

By 3.1

$$\mathrm{Th}(S^*, a) = \mathrm{Th}(S_1, b)_{b \in B} \cup \mathrm{Th}(S, a).$$

This means that $\mathrm{tp}^*(a)$ is axiomatized by $\mathrm{tp}^1(B) \cup \mathrm{tp}(a)$, which implies the lemma. □

Corollary 3.3 *S_1^* is the structure induced by S^* on S_1.* □

Proof of Theorem 1.5: We prove the following claim by induction on k.

1) *For every L–definable X with $\mathrm{MR}\,X \le k$ we have $\mathrm{MRD}^*\,X = \mathrm{MRD}\,X$.*

2) *For all L^*–formulas $\Phi(x,y)$, "$\mathrm{MR}^*\,\Phi(x,b) = k$" is an L^*–elementary property of b.*

Case $k = 0$. Let $\Phi(x,b)$ be of the form $\psi^*(t(x))$, where ψ^* and t are defined from b. Consider t as a map $S \to S_1$. Then $\psi^*(t(x))$ is finite iff the L_1^*–formula $\exists x\,(y \doteq t(x) \land \psi^*(y))$ and all the fibers $t(x) = a$ for $\models \psi^*(a)$ are finite. This can be elementarily expressed since finiteness

---

13 For this we replace T_1 by T_1^{eq}. Actually the sort Σ_1 may be itself a many-sorted structure.

can be expressed in T_1^* and T. This proves 2). 1) is clear.

Case $k + 1$.

1) Assume $\mathrm{MR}\, X \leq k + 1$. If all L^*–definable subsets of X are L–definable, it is clear that $\mathrm{MRD}^* X = \mathrm{MRD}\, X$. So assume that there is an L^*–definable $A \subset X$ which is not L–definable. By Corollary 3.2 there is an L–definable surjection $t : X \to Y \subset S_1^n$ and an L^*–definable $B \subset Y$ such that $A = t^{-1}B$. Since MR is definable in T we can partition Y into finitely many L-definable sets on each of which the ranks of the fibers $t^{-1}y$ have constant rank. The inverse image of this partition is an L–definable partition of X. Since it is enough to prove 1) for each of the sets of the partition, we may assume that all fibers of t have the same rank f. Since A is not L–definable, Y must be infinite. So we have $f = \mathrm{MR}\, X - \mathrm{MR}\, Y \leq k$. By induction all fibers have T^*–rank f. Since, again by induction, all T^*–ranks $\leq k$ are definable, it follows[14] that $\mathrm{MR}^* X = f + \mathrm{MR}^*(Y) = f + \mathrm{MR}\, Y = \mathrm{MR}\, X$.

To prove that $\mathrm{MD}^* X = \mathrm{MD}\, X$, we may assume that $\mathrm{MD}\, X = 1$. We have to show that $\mathrm{MR}^*(X \setminus A) < \mathrm{MR}\, X$ for every L^*–definable $A \subset X$ of T^*–rank $\mathrm{MR}\, X$. This is clear if A is L–definable. If not, we choose Y, t and B as above. Again we may assume that all fibers have rank f. We have then $\mathrm{MD}^* Y = \mathrm{MD}\, Y = 1$. Since $f \leq k$, we have again by induction that $\mathrm{MR}^* B = \mathrm{MR}^* A - f = \mathrm{MR}\, X - f = \mathrm{MR}\, Y = \mathrm{MR}^* Y$. So $\mathrm{MR}^*(X \setminus A) = f + \mathrm{MR}^*(Y \setminus B) < f + \mathrm{MR}^*(Y) = f + \mathrm{MR}\, Y = \mathrm{MR}(X)$.

2) Consider L^*–definable sets $A \subset S^m$. Let N be the T–rank of S^m. $\mathrm{MR}^* X \geq k + 1$ is $\bigwedge$–definable and $\bigvee$–definable , since this is equivalent to *"for all/some L–definable $t : S^m \to S_1^n$ with $A = t^{-1}B$ for $B = t(A)$ there is a number $f \leq N$ such that the T^*–rank of*

$$C_f = \{b \in B \mid \mathrm{MR}(t^{-1}b) = f\}$$

is $\geq k + 1 - f$". Indeed, if there is such a t and f, we have

$$\mathrm{MR}^* A \geq f + \mathrm{MR}^* C_f \geq k + 1.$$

If conversely $\mathrm{MR}^* A \geq k + 1$ and t is such that $A = t^{-1}B$ for $B = t(A)$, there is a C_f such that $\mathrm{MR}^* t^{-1}C_f \geq k + 1$. If $\mathrm{MR}^* C_f \leq k - f$ we would have $f \leq k$ and by definability of T^*–ranks $\leq k$ we have $\mathrm{MR}^* t^{-1}C_f = f + \mathrm{MR}^* C_f \leq k$. So $\mathrm{MR}^* C_f \geq k + 1 - f$. $\qquad\square$

---

14 The reader may consult Lemma 3.11 and (the proof of) Folgerung 4.4 in [7].

Finally let us state an open problem: *Let T be a good theory with two sorts Σ_1 and Σ_2 and T' be a conservative expansion of $T \restriction \Sigma_1$. Does $T' \cup T$ have finite Morley rank?*

References

[1] J. Baldwin and K. Holland, Constructing ω-stable structures: rank 2 fields, *J. Symbolic Logic*, 65(1):371–391, 2000.

[2] A. Baudisch, A. Martin-Pizarro, and M. Ziegler, On fields and colors, *Algebra i Logika*, 45(2), 2006, 159–184, 252; translation in *Algebra Logic* 45 (2006), no. 2, 92–105. (http://arxiv.org/math.LO/0605412).

[3] A. Baudisch, A. Martin-Pizarro, and M. Ziegler, Hrushovski's Fusion, In F. Haug, B. Löwe, and T. Schatz, editors, *Festschrift für Ulrich Felgner zum 65. Geburtstag*, volume 4 of *Studies in Logic*, pages 15–31. College Publications, London, 2007.

[4] Assaf Hasson, Interpreting structures of finite Morley rank in strongly minimal sets, *Ann. Pure Appl. Logic*, 145:96–114, 2007.

[5] Ehud Hrushovski, Strongly minimal expansions of algebraically closed fields, *Israel J. Math.*, 79:129–151, 1992.

[6] Bruno Poizat, Le carré de l'égalité, *J. Symbolic Logic*, 64(3):1338–1355, 1999.

[7] Martin Ziegler, Stabile Gruppen, available at http://home.mathematik. uni-freiburg.de/ziegler/Skripte.html, 1991.

Establishing the o-minimality for expansions of the real field

Jean-Philippe Rolin
Université de Bourgogne

Contents

1 Introduction

A. Grothendieck introduced the notion of "tame geometry" in [8], more precisely in a chapter entitled *"Denunciation of so-called general*

topology, and heuristic reflexions towards a so-called tame topology".
He says there that *general topology* has been *"developed by analysts in*
order to meet the needs of analysis", and *"not for the study of topological*
properties of the various geometrical shapes". Consequently, according
to him, when one tries to work in the technical context of topological
spaces, *"one is confronted at each step with spurious difficulties related*
to wild phenomena".

According to him, the correct answer should be, instead of *"moving*
to contexts which are close to the topological one and less subject to
wildness, such as differentiable manifolds or piecewise linear spaces",
to have an axiomatic approach towards possible foundations for a tame
geometry. He suggests more precisely to extract, among the geometric
properties of the semi-analytic sets in a space $\mathbb{R}^n$, those that make it
possible to use these spaces as *"local models"* for a notion of *"tame*
space". For example, a *triangulability axiom* should be kept, although
it is obviously *"delicate to check".*

It is now widely admitted that the most convenient axiomatic answer
to such a program is the notion of **o-minimal structure**. In particular,
the *tameness axiom* for these structures, which limits the definable sets
of the real line to *finite union of points or intervals*, seems more handy
than a triangulability axiom. Let us say at once that the purpose of
these notes is not to recall why the "nice" geometric properties of semianalytic or sub-analytic subspaces of the euclidean spaces are satisfied by
the definable sets of an o-minimal structure. For these types of results,
the interested reader could refer, for example, to [3].

Grothendieck adds an interesting comment: *"Once the necessary foun-*
dational work will be completed, there will appear, not one tame theory,
but a vast infinity, ranging from the strictest of all, namely the semi-
algebraic subsets of the euclidean spaces, to the one which appears to be
likely the vastest of all, namely using piecewise real analytic spaces (or
semi-analytic spaces)".

Actually, this last sentence contains a double mistake: as we will
show later on, the class of relatively compact semi-analytic or even subanalytic sets is far from the "vastest of all tame classes" which can be
built on the real numbers. Moreover, such a vastest class does not even
exist, for it is known that there exist pairs of o-minimal structures which
do not admit any o-minimal common extensions [16, 17].

These notes are a survey of some methods which have been used to
prove the o-minimality of several expansions of the reals. Of course,
we will not be exhaustive. In particular, the analytic and geometric

methods will be explained to the detriment of the model-theoretic ones. Moreover, our goal is only to give a few central ideas of these proofs, and not all the details. Once again, the interested reader is invited to study the papers listed in the bibliography. Some open questions (probably difficult) will also be raised.

Acknowledgements The author wants to thank the referee for his remarks and careful reading.

2 General notions

2.1 Definitions

Definition 2.1 Let $\mathcal{F}$ be a collection of functions $f : \mathbb{R}^m \to \mathbb{R}$ for various $m \in \mathbb{N}$. A set $S \subset \mathbb{R}^m$ is said to be 0-**definable** in the structure $\mathbb{R}_{\mathcal{F}} = (\mathbb{R}, <, 0, 1, +, -, \cdot, \mathcal{F})$ (called the **expansion** of the ordered real field by $\mathcal{F}$), if S belongs to the smallest collection of subsets of the spaces $\mathbb{R}^p$, $p = 0, 1, \ldots$ which

(1) contains the graphs of addition, multiplication and all functions in $\mathcal{F}$,

(2) contains all diagonals $\{(x_1, \ldots, x_m) : x_i = x_j\} \subset \mathbb{R}^m$, $1 \leqslant i < j \leqslant m$,

(3) is closed under taking cartesian products, finite intersections, complements, and images under projection maps $\mathbb{R}^{m+1} \to \mathbb{R}^m$, $(x_1, \ldots, x_{m+1}) \mapsto (x_1, \ldots, x_m)$.

A set $A \subset \mathbb{R}^n$ is called **definable** in $\mathbb{R}_{\mathcal{F}}$ if there exists a set $S \subset \mathbb{R}^{m+n}$ 0-definable in $\mathbb{R}_{\mathcal{F}}$ and $a \in \mathbb{R}^m$ such that $A = S_a = \{y \in \mathbb{R}^n : (a, y) \in S\}$.

A map $f : S \subset \mathbb{R}^m \to \mathbb{R}^n$ is **definable** in $\mathbb{R}_{\mathcal{F}}$ if its graph is definable in $\mathbb{R}_{\mathcal{F}}$.

Definition 2.2 The structure $\mathbb{R}_{\mathcal{F}}$ is called **model complete** if in the above definition of 0-definable set, the operation of taking complements is superfluous.

The structure $\mathbb{R}_{\mathcal{F}}$ satisfies the property of **quantifier elimination** (in its own language), if in the above definition of 0-definable set, the operation of taking images under projection is superfluous.

The structure $\mathbb{R}_{\mathcal{F}}$ is called **o-minimal** if each definable subset of $\mathbb{R}$ is a finite union of points and intervals.

Remark 2.3 1) The "definable sets", introduced geometrically in the above definition, are exactly the sets which can be defined by a first-order sentence in the language of the ordered field of the real numbers enriched by the elements of $\mathcal{F}$.

2) By contrast to the notion of o-minimality, which expresses a geometric property of a family of sets, the notions of model completeness and quantifier elimination are related to the way these sets are "built" from the elements of $\mathcal{F}$. This means that, although the structure $\mathbb{R}_{\mathcal{F}}$ may not be model complete (or may not satisfy quantifier elimination), there might be another collection of functions $\mathcal{G}$ such that the definable sets of $\mathbb{R}_{\mathcal{G}}$ *are exactly the definable sets of* $\mathbb{R}_{\mathcal{F}}$, and such that the structure $\mathbb{R}_{\mathcal{G}}$ is model complete (or satisfies quantifier elimination).

3) It is known (see for example [3] for a proof) that actually, *any* definable set in an o-minimal structure has finitely many connected components. More precisely, it can be proved that, given any integer $k \in \mathbb{N}$, any definable set in an o-minimal structure admits a (finite) stratification into definable (connected) $\mathcal{C}^k$-manifolds. Consequently, if we refer to the "foundational program" of Grothendieck, it is not necessary to include these kinds of properties in the list of axioms for tameness.

There is a strong relationship between model completeness, quantifier elimination and o-minimality. For example, given a structure $\mathbb{R}_{\mathcal{F}}$, it frequently occurs that the quantifier-free definable subsets of $\mathbb{R}$ are easily shown to have finitely many connected components. Consequently, if $\mathbb{R}_{\mathcal{F}}$ admits the property of quantifier elimination then $\mathbb{R}_{\mathcal{F}}$ is o-minimal.

In the same spirit, although quantifier elimination is not proved (or even not true), it can sometimes be proved that the *quantifier-free* definable sets of $\mathbb{R}_{\mathcal{F}}$ have finitely many connected components. In that case, the model completeness of the structure $\mathbb{R}_{\mathcal{F}}$ immediately implies its o-minimality.

Proposition 2.4 *Consider an o-minimal structure $\mathbb{R}_{\mathcal{F}}$. The germs at $+\infty$ of one variable definable functions form a Hardy field $\mathcal{H}_{\mathcal{F}}$, called the **Hardy field of the structure** $\mathbb{R}_{\mathcal{F}}$.*

Definition 2.5 The o-minimal structure $\mathbb{R}_{\mathcal{F}}$ is called **polynomially bounded** if any element of its Hardy field is bounded by an integer power of the variable.

2.2 Survey of some strategies

In all known examples of o-minimal expansions of the reals, the o-minimality is a consequence of one of two following properties:

1) The elements f of some "sufficiently big" algebra $\mathcal{A}$ of functions generated by the elements of $\mathcal{F}$ can be *characterized* by their asymptotic expansions $\hat{f}$ in some scale. If the expansions $\hat{f}$ are all convergent, it means that $f = \hat{f}$ for any $f \in \mathcal{A}$. In the case where these expansions may diverge, it means that there exists an injective morphism $f \mapsto \hat{f}$ from $\mathcal{A}$ to some algebra of formal power series, a property frequently called **quasianalyticity** of the algebra $\mathcal{A}$. This relationship between o-minimality and quasianalyticity will be developed in section 5.1. Let us just mention briefly that examples of "reasonable" but "not big enough" quasianalytic algebras generated by some collections $\mathcal{F}$ are known, **for which the corresponding structures $\mathbb{R}_{\mathcal{F}}$ are not o-minimal**.

2) The quantifier-free definable sets of $\mathbb{R}_{\mathcal{F}}$ satisfy a uniform finiteness property. In that case, even though model completeness is not known, it may be possible to extend the structure $\mathbb{R}_{\mathcal{F}}$ to a bigger one, say $\mathbb{R}_{\mathcal{F}}^{\infty}$, obtained from $\mathbb{R}_{\mathcal{F}}$ by adding Hausdorff limits of convenient sequences of compact definable sets of $\mathbb{R}_{\mathcal{F}}$. The new structure $\mathbb{R}_{\mathcal{F}}^{\infty}$ will inherit the uniform finiteness property, and will be moreover model complete. It is therefore o-minimal, and hence so is $\mathbb{R}_{\mathcal{F}}$.

This type of proof allows a geometric approach of o-minimality, even though no analytic or formal parametrization of the definable sets is known. It should be noticed nevertheless that such an approach does not necessarily imply the model completeness of the "small" structure $\mathbb{R}_{\mathcal{F}}$.

We focus in these notes on the approach based on asymptotic expansions.

3 The classical examples

In order to introduce in a familiar setting the tools mentioned in section 2, we study in this section the well known classes of semialgebraic and globally subanalytic subsets.

3.1 The semialgebraic sets

Definition 3.1 A **semialgebraic** subset of $\mathbb{R}^n$ is a union of finitely many unions of (basic) subsets of the form:

$$\{x \in \mathbb{R}^n : P(x) = 0, Q_1(x) > 0, \ldots, Q_\ell(x) > 0\}$$

where $\ell \in \mathbb{N}$ and $P, Q_1, \ldots, Q_\ell \in \mathbb{R}[X_1, \ldots, X_n]$.

Theorem 3.2 (Tarski-Seidenberg) *Let A be a semialgebraic subset of $\mathbb{R}^{n+1}$ and $\pi : \mathbb{R}^{n+1} \to \mathbb{R}^n$ the projection onto the first n coordinates. Then $\pi(A)$ is a semialgebraic subset of $\mathbb{R}^n$.*

As an immediate consequence of this famous result, we see that the structure $\mathbb{R}_\emptyset = (\mathbb{R}, <, 0, 1, +, -, \cdot)$ satisfies the property of quantifier elimination, so that its definable sets are exactly the semialgebraic subsets.

Therefore, any definable subset of the real line is a semi-algebraic subset of $\mathbb{R}$, so is equal to a finite union of points and intervals: **the structure $\mathbb{R}_\emptyset$ is o-minimal**.

3.2 The globally subanalytic sets

We follow in this section the Denef and van den Dries approach for subanalytic sets [4]. Let $\mathbb{P}_1$ be the projective real line equipped with its usual analytic structure. The spaces $\mathbb{R}^n$ are naturally embedded in the analytic manifolds $\mathbb{P}_1^n$, so that any subset of $\mathbb{R}^n$ can be seen as a subset of $\mathbb{P}_1^n$.

Definition 3.3 Consider $A \subset \mathbb{R}^n$.

The set A is a **globally semianalytic subset** of $\mathbb{R}^m$ if A is locally defined, in a neighbourhood of any point of $\mathbb{P}_1^n$, by a finite number of analytic equalities and inequalities.

The set A is a **globally subanalytic subset** of $\mathbb{R}^n$ if there exists an integer $m \in \mathbb{N}$ and a globally semianalytic subset $B \subset \mathbb{R}^m \times \mathbb{R}^n$ such that $A = \pi(B)$, where $\pi : \mathbb{R}^m \times \mathbb{R}^n \to \mathbb{R}^m$ is the canonical projection.

Definition 3.4 Let $\mathcal{F}$ be the set of all **restricted analytic functions**, that is to say the functions $f : \mathbb{R}^m \to \mathbb{R}$ whose restriction to $[-1, 1]^m$ is real analytic and which are identically zero outside $[-1, 1]^m$. The structure $\mathbb{R}_{\mathcal{F}}$ is denoted by $\mathbb{R}_{\mathrm{an}}$.

Let us write $^{-1}$ for the function which associates $\frac{1}{x}$ to each nonzero real number x, and 0 to 0. The structure $\mathbb{R}_{\mathcal{F} \cup \{-1\}}$ will be denoted by $\left(\mathbb{R}_{\mathrm{an}}, {}^{-1}\right)$.

Remark 3.5 The notations have changed a little bit between [4] and [20]. In [4], the structure $\mathbb{R}_{\mathrm{an}}$ is generated by the elements f of $\mathcal{F}$ such that $f([-1,1]^m) \subset [-1,1]$, and the division considered there is the **restricted quotient**, defined by $D(x,y) = x/y$ if $|x| \leqslant |y| \leqslant 1$ and $D(x,y) = 0$ otherwise.

It was pointed out by van den Dries in [19] that the fact that $\mathbb{R}_{\mathrm{an}}$ is model complete and o-minimal is a consequence of **Gabrielov's complement theorem**. However, Denef and van den Dries proved in [4] the following stronger quantifier elimination result:

Theorem 3.6 *1) The structure* $\left(\mathbb{R}_{\mathrm{an}}, {}^{-1}\right)$ *admits quantifier elimination.*

2) The structure $\mathbb{R}_{\mathrm{an}}$ *is o-minimal and model complete.*

3) The globally subanalytic subsets are exactly the definable subsets of $\left(\mathbb{R}_{\mathrm{an}}, {}^{-1}\right)$

Remark 3.7 The second of the above statements is an easy consequence of the first one. Indeed, 1) implies that the $\left(\mathbb{R}_{\mathrm{an}}, {}^{-1}\right)$-definable subsets of $\mathbb{R}$ are quantifier-free definable, that is, definable by finitely many equalities and inequalities between one variable terms. But it is clear that these terms are locally equal to analytic functions, or to inverses of analytic functions. The one variable definable subsets of $\left(\mathbb{R}_{\mathrm{an}}, {}^{-1}\right)$ therefore consist of unions of finitely many points and intervals, so the structure $\left(\mathbb{R}_{\mathrm{an}}, {}^{-1}\right)$ is also o-minimal, as well as the structure $\mathbb{R}_{\mathrm{an}}$.

Any quantifier-free definable set of $\left(\mathbb{R}_{\mathrm{an}}, {}^{-1}\right)$ being obviously *existentially definable* in $\mathbb{R}_{\mathrm{an}}$, we conclude that the structure $\mathbb{R}_{\mathrm{an}}$ is model complete.

Proof Let us give a short idea of the proof of the first statement. We first remark that, because of the compactness of the projective spaces, this elimination result is actually a *local process*, up to a finite covering of these spaces by finitely many open polydisks. Then one observes that the inverse symbol appearing inside terms may be replaced by the addition of new variables (which will then have to be eliminated), and new equations.

Consider now a real valued analytic germ f defined in a neighbourhood of $0 \in \mathbb{R}^{m+n}$ by the following convergent power series:

$$f(x,y) = \sum_J f_J(x)y^J, \qquad x = (x_1, \ldots, x_m), \ y = (y_1, \ldots, y_n)$$

where $J = (j_1, \ldots, j_n) \in \mathbb{N}^n$, $y^J = y_1^{j_1} \ldots y_n^{j_n}$. We want to "eliminate the *vertical variables* y" in the equation $f(x,y) = 0$, that is we want to describe the set of *horizontal variables* x for which there exists y with $f(x,y) = 0$. The main problem is to try to apply the Weierstrass preparation theorem, though the function function f is not supposed to be regular in any vertical variable y_j. The classical linear changes of variables, such as $x_i \to x_i + y_j$, which could lead to such a regularity are of course forbidden, because they do not respect the verticality of the y_j. Moreover, we do not want to introduce inverses into these vertical variables, because we want to avoid the addition of new vertical variables and also we want to preserve the classical properties of analytic germs.

The key point of the proof is the following **finiteness property**, which is a first step towards Weierstrass preparation:

There exists an integer $d \in \mathbb{N}$, a real number $r > 0$ and finitely many analytic units U_J, $|J| = j_1 + \cdots + j_n \leqslant d$, such that:

$$f(x,y) = \sum_{|J| \leqslant d} f_J(x) y^J U_J(x,y), \quad x \in I^m, \ y \in I^n, \ I = [-r, r].$$

Let us say a few words about this important result. If we work in the ring of *formal power series*, it is an immediate consequence of noetherianity (consider the ideal generated by the coefficients f_J). However, the formal units we obtain that way are actually infinite sums of germs which do not necessarily have a common domain of convergence.

In order to circumvent this problem, we use a result of commutative algebra. Let $\mathbb{R}\{x\}$ denotes the ring of analytic germs at $0 \in \mathbb{R}^n$, equipped with the topology of *coefficientwise convergence*: a sequence $\left(f^{(j)}\right)$ of elements of $\mathbb{R}\{x\}$ converges to $f \in \mathbb{R}\{x\}$ if for every multi-index $J \in \mathbb{N}^n$, the coefficients of x^J in $f^{(j)}$ converge to the coefficient of x^J in f. Note that the completion of $\mathbb{R}\{x\}$ for this topology is the ring $R[[x]]$ of formal power series. Then for any integer $p \in \mathbb{N}$, it can be proved, by induction on p and n, that any module $M \subset \mathbb{R}\{x\}^p$ is closed for the product topology. An easy consequence of this result is the following: for any germs $f_0, f_1, \ldots, f_q \in \mathbb{R}\{x\}$, if there exist **formal** power series $G_1, \ldots, G_q \in \mathbb{R}[[x]]$ such that $f_0 = \sum_{i=1}^q G_j f_j$, then there exist **convergent** germs $g_1, \ldots, g_q \in \mathbb{R}\{x\}$ such that $f_0 = \sum_{i=1}^q g_j f_j$. Our finiteness property is a immediate application of this property.

Let us now cover the polydisk I^m by the finitely many semianalytic subsets

$$A_J = \{x \in I^m : |f_J(x)| \geqslant |f_L(x)|, \quad \text{for all } L, \ |L| \leqslant d\},$$

so that we can work over one of these sets A_{J_0}. The germ f can be written:

$$f(x,y) = f_{J_0}(x) \sum_{J \neq J_0} \frac{f_J}{f_{J_0}}(x) y^J U_J(x,y) + y^{J_0} U_{J_0}(x,y).$$

Notice the introduction of the bounded quotient applied to the horizontal variables, which can be replaced by new variables V_J bounded by 1, *which do not have to be eliminated in the sequel.* We work consequently with the germ:

$$F(x, V_J, y) = \sum_{J \neq J_0} V_J y^J U_J(x,y) + y^{J_0} U_{J_0}(x,y).$$

Once again, up to covering the unit polydisk in the V_J variables by finitely many polydisks, and up to replacing V_J by $c_J + V_J$, we now work with a germ defined in a neighbourhood of 0 :

$$F(x, V_J, y) = \sum_{J \neq J_0} (c_J + V_J) y^J U_J(x,y) + y^{J_0} U_{J_0}(x,y).$$

It can easily be checked that the *vertical* polynomial change of variables:

$$y \mapsto (y_1 + (y_n)^{d^{n-1}}, y_2 + (y_n)^{d^{n-2}}, \ldots, y_{n-1} + y_n^d, y_n)$$

leads to a germ regular in the last variable: applying Weierstrass preparation, we can suppose that this germ is polynomial in this last variable, which can then be eliminated by Tarski's process.   □

We see that the structures studied in this section are proved to be o-minimal via a quantifier elimination process, up to a "reasonable" enlargement of the language (namely the introduction of the inverse). Nevertheless, the elimination seems deeply related to Weierstrass preparation, which acts as a "bridge" towards the polynomial functions. In the absence of such a preparation result, such a result seems less immediate. Nevertheless some recent results of **A. Rambaud** [14], which will be recalled in a subsequent section, show that this elimination property is not that exceptional.

We will see later what kind of methods are to be developed to obtain, at least, model completeness.

4 The exponential function

Introducing the exponential function into our structures seems natural, from many points of view: it is a very classical transcendental function, it is a morphism, and it is a solution of a simple differential equation. Actually, the notion of o-minimal structure was developed by L. van den Dries in order to establish a correct setting for resolving **Tarski's problem**: *to what extent do the classical results on the semialgebraic subsets remain true after the introduction of the exponential function.*

The tameness properties of the exponential (and logarithm) function, at least in one variable, have been investigated by **Hardy** [9]. More recently, **Khovanskii** proved an n variable result which may be stated in the following way: *the quantifier-free definable sets of the structure $\mathbb{R}exp$ have finitely many connected components* [11].

Consequently, if we refer to what was said in section 2.2, the model completeness of $\mathbb{R}_{\exp}$ would imply its o-minimality. This nice result was obtained by **A. Wilkie** [23]:

Theorem 4.1 *The structure* $\mathbb{R}_{\exp} = (\mathbb{R}, <, 0, 1, +, -, \cdot, \exp)$ *is model complete, and therefore o-minimal.*

As well as the structure $\mathbb{R}_{\mathrm{an}}$, the structure $\mathbb{R}_{\exp}$ does not admit quantifier elimination. It is therefore natural to ask if such a property can be obtained by enlarging the language. Such a result was obtained by **L. van den Dries**, **D. Marker** and **A. Macintyre** in 1994 [20], inspired by ideas of **J.-P. Ressayre** [15]:

Theorem 4.2 *Define the function* log *by its classical definition on the positive numbers, and* $\log(x) = 0$ *if* $x \leqslant 0$.

1) The structure $(\mathbb{R}_{\mathrm{an}}, \exp, \log)$ *admits quantifier elimination.*

2) Any function $f : \mathbb{R}^n \to \mathbb{R}$ *definable in* $(\mathbb{R}_{\mathrm{an}}, \exp)$ *is given piecewise by terms of the language of* $(\mathbb{R}_{\mathrm{an}}, \exp, \log)$.

Remark 4.3 Note that the introduction of inverse is redundant in the above statement, since $\frac{1}{x} = \exp(-\log(x))$ for $x > 0$.

Although these results were perfectly clear for geometers, their proofs were not. The reason is that they were written by specialists in model theory, who used model theoretic criteria for model completeness and quantifier elimination, involving several models of the theory of the exponential function, and not just the field of real numbers.

A geometric proof of these results was given by **J.-M. Lion** and the author in [13]. They were inspired mostly by (*what they could understand of*) the ideas of the model theoretic proof of Theorem 4.2. The key ideas of these geometric proofs are explained in the next section, as well as a possible correspondence between the geometric and the model theoretic approach.

4.1 Quantifier elimination and preparation theorems

4.1.1 Preparation theorem for globally subanalytic functions

A large part of [20] is devoted to the study of theory T_{an} of $\mathbb{R}_{\mathrm{an}}$. One of the main results is the:

Proposition 4.4 *Let $M \subset N$ be models of T_{an}. If $y \in N \setminus M$, let $M\langle y \rangle$ denote the definable closure of $M \cup \{y\}$ in N. Then the value group $v\left(M\langle y \rangle^{\times}\right)$ is the divisible hull of the value group of the field extension $M(y)$.*

It is first proved that, for any model M of the theory T_{an} with value group Γ, there exists an embedding (of analytic structures) of M into the power series field $\mathbb{R}((t^{\Gamma}))$. This is done in two parts, depending on the way the value group of the field M behaves under the addition of the new element y (namely, this value group increases or remains the same).

The problem is to imagine a geometric counterpart to this valuative discussion, and to guess what reasonable geometric meaning can be given to these extensions of models of T_{an}. One possible solution is the following *preparation theorem for globally subanalytic functions*. **From now on, the word *subanalytic* is used for *globally subanalytic*.** For example, a *globally subanalytic function* is a function $f : \mathbb{R}^n \to \mathbb{R}$ whose graph is a globally subanalytic subset of $\mathbb{R}^{n+1}$.

If $B \subset \mathbb{R}^n$ is a subanalytic subset of $\mathbb{R}^n$, a *subanalytic cylinder* $C \subset \mathbb{R}$ with *basis B* is defined in one of the following ways, where φ and ψ are subanalytic functions defined on $\mathbb{R}^n$:

$$C = \{(x,y) : x \in B, \ \varphi(x) < y < \psi(x)\},$$
$$\text{with } \varphi(x) < \psi(x) \text{ on } B,$$
$$C = \{(x,y) : x \in B, \ y < \varphi(x)\},$$
$$C = \{(x,y) : x \in B, \ \varphi(x) < y\},$$
$$C = \{(x,y) : x \in B, \ y = \varphi(x)\}.$$

Theorem 4.5 *Consider a subanalytic function $f : \mathbb{R}^n \times \mathbb{R} \to \mathbb{R}$. Then there exists a finite covering of $\mathbb{R}^n \times \mathbb{R}$ by subanalytic cylinders such that, for each cylinder C of this covering, there exists subanalytic functions A, θ defined on $\mathbb{R}^n$, a rational number r, and a subanalytic unit U defined on C (that is a subanalytic function such that there exist $0 < k < K$ with $k < U(x,y) < K$ for $(x,y) \in C$), with:*

$$f(x,y) = (y - \theta(x))^r A(x) U(x,y)$$

where $\theta \equiv 0$ on B or y is equivalent to $\theta(x)$ on C (that is there exists two positive constants $k_1 < k_2$ such that $k_1 < \theta(x) < k_2$ on B).

Proof Once again, we only give the key points of the proof.

First step. Consider an analytic function f defined by a convergent power series in a neighbourhood of $0 \in \mathbb{R}^n \times \mathbb{R}$. After applying the **finiteness property** of section 3.2 and Weierstrass preparation, we may suppose that:

$$f(x,y) = U(x,y) \left(y^d + a_1(x) y^{d-1} + \cdots + a_d(x) \right), \quad x = (x_1, \ldots, x_n)$$

where the a_i's are obtained by finite composition of analytic functions and restricted quotients. We apply then the classical **Tschirnhausen transformation** $y = y_1 - \frac{a_1(x)}{d}$ (whose interest will appear in a few lines), so that the function becomes

$$f_1(x, y_1) = U_1(x, y_1) \left(y_1^d + b_2(x) y_1^{d-2} + \cdots + b_d(x) \right).$$

We then cover the domain B of the x variable by the following subanalytic sets:

$$B_j = \left\{ x \in B : |b_j(x)|^{1/j} \geqslant |b_i(x)|^{1/i}, \, i = 2, \ldots, d \right\}.$$

Let us work on one these sets, say B_{j_0}. We put $y_1 = |b_{j_0}(x)|^{1/j_0} y_2$, and get

$$
\begin{aligned}
f_2(x, y_2) = \ & U_2(x, y_2) \, |b_{j_0}(x)|^{d/j_0} \left(y_2^d + c_2(x) y_2^{d-2} \right. \\
& \left. + \cdots + y_2^{d-j_0} + \cdots + c_d(x) \right).
\end{aligned}
$$

On the subanalytic sets

$$B_{j_0}^+ = \{ x \in B_{j_0} : b_{j_0}(x) \geq 0 \} \quad \text{and} \quad B_{j_0}^- = \{ x \in B_{j_0}, b_{j_0}(x) < 0 \}$$

the c_j are obtained by finite compositions of analytic functions, restricted quotients and rational powers, and $|c_j| \leqslant 1$, $c_{j_0} \equiv 1$. Notice that the variable y_2 is not bounded any more. But, if y_2 lies outside of a convenient compact interval I, all the previous expression can be factorized by y_2^d, so that the theorem is immediately proved in that case.

Now, up to covering the compact set I by a finite number of open intervals, we may suppose that $y_2 = \alpha + y_3$, $y_3 \in]-\varepsilon, \varepsilon[$. Moreover, replacing the c_j by variables v_j, we may suppose, once again by compactness, that $v_j = c_j^0 + w_j$, $w_j \in]-\varepsilon_j, \varepsilon_j[$. We are then led to:

$$f_3(x, w_2, \ldots, w_d, y_3) = U_3(x, y_3) \left((\alpha + y_3)^d + (c_2^0 + w_2)(\alpha + y_3)^{d-2} \right.$$
$$\left. + \cdots + (c_d^0 + w_d) \right).$$

If $\alpha^d + c_2^0 \alpha^{d-1} + \cdots + c_d^0 \neq 0$, the function f_3 is a unit, and the theorem is proved. Otherwise, the function f_3 is actually a function of order $d_1 < d$ in the variable y_3, because at least one of the coefficients of y_3^k, $k < d$, is a unit. Note that the Tschirnhausen transformation is useful when $\alpha \neq 0$, in order to insure that the order d_1 is equal to $d - 1$. By Weierstrass preparation, the germ f_3 is then the product of a unit and a polynomial (in the variable y_3) of degree $d_1 < d$.

We conclude this step by repeating all the previous process until we get a function of order 1 in the "vertical variable" y.

Second step. The functions c_j and the unit U_3 produced in the first step are obtained by finite compositions of analytic functions, restricted quotient and rational powers. In particular, the subanalytic units produced in this manner have a very precise form: they are *analytic units* applied to finitely many quotients $\frac{(y-\theta(x))^{1/p}}{a(x)}$ or $\frac{b(x)}{(y-\theta(x))^{1/p}}$, where θ, a, b are subanalytic functions and p is a natural number. We now have to prove the same preparation result for these functions. By an easy induction on the "complexity" of such functions, we may assume that we are working with

$$f(x_1, \ldots, x_n, y) = F\left(x_1, \ldots, x_{n-1}, y, \frac{x_n}{y} \right)$$

where x_n/y is a restricted quotient, and F is an analytic function defined by a convergent power series in a neighbourhood of $0 \in \mathbb{R}^{n+1}$

$$F(x_1, \ldots, x_{n-1}, y, t) = \sum_{i,j} F_{i,j}(x_1, \ldots, x_{n-1}) y^i t^j.$$

The preparation theorem is then proved by a reduction to the analytic functions. The above uniform converging sum can now be "split" in two sums as follows:

$$F(x_1, \ldots, x_{n-1}, y, t) = \sum_{i \geqslant j} F_{i,j}(x_1, \ldots, x_{n-1}) y^i t^j$$
$$+ \sum_{i < j} F_{i,j}(x_1, \ldots, x_{n-1}) y^i t^j$$

$$f(x_1, \ldots, x_n, y) = \sum_{i \geqslant j} F_{i,j}(x_1, \ldots, x_{n-1}) y^i \left(\frac{x_n}{y}\right)^j$$
$$+ \sum_{i < j} F_{i,j}(x_1, \ldots, x_{n-1}) y^i \left(\frac{x_n}{y}\right)^j$$
$$= \sum_{i \geqslant j} F_{i,j}(x_1, \ldots, x_{n-1}) y^{i-j} x_n^j$$
$$+ \sum_{i < j} F_{i,j}(x_1, \ldots, x_{n-1}) \left(\frac{x_n}{y}\right)^{j-i} x_n^i.$$

This is the sum of a function F_+ analytic in the variable y and a function F_- analytic in x_n/y. We can then apply the process of the first step to each term of this sum. On cylinders where y is equivalent to some subanalytic function $\theta(x)$, we put $y = \theta(x)(1 + y_1)$, so that the initial function f becomes analytic in the variable y_1.

On the other cylinders, we have

$$f(x_1, \ldots, x_n, y) = y^r A(x) U(x, y) + \left(\frac{x_n}{y}\right)^s B(x) V(x, y)$$

where r, s are positive numbers, A, B are subanalytic and U, V are subanalytic units. We need an extra cylindrical decomposition. On some cylinders, one term of this sum dominates the other one, and the theorem is proved. On the other cylinders, the two terms of this sum are equivalent, which implies that y is equivalent to a subanalytic function $C(x)$, and we reduce to the analytic case by the change of variables: $y = C(x)(1 + y_1)$. □

Remark 4.6 A similar preparation theorem has been proved by **L. van den Dries** and **P. Speissegger** for definable functions of any polynomially bounded o-minimal structure [22]. Observe that this result has not yet been proved by geometric methods.

4.1.2 Preparation theorems for log-analytic functions

In [20], the quantifier elimination result is proved by building log-preserving embeddings of models of T_{an} (exp) into some *saturated* models. The proof is once again achieved by considering how the value group extends by adding new elements to models. The geometric counterpart is another preparation theorem adapted to the introduction of the logarithm and the exponential. The bases of the cylinders involved here are quantifier free definable sets. Their "roofs" and "floors" are graphs of finite composition of logarithm, exponential and global subanalytic functions. The precise statements are given in [13]. We prefer in this note to explain the process of elimination on simple examples.

Consider a simple *log-analytic* function $f(x, y) = F(x, y, \log g(x, y))$, $x = (x_1, \ldots, x_n)$ where F and g are subanalytic. We want to describe the set $D = \{x \in \mathbb{R}^n : \exists y \in \mathbb{R}, f(x, y) = 0\}$. The idea is to cover $\mathbb{R}^{n+1}$ by finitely many *log-exp-cylinders*, such that on each cylinder the function f can be *prepared*. The set D (as well as its complement $\mathbb{R}^n \setminus D$) then appears to be a finite union of bases of some of these cylinders, and hence a *quantifier-free definable set*.

The first step consists in preparing the subanalytic function g on finitely many (subanalytic) cylinders, so that we can suppose $g(x, y) = y_1^r A(x) U(x, y_1)$, $r \in \mathbb{Q}$, A subanalytic, U a subanalytic unit, and $y_1 = y - \theta(x)$, θ subanalytic. Consequently:

$$\log g(x, y) = r \log y + \log A(x) + \log U(x, y).$$

We remark that $\log \circ U$ is a subanalytic function. We are then led to a function:

$$G(x, y_1) = H(x, y_1, \log y_1)$$

where H is subanalytic in the two last variables, and, say, y_1 belongs to a neighbourhood of infinity. Let us prepare H with respect to the last variable $z = \log y_1$. The bases of the cylinders are subanalytic in the variable y_1 and the "roofs" are defined by inequalities of the type $\log y_1 < \varphi(x, y_1)$, where φ is subanalytic in y_1. The prepared form of $H(x, y, \log y_1)$ is

$$H(x, y, \log y_1) = (\log y_1 - \psi(x, y_1))^r B(x, y_1) V(x, y_1, \log y_1)$$

where ψ and B are subanalytic in y_1 and V is a unit. We are then led to preparing functions of the type $\log y_1 - \psi(x, y_1)$, ψ subanalytic in y_1. Naturally, in a first step we prepare ψ with respect to y_1. If, on some cylinders, y_1 is equivalent to some function $C(x)$, we are reduced to the analytic case, as in section 4.1.1, by the change of variables $y_1 = C(x)(1 + y_1)$.

On the other cylinders, we have $\psi(x, y_1) = y_1^s K(x) W(x, y_1)$, $s \in \mathbb{Q}$, K subanalytic in y_1 and W a subanalytic unit in y_1. But, from the relation $\log y_1 \sim \psi(x, y_1)$, we get $\log y_1 \sim y_1^s K(x)$, so that, by applying the function $\log$ to this equivalence, we get that $\log y_1$ is equivalent to some function $R(x)$, so that y_1 is equivalent to some log-exp-function $S(x)$, and we are reduced as before to the analytic case.

4.1.3 A Preparation theorem for log-exp-analytic functions

A *log-exp-analytic function* (LE-function for short) is a finite composition of logarithm, exponential and subanalytic functions. An *LE-cylinder* is defined in the obvious way by LE-functions. The preparation theorem for LE-functions is easily stated:

Theorem 4.7 *Let* $f : \mathbb{R}^n \to \mathbb{R}$ *be an LE-function. There exists a finite covering of* $\mathbb{R}^n$ *by LE-cylinders on which* f *is the product of a log-analytic function and the exponential of an LE-function.*

Note that Theorem 4.2 is an easy corollary of this statement.

In order to give an idea of the proof of Theorem 4.7, consider for example the following simple LE-function:

$$f(x, y) = g(x, y, \exp h(x, y)), \quad x = (x_1, \ldots, x_n)$$

where g is subanalytic and h is a log-analytic function. Let us prepare g in the last variable $z = \exp h(x, y)$. We cover $\mathbb{R}^{n+1}$ by finitely many cylinders, with subanalytic bases, and "roofs" defined by inequalities $\exp h(x, y) < \varphi(x, y)$ (φ being subanalytic), or, equivalently, $h(x, y) < \log \varphi(x, y)$. These inequalities are consequently satisfied by log-analytic functions, which can be prepared on LE-cylinders as explained in the previous section.

On each cylinder, the prepared function has two possible forms:

a) $f(x, y) = \exp(r h(x, y)) A(x, y) U(x, y, \exp h(x, y))$, where $r \in \mathbb{Q}$, A is subanalytic and U is a subanalytic unit. The theorem is therefore proved in this case.

b) $f(x, y) = (\exp h(x, y) - \theta(x, y))^r A(x, y) U(x, y, \exp h(x, y))$, with $\exp h(x, y) \sim \theta(x, y)$. Refining the covering, we can suppose for example that $h(x, y) > K$ for some constant K. On such a cylinder, we can write:

$$\exp h(x, y) = \exp (h(x, y) - \log \theta(x, y)) \, \theta(x, y)$$

where the exponential, applied to bounded arguments, is actually definable in T_{an}, which leads to the desired proof for $f(x,y)$.

4.2 Model completeness of $\mathbb{R}_{\exp}$

We give in this section the ideas used in [12] to derive the model completeness of the structure $\mathbb{R}_{\exp}$ (Theorem 4.1) from the previous preparation results. Note that the original proof of Wilkie is based on a model
theoretic criterion for model completeness, and is independent of [20].

The problem is to prove that the complement of a linear projection of
a quantifier free definable set of $\mathbb{R}_{\exp}$ is a set of the same type. We need
a refinement of the Gabrielov complement theorem, proved by Gabrielov
himself in [6]:

Theorem 4.8 *Let $\mathcal{F}$ be a family of analytic functions defined on open
subspaces of $\mathbb{R}^n$, $n = 0, 1, \ldots$, closed under derivation. Let us denote
by $\mathrm{semi}(\mathcal{F})$ the family of all the globally semi-analytic spaces defined by
elements of $\mathcal{F}$, and by $\mathrm{sub}(\mathcal{F})$ the family of linear projections of elements
of $\mathrm{semi}(\mathcal{F})$.*

1) The family $\mathrm{sub}(\mathcal{F})$ is closed under taking the complements.

*2) Let $Y \subset \mathbb{R}^m \times \mathbb{R}^n$ be an element of $\mathrm{sub}(\mathcal{F})$. There exists an element
Z of $\mathrm{sub}(\mathcal{F})$ included in Y such that the restriction to Z of the canonical
projection $\pi : \mathbb{R}^m \times \mathbb{R}^n \to \mathbb{R}^m$ is a bijection and $\pi(Z) = \pi(Y)$.*

Proof of Theorem 4.1 Consider the family $\mathcal{F}$ of the functions obtained
by finite compositions of (several variable) polynomial functions and the
exponential function. This family is obviously closed under derivation.
Consider a function $f : \mathbb{R}^n \times \mathbb{R} \to \mathbb{R}$ whose graph is an element of $\mathrm{sub}(\mathcal{F})$.
The idea is to apply the previous "explicit" section theorem to the vari-
ous elements existentially introduced by the subanalytic preparation the-
orem 4.5. More precisely, if we fix $M > 1$ and $\kappa = (k, \delta, r)$ with $k \in \mathbb{N}$,
$\delta = (\delta^0, \ldots, \delta^k) \in (0,1)^{k+1}$ and $r = (r^0, \ldots, r^k) \in \mathbb{Q}^{k+1}$, we define the
element Y_κ of $\mathrm{sub}(\mathcal{F})$ consisting of tuples $(x, \varphi^1, \ldots, \varphi^k, \theta^0, \ldots, \theta^k) \in
\mathbb{R}^{n+k+(n+1)k}$ such that:

(1) $\varphi^1 < \ldots < \varphi^k$,
(2) if $\delta^j = 0$ then $\theta^j = 0$, otherwise $|\theta^j|/M < |\varphi^j|$, $|\varphi^{j+1}| < M|\theta^j|$,
(3) if $\varphi^j < y, y' < \varphi^{j+1}$, then

$$|f(x,y)||y' - \theta^j|^{r^j} < M|f(x,y')||y - \theta^j|^{r^j}.$$

Theorem 4.5 says exactly that $\mathbb{R}^n$ is covered by linear projections X_κ of finitely many sets Y_κ (these sets X_κ obviously belong to sub$(\mathcal{F})$). On each X_κ, the explicit section theorem produces functions $\theta^j(x)$ and $\varphi^j(x)$ whose graphs belong to sub$(\mathcal{F})$. We then define

$$\psi_j(x) = (\varphi^j + \varphi^{j+1})(x)/2, \quad A^j(x) = f(x, \psi^j(x))/|\psi^j(x) - \theta^j(x)|^{r^j}.$$

Consequently, on each cylinder $C = \left\{ x \in X_\kappa : \varphi^j(x) < y < \varphi^{j+1}(x) \right\}$, we have:

$$f(x,y) = (y - \theta^j(x))^{r^j} A^j(x) U^j(x,y)$$

where $U^j(x,y) = f(x,y)/(|y - \theta^j(x)|^{r^j} A^j(x))$ is a subanalytic unit bounded from below by $1/M$ and from above by M.

The graphs of all the functions θ^j, A^j, U^j introduced above belong to sub$(\mathcal{F})$, as well as the sets X_κ and Y_κ. Therefore all the functions appearing in the proof of Theorem 4.7 are obtained by finite composition of the exponential and logarithm functions, and functions whose graphs belong to sub$(\mathcal{F})$. Well then the linear projections of quantifier free definable sets, *as well as their complements*, are finite unions of bases of LE-cylinders. They are consequently quantifier-free definable by the same kind of functions, and hence existentially definable in $\mathbb{R}_{\exp}$. This achieves the proof of the complement theorem. $\qquad\square$

5 O-minimality and quasianalyticity

So far we have worked with functions equal to some convergent expansions, in the classical Taylorian scale, or a scale involving the exponential and logarithm functions, sometimes called a scale of *transmonomials*. We refer to [21] for a proof of the o-minimality of the structure generated by the so-called *generalized power series*, *i.e.* convergent power series with positive real exponents and well ordered support.

Nevertheless it is well known that many natural functions of class C^∞ defined, say, in a neighbourhood of the origin, such as solutions of ordinary or partial differential equations, may have a *divergent* Taylor expansion at the origin. We show in this section that o-minimality is a consequence of the fact that there exists a "sufficiently big" algebra of definable functions which can be *characterized* by their Taylor expansion. Such a property is usually called *quasianalyticity*. On the other hand, we will also produce examples of *non o-minimal structures* for which this quasianalyticity property is only satisfied by a "small" algebra.

We end this section by mentioning new results and open questions related to quantifier elimination in this context.

5.1 The general result relating o-minimality and quasi-analyticity

5.1.1 Quasianalyticity and the Λ-Gabrielov property

Let us suppose given, for any compact box $B = [a_1, b_1] \times \cdots \times [a_n, b_n]$ with $a_i < b_i$ for $i = 1, \ldots, n$ and $n \in \mathbb{N}$, an $\mathbb{R}$-algebra $\mathcal{C}_B$ of functions $f : B \to \mathbb{R}$ such that the following holds:

(C1) $\mathcal{C}_B$ contains the functions $(x_1, \ldots, x_n) \mapsto x_i : B \to \mathbb{R}$, and, for every $f \in \mathcal{C}_B$, the restriction of f to $\text{int}(B)$ is C^∞.

(C2) If $B' \subset \mathbb{R}^m$ is a compact box and $g_1, \ldots, g_n \in \mathcal{C}_{B'}$ are such that $g(B') \subset B$, where $g = (g_1, \ldots, g_n)$, then, for every $f \in \mathcal{C}_B$ the function $y \mapsto f(g_1(y), \ldots, g_n(y)) : B' \to \mathbb{R}$ belongs to $\mathcal{C}_{B'}$.

(C3) For every compact box $B' \subset B$ we have $f|B' \in \mathcal{C}_{B'}$ for all $f \in \mathcal{C}_B$, and for every $f \in \mathcal{C}_B$ there is a compact box $B' \subset \mathbb{R}^n$ and $g \in \mathcal{C}_{B'}$ such that $B \subset \text{int}(B')$ and $g|B = f$.

(C4) $\partial f / \partial x_i \in \mathcal{C}_B$ for every $f \in \mathcal{C}_B$ and each $i = 1, \ldots, n$.

(C5) Let us denote by $\mathcal{C}_n$ the $\mathbb{R}$-algebra of germs at the origin of elements of $\mathcal{C}_{I_r}$ for any box $I_r = [-r_1, r_1] \times \cdots \times [-r_n, r_n] \subset R^n$, $r_i > 0$, and by $\hat{\ } : \mathcal{C}_n \to R[[X]] = \mathbb{R}[[X_1, \ldots, X_n]]$ the map that sends each $f \in \mathcal{C}_n$ to its Taylor series $\hat{f}$ at the origin. Then the map $\hat{\ }$ is an isomorphism between $\mathcal{C}_n$ and its image $\hat{\mathcal{C}}_n \subset \mathbb{R}[[X]]$ (**quasianalyticity**).

(C6) If $n > 1$ and $f \in \mathcal{C}_n$ is such that $f(0) = 0$ and $\frac{\partial f}{\partial x_n}(0) \neq 0$, there is an $\alpha \in \mathcal{C}_{n-1}$ and $i \leqslant n$ such that $f(x_1, \ldots, x_{n-1}, \alpha(x_1, \ldots, x_{n-1})) = 0$.

(C7) If $f \in \mathcal{C}_n$ and $i \leqslant n$ are such that $\hat{f}(X) = X_i G(X)$ for some $G \in \mathbb{R}[[X]]$, then $f = x_i g$ for some $g \in \mathcal{C}_n$ such that $G = \hat{g}$.

The closure under composition, derivation and implicit function is exactly what we mean by "sufficiently big". Examples of such large quasianalytic algebras will be given in sections 5.2 and 5.3. The main result obtained in [17] by **P. Speissegger, A. Wilkie** and the author under these hypotheses is:

Theorem 5.1 *For $n \in \mathbb{N}$ let us denote by I^n the unit box $[-1, 1]^n \subset \mathbb{R}^n$. For any function $\tilde{f} \in \mathcal{C}_{I^n}$, consider the function $f : \mathbb{R}^n \to \mathbb{R}$ which coincides with $\tilde{f}$ on I^n and is equal to 0 outside of I^n. Let $\mathcal{F}$ be the family of all these functions f.*

Then the structure $\mathbb{R}_\mathcal{F}$ is model complete, o-minimal and polynomially bounded.

By contrast to the so called *preparation results* of section 4.1, we are not interested here in any *quantifier elimination* property. Consequently, changes of variables which do not preserve the "verticality" of the last variable y are now allowed. The idea is to follow the original proof of Gabrielov's theorem of the complement, by controlling the dimension of the boundary of any linear projection of every quantifier-free definable set.

Definition 5.2 Let $A \subset \mathbb{R}^n$. The set A is called $\mathcal{C}$-*semianalytic* at $a \in \mathbb{R}^n$ if there exists $r \in (0, \infty)^n$ such that $(A - a) \cap I_r$ is a finite union of sets of the following type:

$$\{f = 0, g_1 > 0, \ldots, g_n > 0\}, \quad f, g_1, \ldots, g_n \in \mathcal{C}_{I_r}.$$

The set A is called $\mathcal{C}$-*semianalytic* if A is $\mathcal{C}$-semianalytic at every point $a \in \mathbb{R}^n$.

Moreover, A is called a $\mathcal{C}$-*manifold* if A is a submanifold of $\mathcal{C}_{I_r}$ of dimension $n - k$ and if there exists $f_1, \ldots, f_k \in \mathcal{C}_{I_r}$ vanishing identically on A, such that the gradients $\nabla f_1(x), \ldots, \nabla f_k(x)$ are linearly independent on A.

For $n \in \mathbb{N}$, we denote by Λ_n the family of all $\mathcal{C}$-analytic subsets of I_n. Let $\Lambda = \bigcup_{n \in \mathbb{N}} \Lambda_n$. Let us call Λ-**sets** the elements of Λ, **sub-Λ-sets** the linear projections of Λ-sets, and **sub-Λ-manifolds** the sub-Λ-sets which are in addition smooth manifolds.

A set $A \subset I^n$ satisfies the Λ-**Gabrielov property** if for each $m \leqslant n$ there are connected sub-Λ-manifolds $B_i \subset I^{n+q_i}$, $i = 1, \ldots, k$ and $q_1, \ldots, q_k \geqslant 0$, such that

$$\Pi_m(A) = \Pi_m(B_1) \cup \cdots \cup \Pi_m(B_k)$$

where Π_m means the projection onto the first m co-ordinates and, for each $i = 1, \ldots, k$, we have

(G1) $\bar{B}_i \setminus B_i$ is contained in a closed sub-Λ-set $D_i \subset I^{n+q_i}$ such that D_i has dimension and $\dim(D_i) < \dim(B_i)$;

(G2) $d = \dim(B_i) \leqslant m$, and there is a strictly increasing function $\lambda : \{1, \ldots, d\} \to \{1, \ldots, m\}$ such that the projection $\Pi_\lambda | B_i : B_i \to \mathbb{R}^m$ is an immersion.

It is classical (see [7], [21]) that Theorem 5.1 is a consequence of the following result:

Proposition 5.3 *Every Λ-set has the Λ-Gabrielov property.*

We explain in the following section where this Λ-Gabrielov property comes from.

5.1.2 Quasianalyticity and desingularization

The main tool in the proof of the preparation theorem for subanalytic functions is actually a **Newton polygon** process, applied first to analytic, and then to subanalytic functions. We recall that if we are interested in keeping the "verticality" of the last variable y, the Weierstrass preparation theorem, as well as the so called **finiteness property** (see section 3.2), are used in several places.

But the Weierstrass preparation theorem *is in general not satisfied by quasianalytic algebras* (see [16] for a counterexample). Consequently, the idea is to use the formal information associated to any quasianalytic germ f at the origin (namely, its Taylor expansion), to get a geometric description on the zero set of f. More precisely, instead of solving the equation $f = 0$, a **desingularization process** allows us to parametrize this zero set by a finite union of "blowing-downs" of rectilinear sets. Once again, the Λ-Gabrielov property is an easy and classical consequence of this description.

For example, any nonzero one variable germ $f(x) \in \mathcal{C}_1$ has a nonzero Taylor expansion at the origin $\hat{f}(x) = x^d \hat{u}(x)$, $\hat{u}(0) \neq 0$. By property **(C7)** of section 5.1.1, $\hat{u}$ is the Taylor expansion of a unit $u \in \mathcal{C}_1$ such that $f = x^d u$. It implies in particular that the germ f satisfies the **isolated roots** property.

Let us now call a germ $f(x) \in \mathcal{C}_n$ **normal** if $f(x) = x^r u(x)$, $r \in \mathbb{N}^n$, $u \in \mathcal{C}_n$, $u(0) \neq 0$. An **admissible transformation** is a finite composition of **elementary transformations** of the following type:

1) <u>Ramification.</u> $x_i \mapsto x_i^p$ or $x_i \mapsto -x_i^p$, $p \in \mathbb{N}$
2) <u>Translation.</u> $x_i \mapsto x_i + \alpha(x_1, \ldots, x_{i-1})$, $\alpha(0) = 0$, $\alpha \in \mathcal{C}_{i-1}$
3) <u>Linear transformation</u> $x_i \mapsto x_i + cx_j$, $c \in \mathbb{R}$
4) <u>Regular blowing up</u> $x_i \mapsto x_i(\lambda + x_j)$, $\lambda \in \mathbb{R} \setminus \{0\}$
5) <u>Singular blowing up.</u> $x_i \mapsto x_i x_j$

Proposition 5.4 *Let $f \in \mathcal{C}_n$. Then there exists finitely many $\mathcal{C}$-semi-analytic sets U_i, admissible transformations $\Pi_i : \widetilde{U}_i \to U_i$ and normal germs $\tilde{f}_i$ defined on $\widetilde{U}_i$, $i = 1, \ldots, q$, such that for any i we have*

$$f|U_i \circ \Pi_i = \tilde{f}_i, \qquad i = 1, \ldots, q.$$

Proof The proof is an induction of number n of variables. We will assume that finitely many germs in $\mathcal{C}_n$ can be *simultaneously normalizable*.

Consider now a nonzero germ $f(x,y) \in \mathcal{C}_{n+1}$, where $x = (x_1, \ldots, x_n)$. By quasianalyticity, f admits a nonzero Taylor expansion $\hat{f}$ at the origin, which can be supposed regular in y after a linear change of coordinates. Therefore, f can be supposed regular, with some order $d \in \mathbb{N}$, in the variable y. This implies that the germ $\partial^{d-1} f / \partial y^{d-1}$ is of order 1, so that, applying the property (C6), there exists $\alpha \in \mathcal{C}_n$ such that $\alpha(0) = 0$ and $\partial^{d-1} f / \partial y^{d-1}(x, \alpha(x)) = 0$. Applying Taylor's formula, we obtain

$$
\begin{aligned}
f(x, \alpha(x) + y) &= f(x, \alpha(x)) + \cdots + \frac{1}{(d-2)!} \frac{\partial^{d-2} f}{\partial y^{d-2}}(x, \alpha(x)) y^{d-2} \\
&\quad + y^d U(x, y) \\
&= y^d U(x, y) + a_2(x) y^{d-2} + \cdots + a_d(x)
\end{aligned}
$$

where $U \in \mathcal{C}_{n+1}$ is a unit, and $a_2, \ldots, a_d \in \mathcal{C}_n$, due to properties (C2), (C4) and (C7) of section 5.1.1. This form, which is not strictly speaking a Weierstrass preparation, can be called a **Newton preparation**. Note that the **implicit function theorem** replaces here the **Tschirnhausen transformation** of section 4.1.1.

By the induction hypothesis, all the a_j, $j = 2, \ldots, d$, can be supposed normal, after a convenient admissible transformation. We then follow the sketch of proof of Theorem 4.5 to reduce the order in y of f from d to 1. □

Remark 5.5 We observe that the *Newton prepared* form of f has been obtained after a linear change of coordinates, which in general does not preserve the verticality of the variable y. Therefore the above proof does not lead to a quantifier elimination result. The general preparation result of van den Dries and Speissegger mentioned in Remark 4.6 does not help for this purpose, because the bases of the cylinders they obtain are only *definable sets*, and not *quantifier-free definable sets*.

Remark 5.6 Nevertheless, a new result obtained by **A. Rambaud** in his thesis [14] leads to a quantifier elimination result for quasianalytic algebras satisfying properties (C1) to (C7), up to extending the language by restricted quotients and rational powers. The proof uses a model theoretic criterion for quantifier elimination, and it would be nice to obtain a purely geometric proof.

5.2 The Denjoy-Carleman classes

We mention in this section an example of a quasianalytic algebra for which the previous methods apply.

Let $M = (M_0, M_1, \ldots)$ with $1 \leqslant M_0 \leqslant M_1 \leqslant \ldots$ be a sequence of real numbers and $B = [a_1, b_1] \times \cdots \times [a_n, b_n]$ with $a_i < b_i$ for $i = 1, \ldots, n$. The **Denjoy-Carleman class on B associated to M** is the collection $\mathcal{C}_B^0(M)$ of all functions $f : B \to \mathbb{R}$ for which there exists an open neighbourhood U of B, a C^∞ function $g : U \to \mathbb{R}$ and a constant $A > 0$ such that $f = g|B$ and

$$|g^{(\alpha)}(x)| \leqslant A^{|\alpha|+1} \cdot M_{|\alpha|} \text{ for all } x \in U \text{ and } \alpha \in \mathbb{N}^n.$$

Without loss of generality, the sequence M may be supposed **logarithmically convex** (or **log-convex**), that is $M_i^2 \leqslant M_i M_{i+1}$ for all $i > 0$. It is well known [10, 18] that the class $\mathcal{C}_B^0(M)$ is **quasianalytic** if and only if

$$\sum_{i=0}^{\infty} \frac{M_i}{M_{i+1}} = \infty.$$

In that case, we will say that the **sequence M itself satisfies the property QA**. Moreover it is known that if the sequence M is **strongly log-convex**, *i.e.* $(M_i/i!)$ is log-convex, then the systems of increasing unions $\mathcal{C}_B(M) = \bigcup_{j=0}^{\infty} \mathcal{C}_B^0(M^{(j)})$, where $M^{(j)} = (M_j, M_{j+1}, \ldots)$, for all boxes B, is closed under composition, taking implicit functions and division by monomial terms (and hence by derivation). The results of section 5.1 apply, which tell us that the structure, denoted by $\mathbb{R}_{\mathcal{C}(M)}$ and generated by the elements of all the $\mathcal{C}_B(M)$, is o-minimal, model complete and polynomially bounded.

These examples, combined with some classical facts from analysis, lead to the following interesting theorem [17]:

Theorem 5.7 *1) There are strongly log-convex sequences M and N, each satisfying the property QA, such that the structures $\mathbb{R}_{\mathcal{C}(M)}$ and $\mathbb{R}_{\mathcal{C}(N)}$ are not both reducts of any one o-minimal expansion of the real field.*

2) There is a strongly log-convex sequence M satisfying the property QA such that $\mathbb{R}_{\mathcal{C}(M)}$ does not admit analytic cell decomposition.

Proof The first statement is a consequence of a nice theorem of **S. Mandelbrojt**: for any C^∞ function $f : U \to \mathbb{R}$, where U is a open neighbourhood of $[-1, 1]^n$, $n \in \mathbb{N}$, there exists two strongly log-convex sequences M and N, each satisfying QA, and functions $f_1 \in \mathcal{C}^0_{[-1,1]^n}(M)$, $f_2 \in \mathcal{C}^0_{[-1,1]^n}(N)$, such that $f = f_1 + f_2$ on $[-1,1]^n$. This result being in particular true for an *oscillating* function f, we conclude that the two structures cannot admit any common o-minimal expansion.

Consider now a C^∞ function $f : [-1,1] \to \mathbb{R}$ whose Taylor series at every point is divergent (a Baire category argument shows that such a function exists). By the previous sum result, we can write $f = f_1 + f_2$, with $f_1 \in \mathcal{C}^0_{[-1,1]}(M)$, $f_2 \in \mathcal{C}^0_{[-1,1]}(N)$, where M and N are two strongly log-convex sequences satisfying QA. Therefore, one of the summands must have a divergent Taylor series at every point of some open interval $I \subset [-1,1]$, from which we deduce the second part of the theorem. $\square$

5.3 Quasianalyticity, summability and analytic differential equations

5.3.1 Euler's differential equation

Many solutions of analytic differential equations may have a divergent Taylor expansion at some point. One of the most famous (and easy to produce) examples is given by the following **singular** differential equation

$$x^2 y' = y - x.$$

Besides the obvious constant solution $H(x) = 0$, the other solutions are defined and analytic on $\mathbb{R}^{>0}$ or $\mathbb{R}^{<0}$. The solutions defined for $x > 0$ satisfy $\lim_{x \to 0^+} H(x) = 0$. These functions, extended by $H(0) = 0$, are C^∞ on $(0, +\infty)$, and their common Taylor expansion at the origin is the divergent **Euler power series**

$$\hat{H}(x) = \sum_{i=0}^{\infty} m! x^{m+1}.$$

Only one solution $H_0(x)$ defined on $\mathbb{R}^{<0}$ satisfies $\lim_{x \to 0^-} H_0(x) = 0$, with the same Taylor expansion. The other solutions defined on $\mathbb{R}^{<0}$, which are equal to $H_0(x) + C \cdot \exp(-\frac{1}{x})$, $C \in \mathbb{R} \setminus \{0\}$, have an infinite limit at the origin.

Remark 5.8 All these functions (as well as every non oscillating solution of singular analytic differential equation $x^{p+1} y' = A(x,y)$, $y \in \mathbb{R}$) are definable in the same o-minimal structure, the structure $\mathbb{R}_{\mathrm{Pfaff}}$ generated by the Pfaffian functions, introduced by Wilkie in [24]. We are not going to use this result, because our goal is to recover tameness from the (possibly formal) expansions of the functions. Moreover, given such a solution H, Wilkie's result does not say anything on the possible model completeness of the structure $\mathbb{R}_{an}(H)$ generated by the restricted analytic functions and a "restricted" function which coincides with H on some interval $[0, \varepsilon]$ and vanishes outside this interval.

A natural question is: *is it possible to recover these solutions from the asymptotic expansions?* In other words, does there exist a *summation* process which associates to the formal expansion $\hat{H}(x)$ a function, solution of the differential equation, whose asymptotic expansion at the origin is precisely $\hat{H}(x)$? Basically, the answer is positive, but in general, *the solutions produced by such a process are complex valued.* It is not possible to describe here the complete summation in detail, for any analytic differential equation, but we can explain what happens for Euler's differential equation.

Definition 5.9 1) Let $k > 0$, $A > 0$. A formal power series $\hat{a}(x) = \sum_{m \geqslant 0} a_m x^m \in \mathbb{C}[[x]]$ is **Gevrey of order k and type A** if there exists $C > 0$ and $\alpha > 0$ such that:

$$\forall m \geqslant 0, \ |a_m| \leqslant C A^{m/k} \Gamma(\alpha + m/k).$$

Let us denote by $\mathbb{C}_{k,A}[[x]]$ the algebra of Gevrey series of order k and type A, and $\mathbb{C}_k[[x]] = \bigcup_{A>0} \mathbb{C}_{k,A}[[x]]$.

2) The **(formal) Borel transform of order k** of a formal power series $\hat{a}(x) \in \mathbb{C}[[x]]$ is

$$\mathcal{B}(\hat{a})(\zeta) = \sum_{m \geqslant 0} \frac{1}{\Gamma(1 + m/k)} a_{m+1} \zeta^m.$$

The series $\mathcal{B}(\hat{a})(\zeta)$ converges in a neighbourhood of the origin if and only if $\hat{a}$ is Gevrey of order k.

3) The **Laplace transform of order k** along the ray $d_\theta = \{\arg\zeta = \theta\}$ of a series $\mathbf{H}(\zeta)$ which converges in some neighbourhood of the origin is

$$\mathcal{L}_\theta^k(z) = \int_{d_\theta} \mathrm{e}^{-\zeta^k/z^k} \mathbf{H}(\zeta) \frac{k\zeta^{k-1}}{z^{k-1}} \mathrm{d}\zeta$$

where $\mathbf{H}$ still denotes the analytic continuation of $\mathbf{H}$ along d_θ.

4) The series $\hat{H}$ is **k-summable** if $\mathcal{L}_\theta^k \mathcal{B}(\hat{H})$ is convergent for all but a finite number of θ.

Remark 5.10 The formal power series $\hat{H}(x) = \sum_{m \geqslant 0} m! x^{m+1}$ is Gevrey of order $k = 1$. Its formal Borel transform of order 1 is

$$\mathbf{H}(\zeta) = \sum_{m \geqslant 0} \frac{1}{\Gamma(1+m)} m! \zeta^m = \sum_{m \geqslant 0} \zeta^m = \frac{1}{1-\zeta}.$$

For any ray d_θ with $\theta \neq 0$, the Laplace integral $\mathcal{L}_\theta^1$ is convergent, so that:

$$H_\theta(z) = \mathcal{L}_\theta^1 \mathcal{B}(\hat{H})(z) = \int_{d_\theta} e^{-\zeta/z} \frac{d\zeta}{1-\zeta}.$$

Moreover, this integral defines a holomorphic function on a sector

$$S_\theta = \{|z| < R, \theta - \pi/2 - \delta < \arg z < \theta + \pi/2 + \delta\}$$

with $R > 0$ and some (small) $\delta > 0$. It can be checked that H_θ is a solution of Euler's differential equation. We thus obtain, by analytic continuation, a complex valued solution $H(z)$ of this equation.

Note that for $z \in \mathbb{R}^{>0}$, $H(z) \in \mathbb{C} \setminus \mathbb{R}$, and for $z \in \mathbb{R}^{<0}$, $H(z) \in \mathbb{R}$. Thus H actually coincides on $\mathbb{R}^{<0}$ with the solution H_0 mentioned at the beginning of this section.

Remark 5.11 Suppose we are given a general analytic differential equation $F(x, y, y', \ldots, y^{(d)}) = 0$ and a formal power series solution $\hat{H}(x)$ of this equation. It is known that $\hat{H}$ is **multisummable**: by iterating several Borel transforms followed by several Laplace transforms, we produce a complex-valued solution $H(z)$ whose asymptotic expansion at the origin is precisely $\hat{H}$, and the Laplace integrals converge for almost every θ [2].

5.4 Multisummable real Gevrey functions

The relationship between quasianalyticity and o-minimality was first pointed out by **L. van den Dries** and **P. Speissegger** in [22], in the context of multisummability. They consider the structure $\mathbb{R}_{\mathcal{G}}$ generated by rings of multivariable **real valued functions** defined on the positive real axis, obtained from their Taylor expansion at the origin by the above multi-summation process.

A important property of these rings is their **quasianalyticity**: if the Taylor expansion of an element f of such a ring is equal to zero, then f itself is equal to zero. It is proved in [22] that the structure $\mathbb{R}_{\mathcal{G}}$ is o-minimal, model complete and polynomially bounded. The proof uses blowing up transformations, as in section 5.1.2. The main difference is that, in the multisummable situation, the functions, after convenient blowing-up, become *analytic* in some variables, and that a Weierstrass preparation theorem with respect to these variables can be proved. Actually, the *Newton preparation* of section 5.1.2 could have been used as well, without any reference to analytic properties.

A strong limitation of this theorem is that in general the functions obtained by multi-summation process *are not real-valued on the real axis*. For example, there is only one solution of Euler's equation (namely $H_0(x)$, defined on $\mathbb{R}^{<0}$), which is.

We explain in the next section how it is possible, for convenient solutions of differential equations, to get rid of this restriction. In particular, these methods will be suitable for the other solutions of Euler's equation.

5.5 Quasianalyticity of real solutions of differential equations

5.5.1 Strong quasianalyticity

Three main tools have been used in the study of $\mathbb{R}_{\exp}$ and $(\mathbb{R}_{\mathrm{an}}, \exp)$:

(1) The fact that LE-functions are equal to convergent expansions (in a non Taylorian scale), involving logarithm and exponential functions.

(2) The morphism properties satisfied by the logarithm and the exponential.

(3) The growth properties of these two functions.

Note that the two latter points arise from the differential equation satisfied by these functions. To summarise, the o-minimality results from a **generalized Newton polygon process**, which allows us to solve LE-equations. This process is applied to the functions via (1) above, and a reduction to simpler equations, similar to what is done in the classical analytic case, is a consequence of (2) and (3).

Let us try to generalize these arguments to the more general differential equation

$$(1) \qquad x^{p+1}y' = A(x, y) \qquad y \in \mathbb{R}^r$$

where A is analytic in a neighbourhood of $0 \in \mathbb{R}^{1+r}$ with $A(0) = 0$, and where the *Poincaré rank* p of the equation is a positive integer.

Consider a solution $H : (0, \varepsilon] \to \mathbb{R}^r$ of equation (1), such that $\lim_{x \to 0^+} H(x) = 0$. This solution is analytic on $(0, \varepsilon]$, and in order to get some formal information on $H(x)$ when $x \to 0$, we suppose that $H(x)$

has an *asymptotic expansion* $\hat{H}(x)$ as $x \to 0$. It can easily be proved that under this hypothesis, H can be continued to a C^{∞} function on $[0, \varepsilon]$ and its Taylor series at the origin is equal to $\hat{H}(x)$.

If $\tilde{H} : \mathbb{R} \to \mathbb{R}$ denotes the function which coincides with H on $[0, \varepsilon]$ and vanishes outside, and $\mathcal{F}$ is the union of $\{\tilde{H}\}$ and the family of restricted analytic functions, we denote by $\mathbb{R}_{an}(H)$ the structure $\mathbb{R}_{\mathcal{F}}$.

Definition 5.12 The solution H of (1) is called **quasianalytic** if for any analytic function f defined in a neighbourhood of $0 \in \mathbb{R}^{1+r}$, the Taylor series at the origin of $f(x, H(x))$ is 0 if and only if $f(x, H(x)) \equiv 0$.

Remark 5.13 One important consequence of the quasianalyticity of H is the fact that H is **(analytically) non oscillating**: for any analytic function f defined in a neighbourhood of $0 \in \mathbb{R}^{1+r}$, the graph of H is included in the hypersurface $(f = 0)$ or has finitely many intersection points with this hypersurface. Of course, a solution $H(x)$ may be non oscillating without being quasianalytic: consider $H(x) = \exp(-1/x)$.

We can now ask two natural questions:

(1) Does the **non oscillation** property of H imply the o-minimality of $\mathbb{R}_{an}(H)$? **ANSWER: NO!**
(2) Does the **quasianalyticity property** of H imply the o-minimality of $\mathbb{R}_{an}(H)$? **ANSWER: NO!**

Counterexamples may be found in [16]. These phenomena can be explained as follows: the tameness of a structure means that the *huge* family of definable sets satisfy the finiteness property. We explained in section 5.1 that this tameness is a consequence the quasianalyticity of "sufficiently big" algebras containing H, much "bigger" than the basic algebra of the functions $f(x, H(x))$, for f analytic in a neighbourhood of $0 \in \mathbb{R}^n$. The family of these algebras has to be closed under compositions, monomial division (and hence derivation), and taking implicit functions, and these do not all follow from our quasianalytic assumption.

Nevertheless, it has been proved by **F. Sanz, R. Schäfke** and the author in [16] that o-minimality arises from a slightly stronger assumption:

Definition 5.14 A solution $H = (H_1, \ldots, H_r)$ of (1) is **strongly quasianalytic** if it tends to 0 at the origin, admits an asymptotic expansion $H(x) \sim \hat{H}(x)$ as $x \to 0^+$ and satisfies the following condition:

(SQA) If $k \geqslant 0$, $n \geqslant 1$, an analytic function $f(x, z_{11}, \ldots, z_{rn})$ with

$f(0) = 0$ and polynomials $P_1(x), \ldots, P_n(x)$ with $\mathrm{val}\,(P_l) > 0$ ($\mathrm{val}\,(P_l)$ denotes the *smallest degree* of P_l) and the derivative $P_l^{(\mathrm{val}(P_l))}(0) > 0$ are given, then one has

$$f\left(x, \left\{T_k \hat{H}_j(P_l(x))\right\}_{j,l}\right) \equiv 0 \implies f\left(x, \left\{T_k H_j(P_l(x))\right\}_{j,l}\right) \equiv 0$$

where $T_k H(x)$ means $\frac{H(x) - \hat{H}^k(x)}{x^k}$, $\hat{H}^k(x) = \sum_{i=0}^k \frac{H^{(i)}(0)}{i!} x^i$.

The word "strongly" reflects the fact that the polynomials P_l are allowed as arguments of H. We use the term **simple** for functions of the form $f(x, \{T_k H_j(P_l(x))\}_{j,l})$, as in definition 5.14. These functions form an algebra $\mathcal{S}$.

Theorem 5.15 *If a solution H of (1) is strongly quasianalytic, then the structure $\mathbb{R}_{\mathrm{an}}(H)$ is o-minimal, model complete and polynomially bounded.*

The idea, in order to apply results of section 5.1, is to prove the quasian-alyticity of the smallest algebra $\mathcal{A}$ of functions containing the solution H, the restricted analytic functions, and closed under composition, mono-mial division and taking implicit functions. The following key lemma below allows us to show that the "transcendence" of the functions does not increase under these operations. In some sense, it is a generalization to any solution of a singular analytic equation of the *morphism property* satisfied by the logarithm function.

Lemma 5.16 *For any $L > 0$ there exists a neighbourhood V of $0 \in \mathbb{R}^{1+r}$, $\delta_L > 0$ and an analytic function $B : [-L, L] \times V \to \mathbb{R}$ such that*

$$H(x + x^{p+1} z) = B(z, x, H(x)), \quad \text{for } |z| \leqslant L \text{ and } 0 < x < \delta_L.$$

Proof Just observe that the function $G_x : z \mapsto H(x + x^{p+1} z)$ is a solution of a **regular** differential equation, which depends analytically of the parameter x, and that $G_x(0) = H(x)$. The lemma is an immediate consequence of the analytic dependence of the solution upon parameters (namely x) and initial conditions (namely $H(x)$). $\qquad\qquad\square$

Proof of Theorem 5.15 We just give some examples of why some func-tions of one variable belonging to the algebra $\mathcal{A}$ are actually **simple**. Let us suppose that $H'(0) \neq 0$.

1) $\underline{f(x) = H(H(x))}$. We write $H(x) = \hat{H}^{p+1}(x) + x^{p+1}T_{p+1}H(x)$ and $\hat{H}^{p+1}(x) = xu(x)$ so that, according to Lemma 5.16:

$$
\begin{aligned}
H(H(x)) &= H\left(\hat{H}^{p+1}(x) + x^{p+1}T_{p+1}H(x)\right) \\
&= H\left(xu(x) + (xu(x))^{p+1}T_{p+1}H(x)/u(x)^{p+1}\right) \\
&= B\left(T_{p+1}H(x)/u(x)^{p+1}, xu(x), H(xu(x))\right) \\
&= C\left(x, H\left(\hat{H}^{p+1}(x)\right), T_{p+1}H(x)\right) \in \mathcal{S}.
\end{aligned}
$$

2) $\underline{F(x, f(x), H(f(x))) = 0}$. Here we mean that $f(x)$ is a solution of the implicit equation $F(x, y, H(y)) = 0$, where F is an analytic function. We then have

$$
\begin{aligned}
& F(x, f(x), H(f(x))) \\
&= F\left(x, \hat{f}^{p+1}(x) + x^{p+1}T_{p+1}f(x), H(\hat{f}^{p+1}(x) + x^{p+1}T_{p+1}f(x))\right)
\end{aligned}
$$

which shows that $T_{p+1}f(x)$ is a solution of the implicit equation

$$
F\left(x, \hat{f}^{p+1}(x) + x^{p+1}z, H(\hat{f}^{p+1}(x) + x^{p+1}z)\right) = 0.
$$

But this equation, according to lemma 5.16, can be written

$$
G\left(x, H(\hat{f}^{p+1}(x)), z\right) = 0
$$

where G is analytic. Its solution is an analytic function of x and of $H\left(\hat{f}^{p+1}(x)\right)$, so is a simple function. $\qquad\square$

5.5.2 Results and counterexamples

We conclude this section by an application of the previous results, which gives rise to some interesting examples (and problems).

Consider the differential system

$$
(2) \qquad
\begin{cases}
x^2 \dfrac{dy_1}{dx} &= y_1 + y_2 - x \\[2mm]
x^2 \dfrac{dy_2}{dx} &= -y_1 + y_2
\end{cases}
$$

It is proved in [16] that every solution $H(x) = (H_1(x), H_2(x))$ of (2) defined for $x > 0$ verifies $\lim_{x \to 0^+} H(x) = 0$ and is **strongly quasianalytic**. Therefore, according to theorem 5.15, the structure $\mathbb{R}_{an}(H)$ is o-minimal and model complete. Each solution of (2) defined for $x > 0$ is in particular *non oscillating*. Nevertheless it can be proved by simple computations, that, given two solutions H and G of (2) defined for $x > 0$, the argument of the vector $H(x) - G(x) \to \infty$ as $x \to 0$: the two solutions are *asymptotically linked*. As a consequence, the structures $\mathbb{R}_{an}(H)$ and $\mathbb{R}_{an}(G)$ do not admit any common o-minimal extension. This simple example thus gives, in a much simpler way than in section 5.2, an *explicit infinite family of "mutually incompatible" o-minimal structures*.

Moreover, for any pair $(H(x), G(x))$ of solutions of (2) defined for $x > 0$, let us define the function $H^*(x) = (H(x), G(2x))$, which is a solution of a singular analytic differential equation in $\mathbb{R}^4$. It is proved in [16] that this solution is non-oscillating. On the other hand, the curve $x \mapsto H^*(x)$ intersects infinitely many times the semialgebraic set $\{(v_1, v_2) \in \mathbb{R}^4 / v_1 - v_2 = 0\}$, and hence H^* is not definable in an o-minimal expansion of the reals. Therefore:

There exist non oscillating solutions of singular analytic differential equations which are not definable in any o-minimal expansion of the reals.

6 Some open questions

Let us state in this last section a few open questions about o-minimal expansions of the reals which should be of some interest.

6.1 Analytic vector fields in dimension 3

Consider three analytic functions a, b, c defined on a open neighbourhood U of the origin $0 \in \mathbb{R}^3$, and the *vector field* $X = (a, b, c)$. A *trajectory* of X is a differentiable curve $\gamma : (0, \varepsilon) \to U$ such that $\dot{\gamma}(t) = X(\gamma(t))$ for every $t \in (0, \varepsilon)$. The trajectory γ is called *non oscillating* if, for every analytic function $f : U \to \mathbb{R}$, the *support* $|\gamma| = \{\gamma(t) : t \in (0, \varepsilon)\}$ is either included in $f^{-1}(0)$ or else intersects it finitely many times. Let us denote by $\mathbb{R}_{an}(\gamma)$ the structure generated by the restricted analytic functions and the components of the γ. Some recent results [16] suggest the following question:

Suppose we are given a non oscillating trajectory γ of an analytic vector field in $\mathbb{R}^3$. Is the structure $\mathbb{R}_{an}(\gamma)$ o-minimal (and model complete)?

The o-minimality of $\mathbb{R}_{an}(\gamma)$ for non oscillating solutions of analytic vector fields in $\mathbb{R}^2$ is a consequence of Wilkie's theorem on Pfaffian functions [24]. It can be proved that the structures $\mathbb{R}_{an}(\gamma, \exp)$ are model complete for any non oscillating γ. Nevertheless, this last result is not known (*and probably not true!*) for many structures $\mathbb{R}_{an}(\gamma)$. For example, if $H(x) = H_0(x) + \exp(-1/x)$, $x < 0$, is a non bounded solution of Euler's equation, the question of model completeness of $\mathbb{R}_{an}(H)$ is still unsolved.

6.2 O-minimal expansions of class C^k

Suppose we are given an o-minimal expansion $\mathbb{R}_\mathcal{F}$ of the reals and an integer $k > 0$. It is known that any definable set of $\mathbb{R}_\mathcal{F}$ admits a finite stratification in definable manifolds of class C^k. We say that the structure $\mathbb{R}_\mathcal{F}$ *admits* C^k *stratifications*. Actually, most of the known o-minimal expansions of the reals admit *analytic* stratifications. The only examples produced so far of o-minimal expansions which admit C^∞ stratifications and no analytic stratifications are those of section 5.2. The question of existence of o-minimal structures which do not admit C^∞ stratifications is still open[1].

6.3 Transexponential o-minimal structures

All the known examples of o-minimal structures have the property of *exponential growth*: for any element f of the Hardy field of an o-minimal expansion of the reals, there exists an integer $n \in \mathbb{N}$ such that $f(x) < \exp_n(x)$ in a neighbourhood of $+\infty$, where $\exp_0 = \mathrm{Id}$, and $\exp_n = \exp \circ \exp_{n-1}$. Let us call *transexponential* a germ f of a real function at $+\infty$, such that $f > \exp_n$ for every $n \in \mathbb{N}$. A natural question concerns the existence of *transexponential* o-minimal structures, that is which admit transexponential definable functions.

A possible idea is to consider solutions E of the functional equation:

$$(3) \qquad\qquad \exp(E(x)) = E(x + 1).$$

Such a function E, which conjugates the exponential and the translation by 1, is called by **J. Ecalle** [5] an *iterator* of the exponential function: it allows us to define *rational iterates* of the exponential, as in

$$e_{1/2} = E \circ T_{1/2} \circ E^{-1}.$$

M. Boshernitzan builds in [1] an analytic solution E of (3) which *belongs to a Hardy field* and such that $\exp_n < E$ for all integer $n \in \mathbb{N}$. We should point out that such a solution is not unique. The question

---

1 Added in print: This question has since been solved positively, see O. Legal, J.-P. Rolin, Une structure o-minimale sans décomposition cellulaire C^∞, note submitted to C. R. Acad. Sci. Paris Sér. I Math., 2007.

is: does there exist a solution E of (3) such that the structure $\mathbb{R}_{an}(E)$ is o-minimal?

References

[1] M. Boshernitzan, Hardy fields and existence of transexponential functions, *Aequationes Math.*, 30 (1986), pp. 258–280.

[2] B. L. J. Braaksma, Multisummability of formal power series solutions of nonlinear meromorphic differential equations, *Ann. Inst. Fourier (Grenoble)*, 42 (1992), pp. 517–540.

[3] M. Coste, *An introduction to o-minimal geometry*, Dipartimento di Matematica Dell'Universita di Pisa, 1975.

[4] J. Denef and L. van den Dries, *p*-adic and real subanalytic sets, *Ann. of Math.* (2), 128 (1988), pp. 79–138.

[5] J. Écalle, *Introduction aux fonctions analysables et preuve constructive de la conjecture de Dulac*, Actualités Mathématiques, Hermann, Paris, 1992.

[6] A. Gabrielov, Complements of subanalytic sets and existential formulas for analytic functions, *Invent. Math.*, 125 (1996), pp. 1–12.

[7] A. M. Gabrielov, Projections of semianalytic sets, *Funkcional. Anal. i Priložen.*, 2 (1968), pp. 18–30.

[8] A. Grothendieck, Esquisse d'un programme, in *Geometric Galois actions, 1*, vol. 242 of London Math. Soc. Lecture Note Ser., Cambridge Univ. Press, Cambridge, 1997, pp. 5–48. With an English translation on pp. 243–283.

[9] G. H. Hardy, *Orders of infinity. The Infinitärcalcül of Paul du Bois-Reymond*, Hafner Publishing Co., New York, 1971. Reprint of the 1910 edition, Cambridge Tracts in Mathematics and Mathematical Physics, No. 12.

[10] Y. Katznelson, *An introduction to harmonic analysis*, Cambridge Mathematical Library, Cambridge University Press, Cambridge, third ed., 2004.

[11] A. G. Khovanskiĭ, Real analytic manifolds with the property of finiteness, and complex abelian integrals, *Funktsional. Anal. i Prilozhen.*, 18 (1984), pp. 40–50.

[12] J.-M. Lion and J.-P. Rolin, Théorème de Gabrielov et fonctions log-exp-algébriques, *C. R. Acad. Sci. Paris Sér. I Math.*, 324 (1997), pp. 1027–1030.

[13] J.-M. Lion and J.-P. Rolin, Théorème de préparation pour les fonctions logarithmico-exponentielles, *Ann. Inst. Fourier (Grenoble)*, 47 (1997), pp. 859–884.

[14] A. Rambaud, *Quasianalyticité, o-minimalité et élimination des quantificateurs*, PhD thesis, Université Paris 7, 2005.

[15] J.-P. Ressayre, Integer parts of real closed exponential fields, in *Arithmetic, Proof Theory and Computational Complexity*, P. Clote and J. Krajicek, eds., Oxford University Press, 1993, pp. 278–288.

[16] J.-P. Rolin, F. Sanz, and R. Schäfke, Quasi-analytic solutions of analytic ordinary differential equations and o-minimal structures, *Proc. London Math. Soc.* 95 Nr 2 (2007), pp. 413–442.

[17] J.-P. Rolin, P. Speissegger, and A. J. Wilkie, Quasianalytic Denjoy-Carleman classes and o-minimality, *J. Amer. Math. Soc.*, 16 (2003), pp. 751–777 (electronic).

[18] W. Rudin, *Real and complex analysis*, McGraw-Hill Book Co., New York, third ed., 1987.

[19] L. van den Dries, *Tame topology and o-minimal structures*, Cambridge University Press, Cambridge, 1998.

[20] L. van den Dries, A. Macintyre, and D. Marker, The elementary theory of restricted analytic fields with exponentiation, *Ann. of Math.* (2), 140 (1994), pp. 183–205.

[21] L. van den Dries and P. Speissegger, The real field with convergent generalized power series, *Trans. Amer. Math. Soc.*, 350 (1998), pp. 4377–4421.

[22] ———, The field of reals with multisummable series and the exponential function, *Proc. London Math. Soc.* (3), 81 (2000), pp. 513–565.

[23] A. J. Wilkie, Model completeness results for expansions of the ordered field of real numbers by restricted Pfaffian functions and the exponential function, *J. Amer. Math. Soc.*, 9 (1996), pp. 1051–1094.

[24] ———, A theorem of the complement and some new o-minimal structures, *Selecta Math. (N.S.)*, 5 (1999), pp. 397–421.

On the tomography theorem by P. Schapira

Sergei Starchenko[†]
University of Notre Dame

Summary

In this note we present the Tomography Theorem of P. Schapira [3] (Section 5) adapted to o-minimal structures. This result is well-known, and there are many different proofs of it. For example, in [1] it is proved by the induction on dimension. The main purpose of this note is to demonstrate how the original "chasing diagrams" method of P. Schapira works nicely in the o-minimal setting. For more on applications of constructable functions and a list of references we refer to [1].

1 Introduction

We work in o-minimal expansion $\mathcal{M} = \langle \mathbf{R}, +, \cdot, <, \ldots \rangle$ of a real closed field $\mathbf{R}$. Out of many "tame" properties of o-minimal structures we will need only the definable trivialization theorem and the existence of the o-minimal Euler characteristic. We state them below for the sake of completeness and refer to [2] for more details.

Recall that a definable map $f \colon A \to B$ is called *trivial* if there is a definable set F and a definable bijection $h \colon A \to B \times F$ such that the following diagram is commutative, where π_B is the projection onto B.

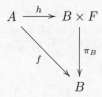

† The author was partially supported by NSF

Let $f\colon A \to B$ be a definable map and $B' \subseteq B$ a definable subset. The map f is called *definably trivial over B'* is the restriction $f \restriction f^{-1}(B')\colon f^{-1}(B') \to B'$ is definably trivial.

Definable Trivialization Theorem. *Let $f\colon A \to B$ be a definable map. Then there is a finite partition $B = B_1 \cup \ldots \cup B_k$ such that every B_i is definable and f is trivial over B_i.*

Existence of Euler Characteristic. *There is a unique function χ that assigns to every definable set A an integer $\chi(A)$ satisfying the following properties:*

(EC1) $\chi(\varnothing) = 0$ and $\chi(\mathbf{R}^n) = (-1)^n$ for any $n \in \mathbb{N}$;
(EC2) $\chi(A \cup B) = \chi(A) + \chi(B) - \chi(A \cap B)$ for all definable A, B;
(EC3) $\chi(A \times B) = \chi(A)\chi(B)$ for all definable A, B;
(EC4) If $f\colon A \to B$ is a definable bijection between definable sets then
$\chi(A) = \chi(B)$.

For a definable set A the number $\chi(A)$ from the previous fact is called *the o-minimal Euler characteristic (or just the Euler characteristic) of A.*

Example 1.1 (1) Since $\chi(\mathbf{R}^0) = 1$, we have $\chi(P) = 1$ for every one-element set P.

(2) If S_n is a unit n-dimensional sphere in $\mathbf{R}^{n+1}$ then S_n is definably bijective to the disjoint union of $\mathbf{R}^n$ and a point. Hence

$$\chi(S_n) = 1 + (-1)^n = \begin{cases} 0 & \text{if } n \text{ is odd} \\ 2 & \text{if } n \text{ is even.} \end{cases}$$

Notation. For a definable set A by $\mathbb{1}_A$ we will denote the characteristic function of A.

1.1 Grassmanians

For $n \in \mathbb{N}^{>0}$, as usual, we will denote by $\mathbb{P}^n$ the projective space of dimension n over $\mathbf{R}$, i.e. the set of all lines in $\mathbf{R}^{n+1}$ through the origin. It is not hard to find a definable set A and a bijection $h\colon A \to \mathbb{P}^n$ such that the incidence relation

$$I = \{(x, a) \in \mathbf{R}^{n+1} \times A\colon x \in h(a)\}$$

is a definable subset of $\mathbf{R}^{n+1} \times A$. Identifying $\mathbb{P}^n$ with such A we will always view each $\mathbb{P}^n$ as a definable set with the incidence relation $I = \{(x,l) \in \mathbf{R}^{n+1} \times \mathbb{P}^n : x \in l\}$ also definable.

Since $\mathbf{R}^{n+1} \setminus \{0\}$ is definably bijective to $\mathbb{P}^n \times (\mathbf{R} \setminus \{0\})$, we have $(-1)^{n+1} - 1 = \chi(\mathbb{P}^n) \times (-2)$, and

$$\chi(\mathbb{P}^n) = \frac{1 + (-1)^n}{2} = \begin{cases} 0 & \text{if } n \text{ is odd} \\ 1 & \text{if } n \text{ is even.} \end{cases}$$

Also, for $n \in \mathbb{N}^{>0}$, we will denote by $\mathrm{Gr}(n, n-1)$ the set of all affine hyperplanes in $\mathbf{R}^n$.

As in the case of projective spaces, we will view $\mathrm{Gr}(n, n-1)$ as a definable set with the incidence relation

$$\mathcal{I} = \{(x, \Pi) \in \mathbf{R}^n \times \mathrm{Gr}(n, n-1) : x \in \Pi\}$$

also definable.

Our main goal is to prove the following theorem of P. Schapira.

Theorem 1.2 (Schapira's Tomography Theorem) *Let $n > 1$ be an odd integer. For a definable set $A \subseteq \mathbf{R}^n$ let $\Xi_A \colon \mathrm{Gr}(n, n-1) \to \mathbb{Z}$ be the function that assigns to each $\Pi \in \mathrm{Gr}(n, n-1)$ the number $\chi(\Pi \cap A)$. Then the map $A \mapsto \Xi_A$ is injective.*

2 Constructible functions

Let $A \subseteq \mathbf{R}^n$ be a definable set. *A constructible function on A* is a definable function $\sigma \colon A \to \mathbb{Z}$. In other words, a function $\sigma \colon A \to \mathbb{Z}$ is constructible if the range of σ is finite and for every $i \in \mathbb{Z}$, the set $\sigma^{-1}(i)$ is a definable subset of A. We will denote the set of all constructible functions on A by $\mathcal{CF}(A)$. The set $\mathcal{CF}(A)$ is closed under point-wise addition and has a natural structure of a $\mathbb{Z}$-module.

Example 2.1 (1) Let A be a definable set. If B is a definable subset of A then $\mathbb{1}_B$ is a constructible functions on A.

(2) Let $f \colon A \to B$ be a definable map. By the Definable Trivialization Theorem, the function that assigns to each $b \in B$ the Euler characteristic of $f^{-1}(b)$ is constructible. We will denote this function by f^χ. Thus for every definable map $f \colon A \to B$ we have a constructible function $f^\chi \in \mathcal{CF}(B)$.

According to the definition, every $\sigma \in \mathcal{CF}(A)$ can be written as a finite sum $\sum n_i \mathbb{1}_{A_i}$, where $n_i \in \mathbb{Z}$ and every A_i is a definable subset of A. Thus $\mathcal{CF}(A)$ is generated, as a $\mathbb{Z}$-module, by $\{\mathbb{1}_B : B \subseteq A \text{ is definable}\}$.

Constructible functions can be integrated with respect to the o-minimal Euler characteristic. Namely, for $\sigma \in \mathcal{CF}(A)$ and a definable $B \subseteq A$ we set

$$\int_B \sigma d\chi = \sum_{i \in \mathbb{Z}} i\chi\big(\sigma^{-1}(i) \cap B\big).$$

Since, by (EC2), χ is additive, this integration is $\mathbb{Z}$-linear, and

$$\int_A \left(\sum n_i \mathbb{1}_{A_i}\right) d\chi = \sum n_i \chi(A_i).$$

2.1 Operations on Constructible functions

2.1.1 Multiplication

Let A be a definable set. Clearly, $\mathcal{CF}(A)$ is closed under point-wise multiplication. Hence every $\alpha \in \mathcal{CF}(A)$ defines a $\mathbb{Z}$-linear map $\Lambda_\alpha \colon \mathcal{CF}(A) \to \mathcal{CF}(A)$ given by $\Lambda_\alpha \colon \sigma \mapsto \alpha\sigma$.

2.1.2 Pull-backs

Let $f \colon A \to B$ be a definable map. For every $\sigma \in \mathcal{CF}(B)$ the composition $\sigma \circ f$ is a constructible function on A. We will denote this function by $f^*\sigma$ and call it *the pull-back of σ under f.* Thus for every definable map $f \colon A \to B$ we have $f^* \colon \mathcal{CF}(B) \to \mathcal{CF}(A)$. Obviously, f^* is $\mathbb{Z}$-linear and $(f \circ g)^* = g^* \circ f^*$.

Example 2.2 $f^*(\mathbb{1}_B) = \mathbb{1}_A$ for any definable map $f \colon A \to B$.

2.1.3 Push-forwards

Let $f \colon A \to B$ be a definable map. For $\sigma \in \mathcal{CF}(A)$ let $f_! \sigma$ be the function from B to $\mathbb{Z}$ given by

$$f_! \sigma \colon b \mapsto \int_{f^{-1}(b)} \sigma d\chi.$$

Example 2.3 If $f \colon A \to B$ is a definable function then $f_! \mathbb{1}_A = f^\chi$.

If $\sigma = \sum n_i \mathbb{1}_{A_i} \in \mathcal{CF}(A)$ then

$$f_! \sigma = f_! \left(\sum n_i \mathbb{1}_{A_i}\right) = \sum n_i f_!(\mathbb{1}_{A_i}) = \sum n_i (f \restriction A_i)^\chi.$$

Hence $f_!\sigma \in C\mathcal{F}(B)$.

Thus $f_!$ is a map from $C\mathcal{F}(A)$ into $C\mathcal{F}(B)$, and it is easy to see that $f_!$ is $\mathbb{Z}$-linear. For $\sigma \in C\mathcal{F}(A)$, the constructible function $f_!\sigma$ is called *the push-forward of σ under f*.

Example 2.4 If $f: A \to B$ is a definable map that is trivial over B, then all fibers $f^{-1}(b)$ have the same Euler characteristic. In this case $f_!\mathbb{1}_A = n\mathbb{1}_B$, where n is the Euler characteristic of any fiber $f^{-1}(b)$.

Claim 2.5 *If $f: A \to B$ and $g: B \to C$ are definable maps then*

$$(g \circ f)_! = (g_!) \circ (f_!).$$

Proof Using the linearity of push-forwards and the Trivialization Theorem, we only need to check that $(f \circ g)_!\mathbb{1}_A = f_!(g_!\mathbb{1}_A)$ in the case when both f and g are definably trivial. From the example above it follows that on both sides we get the same function $mn\mathbb{1}_C$, where m and n are the Euler characteristics of fibers $f^{-1}(b), b \in B$, and $g^{-1}(c), c \in C$, respectively. □

2.2 Cartesian squares

Recall that if Y, Y', X are sets and $f: Y \to X$, $f': Y' \to X$ are maps then *the fiber product $Y \times_X Y'$* is defined to be the set

$$\{(y, y') \in Y \times Y': f(y) = f(y')\}$$

together with the natural projections $\pi_Y: Y \times_X Y' \to Y$, $\pi_{Y'}: Y \times_X Y' \to Y'$. The following diagram is commutative.

$$
\begin{array}{ccc}
Y \times_X Y' & \xrightarrow{\pi_{Y'}} & Y' \\
\downarrow{\scriptstyle \pi_Y} & & \downarrow{\scriptstyle f'} \\
Y & \xrightarrow{f} & X
\end{array}
$$

Obviously, $Y \times_X Y'$ and the projections are definable in the case when f and f' are definable. In this case we call $Y \times_X Y'$ *a definable fiber product*.

Theorem 2.6 (Base Change Formula) *If*

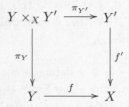

is a definable fiber product then $\pi_{Y'!} \circ \pi_Y^* = f'^* \circ f_!$

Proof We need to show that $(\pi_{Y'!} \circ \pi_Y^*)\sigma = (f'^* \circ f_!)\sigma$ for any $\sigma \in \mathcal{CF}(Y)$. Using linearity of pull-backs and push-forwards we need to consider only the case when $\sigma = \mathbb{1}_A$ for some definable $A \subseteq Y$. Replacing Y by A and restricting f to A if needed, we may assume that $\sigma = \mathbb{1}_Y$.

Let $b \in Y'$. Since $\pi_Y^* \mathbb{1}_Y = \mathbb{1}_{Y \times_X Y'}$, the value of $(\pi_{Y'!} \circ \pi_Y^*)\mathbb{1}_Y$ at b is the Euler characteristic of the set $\{\langle y, b\rangle \in Y \times_X Y'\}$. It is not hard to see that $(f'^* \circ f_!)\mathbb{1}_Y$ assigns to b the Euler characteristic of the set $\{y \in Y\colon f(y) = f'(b)\}$. These two sets are definably bijective, hence the Euler characteristics coincide.   □

2.3 Radon Transform

Let A, B be two definable sets and S a definable subset of $A \times B$. We denote by π_A and π_B the corresponding projections from S:

$$A \xleftarrow{\ \pi_A\ } S \xrightarrow{\ \pi_B\ } B \ .$$

For $\sigma \in \mathcal{CF}(A)$ *the Radon transform of* σ *with respect to* S is the constructible function on B given by $(\pi_{B!} \circ \pi_A^*)\sigma$. We will denote this function by $\mathcal{R}_S(\sigma)$. Clearly, $\mathcal{R}_S$ is a $\mathbb{Z}$-linear map from $\mathcal{CF}(A)$ to $\mathcal{CF}(B)$.

Example 2.7 Let $\mathcal{I} \subseteq \mathbf{R}^n \times \mathrm{Gr}(n, n-1)$ be the incidence relation and $A \subseteq \mathbf{R}^n$ a definable set. Then $\mathcal{R}_{\mathcal{I}}(\mathbb{1}_A) = \Xi_A$, where $\Xi_A\colon \mathrm{Gr}(n, n-1) \to \mathbb{Z}$ is the function from Theorem 1.2.

Let $S' \subseteq B \times A$ be the inverse of S, i.e. $S' = \{(b, a)\colon (a, b) \in S\}$. The composition $\mathbf{R}_{S'} \circ \mathbf{R}_S$ is a $\mathbb{Z}$-linear map from $\mathcal{CF}(A)$ into itself, and in a special case it has a very simple form.

Theorem 2.8 *Let* A, B *be definable sets and* $S \subseteq A \times B$ *a definable subset. Assume that there are integers* ξ, η *such that*

(1) *for all $a \in A$ the Euler characteristic of the set $S_a = \{b \in B : (a, b) \in S\}$ equals ξ;*

(2) *for all $a \neq a' \in A$ the Euler charactersitc of the set $S_a \cap S_{a'}$ equals η.*

Then

$$\mathcal{R}_{S'} \circ \mathcal{R}_S \colon \mathbb{1}_X \mapsto (\xi - \eta)\mathbb{1}_X + \eta \mathcal{X}(X)\mathbb{1}_A$$

for every definable $X \subseteq A$.

Proof Consider the diagram

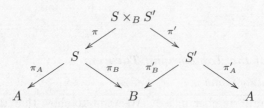

with natural projections. According to the definition of the Radon transform, we have

$$\mathcal{R}_{S'} \circ \mathcal{R}_S = \pi'_{A!} \circ \pi'^*_B \circ \pi_{B!} \circ \pi^*_A$$

By the Base Change Formula, $\pi'^*_B \circ \pi_{B!} = \pi'_! \circ \pi^*$, hence

$$\mathcal{R}_{S'} \circ \mathcal{R}_S = \pi'_{A!} \circ \pi'_! \circ \pi^* \circ \pi^*_A = (\pi'_A \circ \pi')_! \circ (\pi_A \circ \pi)^*$$

Let $r \colon S \times_B S' \to A \times A$ be the map $r \colon \big((a, b), (b, a')\big) \mapsto (a, a')$. Consider the commutative diagram

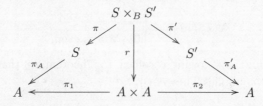

Since $\pi_A \circ \pi = \pi_1 \circ r$, $\pi'_A \circ \pi' = \pi_2 \circ r$ and $r_! \circ r^* = \Lambda_{r\mathcal{X}}$ (see Example 2.1), we obtain

$$\mathcal{R}_{S'} \circ \mathcal{R}_S = (\pi_2 \circ r)_! \circ (\pi_1 \circ r)^* = \pi_{2!} \circ r_! \circ r^* \circ \pi^*_1 = \pi_{2!} \circ \Lambda_{r\mathcal{X}} \circ \pi^*_1.$$

According to the assumptions, $r\mathcal{X} = \eta\Lambda_{\mathbb{1}_{A \times A}} + (\xi - \eta)\Lambda_{\mathbb{1}_\Delta}$, where Δ is the diagonal of $A \times A$. We have then

$$
\begin{aligned}
\mathcal{R}_{S'} \circ \mathcal{R}_S &= \pi_{2!} \circ \left(\eta \Lambda_{\mathbb{1}_{A \times A}} + (\xi - \eta) \Lambda_{\mathbb{1}_\Delta} \right) \circ \pi_1^* \\
&= \eta \, \pi_{2!} \circ \Lambda_{\mathbb{1}_{A \times A}} \circ \pi_1^* + (\xi - \eta) \, \pi_{2!} \circ \Lambda_{\mathbb{1}_\Delta} \circ \pi_1^* \\
&= \eta \, \pi_{2!} \circ \pi_1^* + (\xi - \eta) \, \pi_{2!} \circ \Lambda_{\mathbb{1}_\Delta} \circ \pi_1^*.
\end{aligned}
$$

If X is a definable subset of A, then $\pi_1^*(\mathbb{1}_X) = \mathbb{1}_{X \times A}$, $\Lambda_{\mathbb{1}_\Delta}(\mathbb{1}_{X \times A}) = \mathbb{1}_{\Delta \cap X \times X}$, $\pi_{2!}(\mathbb{1}_{\Delta \cap X \times X}) = \mathbb{1}_X$, and $\pi_{2!}(\mathbb{1}_{X \times A}) = \mathcal{X}(X)\mathbb{1}_A$.
Thus

$$
\mathcal{R}_{S'} \circ \mathcal{R}_S(\mathbb{1}_X) = (\xi - \eta)\mathbb{1}_X + \eta \mathcal{X}(X)\mathbb{1}_A.
$$

□

2.4 Proof of the Tomography Theorem

2.4.1 Computing ξ and η

We fix a natural number $n > 1$. We want to show that the incidence relation $\mathcal{I} \subseteq \mathbf{R}^n \times \mathrm{Gr}(n, n-1)$ satisfies the assumptions of Theorem 2.8 and to compute ξ and η in this case.

Let $a \in \mathbf{R}^n$. Obviously the set $\mathcal{I}_a = \{\Pi \in \mathrm{Gr}(n, n-1) \colon a \in \Pi\}$ is definably bijective to $\mathbb{P}^n$. Hence

$$
\mathcal{X}(\mathcal{I}_a) = \begin{cases} 0 & \text{if } n \text{ is odd} \\ 1 & \text{if } n \text{ is even.} \end{cases}
$$

Let $a \neq b \in \mathbf{R}^n$. It is not hard to see that the set $\mathcal{I}_a \cap \mathcal{I}_b$ is definably bijective to $\mathbb{P}^{n-1}$ and

$$
\mathcal{X}(\mathcal{I}_a \cap \mathcal{I}_b) = \begin{cases} 1 & \text{if } n \text{ is odd} \\ 0 & \text{if } n \text{ is even.} \end{cases}
$$

Thus we obtain the following corollary to Theorem 2.8 that, according to Example 2.7, implies Theorem 1.2.

Corollary 2.9 *Let $n > 1$ be an integer and $\mathcal{I} \subseteq \mathbf{R}^n \times \mathrm{Gr}(n, n-1)$ be the incidence relation. Then for any definable $A \subseteq \mathbf{R}^n$*

$$
\mathcal{R}_{\mathcal{I}'} \circ \mathcal{R}_{\mathcal{I}} \colon \mathbb{1}_A \mapsto \begin{cases} \mathbb{1}_A & \text{if } n \text{ is odd} \\ \mathcal{X}(A)\mathbb{1}_{\mathbf{R}^n} - \mathbb{1}_A & \text{if } n \text{ is even.} \end{cases}
$$

Remark 2.10 1. Since the $\mathbb{Z}$-module of constructible functions is generated by characteristic functions of definable subset, it follows that for any odd integer $n > 1$ the Radon transform $\mathcal{R}_\mathcal{I}$ is an isomorphism from $\mathcal{CF}(\mathbf{R}^n)$ onto $\mathcal{CF}(\mathrm{Gr}(n, n-1))$ whose inverse is $\mathcal{R}_{\mathcal{I}'}$.

2. In the case when n is an even integer the map $A \mapsto \mathcal{R}_\mathcal{I}(\mathbb{1}_A)$ is still injective. Indeed, it is not hard to see that the map $A \mapsto \mathcal{R}_{\mathcal{I}'} \circ \mathcal{R}_\mathcal{I}(\mathbb{1}_A)$ is injective on the family of *non-empty* definable subsets $A \subseteq \mathbf{R}^n$, and

$$\mathcal{R}_{\mathcal{I}'} \circ \mathcal{R}_\mathcal{I}(\mathbb{1}_\varnothing) = 0 = \mathcal{R}_{\mathcal{I}'} \circ \mathcal{R}_\mathcal{I}(\mathbb{1}_{\mathbf{R}^n}).$$

Thus the map $A \mapsto \mathcal{R}_\mathcal{I}\mathbb{1}_A$ is also injective on the family of non-empty definable subsets $A \subseteq \mathbf{R}^n$, and we need to check only that $\mathcal{R}_\mathcal{I}(\mathbb{1}_\varnothing) \neq \mathcal{R}_\mathcal{I}(\mathbb{1}_{\mathbf{R}^n})$. By direct computations,

$$\mathcal{R}_\mathcal{I}(\mathbb{1}_\varnothing) = 0 \quad \text{and} \quad \mathcal{R}_\mathcal{I}(\mathbb{1}_{\mathbf{R}^n}) = -\mathbb{1}_{\mathrm{Gr}(n,n-1)}.$$

References

[1] Ludwig Bröcker, Euler integration and Euler multiplication, *Adv. Geom.*, **5**, 2005, no. 1, 145 – 169.

[2] Lou van den Dries, *Tame topology and o-minimal structures*, London Mathematical Society Lecture Note Series, vol. 248, Cambridge University Press, Cambridge, 1998.

[3] P. Schapira, Tomography of constructible functions, in: *Applied algebra, algebraic algorithms and error-correcting codes* (Paris, 1995), Lecture Notes in Comput. Sci., vol. 948, Springer, Berlin, 1995, 427 – 435

A class of quantum Zariski geometries

Boris Zilber

Oxford University

1 Introduction

This paper is an attempt to understand the nature of *non-classical* Zariski geometries. Examples of such structures were first discovered in [HZ].

These examples showed that contrary to some expectations, one-dimensional Zariski geometries are not necessarily algebraic curves. Given a smooth algebraic curve C with a big enough group of regular automorphisms, one can produce a "smooth" Zariski curve $\tilde{C}$ along with a finite cover $p : \tilde{C} \to C$. $\tilde{C}$ cannot be identified with any algebraic curve because the construction produces an unusual subgroup of the group of regular automorphisms of $\tilde{C}$ ([HZ], section 10). The main theorem of [HZ] states that every Zariski curve has the form $\tilde{C}$, for some algebraic C. So, only in the limit case, when p is bijective, is the curve algebraic.

A typical example of an unusual subgroup of the automorphism group of such a $\tilde{C}$ is the nilpotent group of two generators $\mathbf{U}$ and $\mathbf{V}$ with the central commutator $\epsilon = [\mathbf{U}, \mathbf{V}]$ of finite order N. So, the defining relations are

$$\mathbf{U}\mathbf{V} = \epsilon\,\mathbf{V}\mathbf{U}, \quad \epsilon^N = 1.$$

This, of course, hints towards the known object of non-commutative geometry, the non-commutative (quantum) torus at the Nth root of unity. This observation encourages us to look for systematic links between non-commutative geometry and model theory. More specifically, we would like to give arguments towards the thesis that any non-classical Zariski geometry is in some way an object of non-commutative (quantum) geometry and the classical ones are just the limit cases of the general situation. Towards this end we carried out some analysis of the above examples in [Z2].

In this paper we attempt to give a general method which associates a "geometric object" to a typical *quantum algebra*. Note that this is in fact an open question. Yu. Manin mentions this foundational problem in [Man] I.1.4. Indeed, in general non-commutative geometry does not assume that one has (as is the case in commutative geometry) a procedure of getting a manifold-like structure from the algebra of "observables", yet it is desirable both for technical and conceptual reasons. See also the survey paper [Sk]. The approach in [RVW] looks quite similar to what our paper suggests.

More specifically, we restrict ourselves with quantum algebras *at roots of unity*.

Strictly speaking the general notion of a *quantum algebra* does not exist, and we have to start our construction by introducing algebraic assumptions on A which make the desired theorem feasible.

The next step, after proving that the geometric object we obtain has the right properties, would be to check if our assumptions cover all interesting cases. If it were the case our assumptions would deserve the status of a definition of a quantum algebra.

Our construction always produces a Zariski geometry and when the algebra in question is big enough the structure is provably non-classical, that is not an object of (commutative) algebraic geometry. This might be seen as a good criterion for the adequacy of the construction. Among the structures which satisfy our assumptions is the quantum group $U_\epsilon(\mathfrak{sl}_2)$, but we couldn't check it for higher-dimensional objects because of algebraic difficulties.

In more detail, we consider F-algebras A over an algebraically closed field F. Our assumptions imply that a typical irreducible A-module is of finite dimension over F.

We introduce the structure associated with A as a two-sorted structure $(\tilde{V}, F)$ where F is given with the usual field structure and $\tilde{V}$ is the bundle over an affine variety V of A-modules of a fixed finite F-dimension N. Again by the assumptions the isomorphism types of N-dimensional A-modules are determined by points in V. "Inserting" a module M_m of the corresponding type in each point m of V we get

$$\tilde{V} = \coprod_{m \in V} M_m.$$

In fact, for any m belonging to an open subset of V, the module M_m is irreducible.

Our language contains a function symbol $\mathbf{U}_i$ acting on each M_m (and so on the sort $\tilde{\mathbf{V}}$) for each generator $\mathbf{U}_i$ of the algebra A. We also have the binary function symbol for the action of F by scalar multiplication on the modules. Since M_m may be considered an $\mathrm{A}/\mathrm{Ann}\,M_m$-module we have the bundle of finite-dimensional algebras $\mathrm{A}/\mathrm{Ann}\,M_m$, $m \in \mathrm{V}$, represented in $\tilde{\mathbf{V}}$. In typical cases the intersection of all such annihilators is 0. As a consequence of this, the algebra A is faithfully represented by its action on the bundle of modules. In fact the whole construction of the structure is aiming to present the category of all finite dimensional A-modules.

Note that in the case when all the M_m are irreducible our structure is a groupoid in the same sense as in Hrushovski's paper [H]. But in general, e.g. in the case of the quantum group $U_\epsilon(\mathfrak{sl}_2)$ the structure is not a groupoid and this is one of the features that makes it richer and more interesting.

We write down our description of $\tilde{\mathbf{V}}$ as the set of first-order axioms Th(A-mod).

We prove two main theorems.

Theorem A (Sections 2.4 and 3.2) *The theory* Th(A-mod) *is categorical in uncountable cardinals and model complete.*

Theorem B (Section 4.3) $\tilde{\mathbf{V}}$ *is a Zariski geometry in both sorts.*

Theorem A is rather easy to prove, and in fact the proof uses not all of the assumptions on A we assumed. Yet despite the apparent simplicity of the construction, for certain A, $\tilde{\mathbf{V}}$ is not definable in an algebraically closed field, that is, $\tilde{\mathbf{V}}(\mathrm{A})$ *is not classical.*

Theorem B requires much more work, mainly the analysis of definable sets. This is due to the fact that the theory of $\tilde{\mathbf{V}}$, unlike the case of Zariski geometries coming from algebraic geometry, does not have quantifier elimination in the natural algebraic language. We hope that this technical analysis will be instrumental in practical applications to noncommutative geometry.

Acknowledgement I would like to thank Shahn Majid for very helpful discussions.

I started to work on the paper while I was a member of *Model Theory*

and Applications to Algebra and Analysis programme at the Isaac Newton Institute for Mathematical Sciences, Cambridge, UK. I am grateful
to the organisers of the program, the staff of the Institute, and the participants.

2 From algebras to structures

2.1 We fix below until the end of the paper an F-algebra A, satisfying
the following

Assumptions.

1. We assume that F is an algebraically closed field and A is an as-
   sociative unital affine F-algebra with generators $\mathbf{U}_1, \ldots, \mathbf{U}_d$ and
   defining relations with parameters in a finite $C \subset$ F. We also
   assume that A is a finite dimensional module over its central
   subalgebra Z_0.
2. Z_0 is a unital finitely generated commutative F-algebra without
   zero divisors, so Max Z_0, the space of maximal ideals of Z_0, can
   be identified with the F-points of an irreducible affine algebraic
   variety V over C.
3. There is a positive integer N such that to every $m \in$ Max Z_0 we
   can put in correspondence with m an A-module M_m of dimension
   N over F with the property that the maximal ideal m annihilates
   M_m.

   The isomorphism type of the module M_m is determined uniformly by a solution to a system of polynomial equations P^A in
   variables $t_{ijk} \in$ F and $m \in$ V such that:
   for every $m \in$ V there exists $t = \{t_{ijk} : i \leq d, \ j, k \leq N\}$ satisfying
   $P^A(t, m) = 0$ and for each such t there is a basis $e(1), \ldots, e(N)$
   of the F-vector space on M_m with

$$\bigwedge_{i \leq d, \ j \leq N} \mathbf{U}_i e(j) = \sum_{k=1}^{N} t_{ijk} e(k).$$

   We call any such basis $e(1), \ldots, e(N)$ **canonical**.
4. There is a finite group Γ and a (partial) map $g : V \times \Gamma \to \mathrm{GL}_N(\mathrm{F})$
   such that,

   - for each $\gamma \in \Gamma$, the map $g(\cdot, \gamma) : V \to \mathrm{GL}_N(\mathrm{F})$ is rational
     C-definable (defined on an open subset of V)

- for any $m \in V$, the (partial) map $g(m, \cdot) : \Gamma \to GL_N(F)$ has for domain a subgroup Dom_m of Γ, is injective on its domain, and for any two canonical bases $e(1), \ldots, e(N)$ and $e'(1), \ldots, e'(N)$ of M_m there is $\lambda \in F^*$ and $\gamma \in \mathrm{Dom}_m$ such that

$$e'(i) = \lambda \sum_{1 \leq j \leq N} g_{ij}(m, \gamma) e(j), \quad i = 1, \ldots, N.$$

We denote

$$\Gamma_m := g(m, \mathrm{Dom}_m).$$

Remark The correspondence $m \mapsto M_m$ between points in V and the isomorphism types of modules is bijective by the assumption 2.1. Indeed, for distinct $m_1, m_2 \in \mathrm{Max}\, Z_0$ the modules M_{m_1} and M_{m_2} are not isomorphic, for otherwise the module will be annihilated by Z_0.

2.2 The structure

Recall that $V(A)$ or simply V stands for the F-points of the algebraic variety $\mathrm{Max}\, Z_0$. By assumption 2.1.1 this can be viewed as the set of A-modules M_m, $m \in \mathrm{Max}\, Z_0$.

Consider the set $\tilde{V}$ as the disjoint union

$$\tilde{V} = \coprod_{m \in V} M_m.$$

We also pick up arbitrarily for each $m \in V$ a canonical basis $e = \{e(1), \ldots, e(N)\}$ in M_m and all the other canonical bases conjugated to e by Γ_m. We denote the set of bases for each $m \in V$ as

$$E_m := \Gamma_m e = \{(e'(1), \ldots, e'(N)) : e'(i) = \sum_{1 \leq j \leq N} \gamma_{ij} e(j), \; \gamma \in \Gamma_m\}.$$

Consider, along with the sort $\tilde{V}$ also the field sort F, the sort V identified with the corresponding affine subvariety $V \subseteq F^k$, some k, and the projection map

$$\pi : x \mapsto m \text{ if } x \in M_m, \text{ from } \tilde{V} \text{ to } V.$$

We assume the *full language of* $\tilde{V}$ contains:

1. the ternary relation $S(x, y, z)$ which holds if and only if there is $m \in V$ such that $x, y, z \in M_m$ and $x + y = z$ in the module;
2. the ternary relation $a \cdot x = y$ which for $a \in F$ and $x, y \in M_m$ is interpreted as the multiplication by the scalar a in the module M_m;

3. the binary relations $\mathbf{U}_i x = y$, $(i = 1, \ldots, d)$ which for $x, y \in M_m$ are interpreted as the actions by the corresponding operators in the module M_m;

4. the relations $E \subseteq \mathrm{V} \times \tilde{\mathrm{V}}^N$ with $E(m, e)$ interpreted as $e \in E_m$.

The *weak language* is the sublanguage of the full one which includes 1–3 above only.

Finally, denote $\tilde{\mathrm{V}}$ the 3-sorted structure $(\tilde{\mathrm{V}}, \mathrm{V}, \mathrm{F})$ described above, with V endowed with the usual Zariski language as the algebraic variety.

Remarks 1. Notice that the sorts V and F are bi-interpretable over C.

2. The map $g : \mathrm{V} \times \Gamma \to \mathrm{GL}_N(\mathrm{F})$ being rational is definable in the weak language of $\tilde{\mathrm{V}}$.

Now we introduce the first order theory Th(A-mod) describing $(\tilde{\mathrm{V}}, \mathrm{V}, \mathrm{F})$. It consists of axioms:

Ax 1. F is an algebraically closed field of characteristic p and V is the Zariski structure on the F-points of the variety Max Z_0.

Ax 2. For each $m \in \mathrm{V}$, the action of the scalars of F and of the operators $\mathbf{U}_1, \ldots, \mathbf{U}_d$ defines on $\pi^{-1}(m)$ the structure of an A-module of dimension N.

Ax 3. Assumption 2.1.3 holds for the given P^A.

Ax 4. For the $g : \mathrm{V} \times \Gamma \to \mathrm{GL}_N(\mathrm{F})$ given by the assumption 2.1.4, for any $e, e' \in E_m$ there exists $\gamma \in \Gamma$ such that

$$e'(i) = \sum_{1 \leq j \leq N} g_{ij}(m, \gamma) e(j), \quad i = 1, \ldots, N.$$

Moreover, E_m is an orbit under the action of Γ_m.

Remark The referee of the paper notes that if M_m is irreducible then associated to a particular collection of coefficients t_{kij} there is a unique (up to scalar multiplication) canonical base for M_m (as in 2.1.3). It follows that the only possible automorphisms of $\tilde{\mathrm{V}}$ which fix all of F are induced by multiplication by scalars in each module (the scalars do not have to be the same for each fibre, and typically are not). So the 'projective' bundle $\coprod_{m \in \mathrm{V}}(M_m/\text{scalars})$ is definable in the field F, but the original $\tilde{\mathrm{V}}$ in general is not (see subsection 2.5).

2.3 Examples We assume below that $\epsilon \in \mathrm{F}$ is a primitive root of 1 of order ℓ, and ℓ is not divisible by the characteristic of F.

1. Let A be generated by $\mathbf{U}$, $\mathbf{V}$, $\mathbf{U}^{-1}$, $\mathbf{V}^{-1}$ satisfying the relations

$$\mathbf{U}\mathbf{U}^{-1} = 1 = \mathbf{V}\mathbf{V}^{-1}, \qquad \mathbf{U}\mathbf{V} = \epsilon\mathbf{V}\mathbf{U}.$$

We denote this algebra T_ϵ^2 (equivalent to $\mathcal{O}_\epsilon((\mathrm{F}^\times)^2)$ in the notations of [BG]).

The centre $Z = Z_0$ of T_ϵ^2 is the subalgebra generated by $\mathbf{U}^\ell$, $\mathbf{U}^{-\ell}$, $\mathbf{V}^\ell$, $\mathbf{V}^{-\ell}$. The variety $\mathrm{Max}\, Z$ is isomorphic to the 2-dimensional torus $\mathrm{F}^* \times \mathrm{F}^*$.

Any irreducible T_ϵ^2-module M is an F-vector space of dimension $N = \ell$. It has a basis $\{e_0, \ldots, e_{\ell-1}\}$ of the space consisting of $\mathbf{U}$-eigenvectors and satisfying, for an eigenvalue μ of $\mathbf{U}$ and an eigenvalue ν of $\mathbf{V}$,

$$\mathbf{U}e_i = \mu\epsilon^i e_i$$
$$\mathbf{V}e_i = \begin{cases} \nu e_{i+1}, & i < \ell - 1, \\ \nu e_0, & i = \ell - 1. \end{cases}$$

We also have a basis of $\mathbf{V}$-eigenvectors $\{g_0, \ldots, g_{\ell-1}\}$ satisfying

$$g_i = e_0 + \epsilon^i e_1 + \cdots + \epsilon^{i(\ell-1)}e_{\ell-1}$$

and so

$$\mathbf{V}g_i = \nu\epsilon^i g_i$$
$$\mathbf{U}g_i = \begin{cases} \mu g_{i+1}, & i < \ell - 1, \\ \mu g_0, & i = \ell - 1. \end{cases}$$

For $\mu^\ell = a \in \mathrm{F}^*$ and $\nu^\ell = b \in \mathrm{F}^*$, $(\mathbf{U}^\ell - a), (\mathbf{V}^\ell - b)$ are generators of $\mathrm{Ann}(M)$. The module is determined uniquely once the values of a and b are given. So, V is isomorphic to the 2-dimensional torus $\mathrm{F}^* \times \mathrm{F}^*$.

The coefficients t_{ijk} in this example are determined by μ and ν, which satisfy the polynomial equations $\mu^\ell = a, \nu^\ell = b$.

$\Gamma_m = \Gamma$ is the fixed nilpotent group of order ℓ^3 generated by the matrices

$$\begin{pmatrix} 0 & 1 & 0 & \ldots & 0 \\ 0 & 0 & 1 & \ldots & 0 \\ \ldots & \ldots & & & \\ 1 & 0 & & \ldots & 0 \end{pmatrix} \text{ and } \begin{pmatrix} 1 & 0 & 0 & \ldots & 0 \\ 0 & \epsilon & 0 & \ldots & 0 \\ \ldots & \ldots & & & \\ 0 & 0 & & \ldots & \epsilon^{\ell-1} \end{pmatrix}$$

2. Similarly, the d-dimensional quantum torus $T_{\epsilon,\theta}^d$ which is generated

by $\mathbf{U}_1, \ldots, \mathbf{U}_d, \mathbf{U}_1^{-1} \ldots, \mathbf{U}_d^{-1}$ satisfying

$$\mathbf{U}_i \mathbf{U}_i^{-1} = 1, \quad \mathbf{U}_i \mathbf{U}_j = \epsilon^{\theta_{ij}} \mathbf{U}_j \mathbf{U}_i, \quad 1 \le i, j \le d,$$

where θ is an antisymmetric integer matrix, g.c.d.$\{\theta_{ij} : 1 \le j \le d\}) = 1$ for some $i \le d$.

There is a simple description of the bundle of irreducible modules all of which are of the same dimension $N = \ell$.

$T_{\epsilon,\theta}^d$ satisfies all the assumptions.

3. $A = U_\epsilon(\mathfrak{sl}_2)$, the quantum universal enveloping algebra of $\mathfrak{sl}_2(F)$. It is given by generators K, K^{-1}, E, F satisfying the defining relations

$$KK^{-1} = 1, \ KEK^{-1} = \epsilon^2 E, \ KFK^{-1} = \epsilon^{-2}F, \ EF - FE = \frac{K - K^{-1}}{\epsilon - \epsilon^{-1}}.$$

The centre Z of $U_\epsilon(\mathfrak{sl}_2)$ is generated by K^ℓ, E^ℓ, F^ℓ and the element

$$C = FE + \frac{K\epsilon + K^{-1}\epsilon^{-1}}{(\epsilon - \epsilon^{-1})^2}.$$

We use [BG], Chapter III.2, to describe $\tilde{V}$. We assume $\ell \ge 3$ odd.

Let $Z_0 = Z$ and so $V = \text{Max}\, Z$ is an algebraic extension of degree ℓ of the commutative affine algebra $K^\ell, K^{-\ell}, E^\ell, F^\ell$.

To every point $m = (a, b, c, d) \in V$ corresponds the unique, up to isomorphism, module with a canonical basis $e_0, \ldots, e_{\ell-1}$ satisfying

$$Ke_i = \mu\epsilon^{-2i}e_i,$$

$$Fe_i = \begin{cases} e_{i+1}, & i < \ell - 1, \\ be_0, & i = \ell - 1. \end{cases}$$

$$Ee_i = \begin{cases} \rho e_{\ell-1}, & i = 0, \\ \left(\rho b + \frac{(\epsilon^i - \epsilon^{-i})(\mu\epsilon^{1-i} - \mu^{-1}\epsilon^{i-1})}{(\epsilon - \epsilon^{-1})^2}\right)e_{i-1}, & i > 0. \end{cases}$$

where μ, ρ satisfy the polynomial equations

$$(1) \qquad \mu^\ell = a, \qquad \rho b + \frac{\mu\epsilon + \mu^{-1}\epsilon^{-1}}{(\epsilon - \epsilon^{-1})^2} = d$$

and

$$(2) \qquad \rho \prod_{i=1}^{\ell-1} \left(\rho b + \frac{(\epsilon^i - \epsilon^{-i})(\mu\epsilon^{1-i} - \mu^{-1}\epsilon^{i-1})}{(\epsilon - \epsilon^{-1})^2}\right) = c.$$

We may characterise V as

$$V = \{(a, b, c, d) \in F^4 : \exists \rho, \mu \ (1) \text{ and } (2) \text{ hold}\}$$

In fact, the map $(a, b, c, d) \mapsto (a, b, c)$ is a cover of the affine variety $A^3 \cap \{a \neq 0\}$ of order ℓ.

In almost all points of V, except for the points of the form $(1, 0, 0, d_+)$ and $(-1, 0, 0, d_-)$, the module is irreducible. In the exceptional cases, for each $i \in \{0, \ldots, \ell - 1\}$ we have exactly one ℓ-dimensional module (denoted $Z_0(\epsilon^i)$ or $Z_0(-\epsilon^i)$ in [BG], depending on the sign) which satisfies the above description with $\mu = \epsilon^i$ or $-\epsilon^i$. The Casimir invariant is

$$d_+ = \frac{\epsilon^{i+1} + \epsilon^{-i-1}}{(\epsilon - \epsilon^{-1})^2} \quad \text{or} \quad d_- = -\frac{\epsilon^{i+1} + \epsilon^{-i-1}}{(\epsilon - \epsilon^{-1})^2}$$

and the module, for $i < \ell - 1$, has the unique proper irreducible submodule of dimension $\ell - i - 1$ spanned by $e(i+1), \ldots, e(\ell - 1)$. For $i = \ell - 1$ the module is irreducible. According to [BG], III.2, all the irreducible modules of A have been listed above, either as M_m or as submodules of M_m for the exceptional $m \in V$.

To describe Γ_m consider two canonical bases e and e' in M_m. If e' is not of the form λe, then necessarily $e'_0 = \lambda e_k$, for some $k \leq \ell - 1$, $b \neq 0$ and

$$e'_i = \begin{cases} \lambda e_{i+k}, & 0 \leq i < \ell - k, \\ \lambda b e_{i+k}, & \ell - 1 \geq i \geq \ell - k, \end{cases}$$

If we put $\lambda = \lambda_k = \nu^{-k}$, for $\nu^\ell = b$, we get a finite order transformation. So we can take $\Gamma_{(a,b,c,d)}$, for $b \neq 0$, to be the Abelian group of order ℓ^2 generated by the matrices

$$\begin{pmatrix} 0 \ \nu^{-1} \ 0 \ldots 0 \\ 0 \ 0 \ \nu^{-1} \ldots 0 \\ \ldots \ \ldots \ \nu^{-1} \\ \nu^{\ell-1} \ 0 \ \ldots 0 \end{pmatrix} \quad \text{and} \quad \begin{pmatrix} \epsilon \ 0 \ 0 \ldots 0 \\ 0 \ \epsilon \ 0 \ldots 0 \\ \ldots \ \ldots \\ 0 \ 0 \ \ldots \epsilon \end{pmatrix}$$

where ν is defined by

$$\nu^\ell = b.$$

When $b = 0$ the group $\Gamma_{(a,0,c,d)}$ is just the cyclic group generated by the scalar matrix with ϵ on the diagonal.

The isomorphism type of the module depends on $\langle a, b, c, d \rangle$ only. This basis satisfies all the assumptions 2.1.1–4.

$U_\epsilon(\mathfrak{sl}_2)$ is one of the simplest examples of a *quantum group*. Quantum groups, as all bi-algebras, have the following crucial property: *the* *tensor product $M_1 \otimes M_2$ of any two* A-*modules is well-defined and is an* A-*module.* So, the tensor product of two modules in $\tilde{V}$ produces a $U_\epsilon(\mathfrak{sl}_2)$-module of dimension ℓ^2, definable in the structure, and which 'contains' finitely many modules in $\tilde{V}$. This defines a multivalued operation on V (or on an open subset of V, in the second case).

More examples and the most general known cases $U_\epsilon(\mathfrak{g})$, for $\mathfrak{g}$ a semisimple complex Lie algebra, and $\mathcal{O}_\epsilon(G)$, the quantised group G, for G a connected simply connected semisimple complex Lie group, are shown to have properties 1 and 2 for the central algebra Z_0 generated by the corresponding U_i^ℓ, $i = 1, \ldots, d$.

The rest of the assumptions are harder to check. We leave this open.

4. $A = \mathcal{O}_\epsilon(\mathrm{F}^2)$, Manin's quantum plane is given by generators $\mathbf{U}$ and $\mathbf{V}$ and defining relations $\mathbf{UV} = \epsilon\mathbf{VU}$. The centre Z is again generated by $\mathbf{U}^\ell$ and $\mathbf{V}^\ell$ and the maximal ideals of Z in this case are of the form $\langle(\mathbf{U}^\ell - a), (\mathbf{V}^\ell - b)\rangle$ with $\langle a, b\rangle \in \mathrm{F}^2$.

This example, though very easy to understand algebraically, does not quite fit into our construction. Namely, the assumption 3 is satisfied only in generic points of $V = \mathrm{Max}\, Z$. But the main statement still holds true for this case as well. We just have to construct $\tilde{V}$ by gluing two Zariski spaces, each corresponding to a localisation of the algebra A.

To each maximal ideal with $a \neq 0$ we put in correspondence the module of dimension ℓ given in a basis $e_0, \ldots, e_{\ell-1}$ by

$$\mathbf{U}e_i = \mu\epsilon^i e_i$$
$$\mathbf{V}e_i = \begin{cases} e_{i+1}, & i < \ell - 1, \\ be_0, & i = \ell - 1, \end{cases}$$

for μ satisfying $\mu^\ell = a$.

To each maximal ideal with $b \neq 0$ we put in correspondence the module of dimension ℓ given in a basis $g_0, \ldots, g_{\ell-1}$ by

$$\mathbf{V}g_i = \nu\epsilon^i g_i$$
$$\mathbf{U}g_i = \begin{cases} g_{i+1}, & i < \ell - 1, \\ ae_0, & i = \ell - 1, \end{cases}$$

for ν satisfying $\nu^\ell = b$.

When both $a \neq 0$ and $b \neq 0$ we identify the two representations of the same module by choosing g (given e and ν) so that

$$g_i = e_0 + \nu^{-1}\epsilon^i e_1 + \cdots + \nu^{-k}\epsilon^{ik}e_k + \cdots + \nu^{-(\ell-1)}\epsilon^{i(\ell-1)}e_{\ell-1}.$$

This induces a definable isomorphism between modules and defines a gluing between $\tilde{V}_{a\neq 0}$ and $\tilde{V}_{b\neq 0}$. In fact $\tilde{V}_{a\neq 0}$ corresponds to the algebra given by three generators $\mathbf{U}, \mathbf{U}^{-1}$ and $\mathbf{V}$ with relations $\mathbf{UV} = \epsilon\mathbf{VU}$ and $\mathbf{UU}^{-1} = 1$, a localisation of $\mathcal{O}_\epsilon(\mathbf{F}^2)$, and $\tilde{V}_{b\neq 0}$ corresponds to the localisation by $\mathbf{V}^{-1}$.

2.4 Categoricity

Lemma (i) Let $\tilde{V}_1$ and $\tilde{V}_2$ be two structures in the weak language satisfying 2.1.1–3 and 2.2.1–3 with the same P^A over the same algebraically closed field F. Then the natural isomorphism $i : V_1 \cup F \to V_2 \cup F$ over C can be lifted to an isomorphism

$$i : \tilde{V}_1 \to \tilde{V}_2.$$

(ii) Let $\tilde{V}_1$ and $\tilde{V}_2$ be two structures in the full language satisfying 2.1.1–4 and 2.2.1–4 with the same P^A over the same algebraically closed field F. Then the natural isomorphism $i : V_1 \cup F \to V_2 \cup F$ over C can be lifted to an isomorphism

$$i : \tilde{V}_1 \to \tilde{V}_2.$$

Proof. We may assume that i is the identity on V and on the sort F.

The assumptions 2.1 and the description 2.2 imply that in both structures $\pi^{-1}(m)$, for $m \in V$, has the structure of a module. Denote these $\pi_1^{-1}(m)$ and $\pi_2^{-1}(m)$ in the first and second structure correspondingly.

For each $m \in V$ the modules $\pi_1^{-1}(m)$ and $\pi_2^{-1}(m)$ are isomorphic.

Indeed, using 2.1.3 choose t_{ijk} satisfying P^A for m and find bases e in $\pi_1^{-1}(m)$ and e' in $\pi_2^{-1}(m)$ with the $\mathbf{U}_i$'s represented by the matrices $\{t_{ijk} : k, j = 1, \ldots, N\}$ in both modules. It follows that the map

$$i_m : \sum z_j e(j) \mapsto \sum z_j e'(j), \quad z_1, \ldots, z_N \in F$$

is an isomorphism of the A-modules

$$i_m : \pi_1^{-1}(m) \to \pi_2^{-1}(m).$$

Hence, the union

$$\mathbf{i} = \bigcup_{m \in V} i_m, \quad \mathbf{i} : \tilde{V}_1 \to \tilde{V}_2,$$

is an isomorphism. This proves (i).

In order to prove (ii) choose, using 2.1.4, e and e' in E_m in $\pi_1^{-1}(m)$

and $\pi_2^{-1}(m)$ correspondingly. Then the map i_m by the same assumption also preserves E_m, and so **i** is an isomorphism in the full language. □

As an immediate corollary we get

Theorem Th(A-mod) *is categorical in uncountable cardinals both in the full and the weak languages.*

Remark 1 The above Lemma is a special case of Lemma 3.2.

Remark 2 It is not difficult to see that in the general case the theory Th(A-mod) is not almost strongly minimal in the weak language but is always almost strongly minimal in the full language.

2.5 We prove in this subsection that despite the simplicity of the construction and the proof of categoricity the structures obtained from algebras A in our list of examples are nonclassical.

Assume for simplicity that char $F = 0$. The statements in this subsection are in their strongest form when we choose the weak language for the structures.

Proposition $\tilde{V}(T_\epsilon^n)$ *is not definable in an algebraically closed field, for* $n \geq 2$.

Proof. We write A for T_ϵ^2. We consider the structure in the weak language.

Suppose towards the contradiction that $\tilde{V}(A)$ is definable in some F'. Then F is also definable in this algebraically closed field. But, as is well-known, the only infinite field definable in an algebraically closed field is the field itself. So, $F' = F$ and so we have to assume that $\tilde{V}$ is definable in F.

Given $\mathbf{W} \in A$, $v \in \tilde{V}$, $x \in F$ and $m \in V$, denote $\text{Eig}(\mathbf{W}; v, x, m)$ the statement:

v is an eigenvector of $\mathbf{W}$ *in* $\pi^{-1}(m)$ *(or simply in* M_m*) with the eigenvalue x.*

For any given $\mathbf{W}$ the ternary relation $\text{Eig}(\mathbf{W}; v, x, m)$ is definable in $\tilde{V}$ by 2.2.

Let $m \in V$ be such that μ is an **U**-eigenvalue and ν is a **V**-eigenvalue in the module M_m. $\langle \mu^\ell, \nu^\ell \rangle$ determines the isomorphism type of M_m (see 2.3), in fact $m = \langle \mu^\ell, \nu^\ell \rangle$.

Consider the definable set

$$\mathrm{Eig}(\mathbf{U}) = \{v \in \tilde{\mathrm{V}} : \exists \mu, m \; \mathrm{Eig}(\mathbf{U}; v, \mu, m)\}.$$

By our assumption and elimination of imaginaries in ACF this is in a definable bijection with an algebraic subset S of F^n, some n, defined over some finite C'. We may assume that $C' = C$. Moreover the relations and functions induced from $\tilde{\mathrm{V}}$ on $\mathrm{Eig}(\mathbf{U})$ are algebraic relations definable in F over C.

Consider μ and ν as variables running in F and let $\tilde{\mathrm{F}} = \mathrm{F}\{\mu, \nu\}$ be the field of Puiseux series in variables μ, ν. Since $S(\tilde{\mathrm{F}})$ as a structure is an elementary extension of $\mathrm{Eig}(\mathbf{U})$ there is a tuple, say e_μ, in $S(\tilde{\mathrm{F}})$ which is an $\mathbf{U}$-eigenvector with the eigenvalue μ.

By definition the coordinates of e_μ are Laurent series in the variables $\mu^{\frac{1}{k}}$ and $\nu^{\frac{1}{k}}$, for some positive integer k. Let K be the subfield of $\tilde{\mathrm{F}}$ consisting of all Laurent series in variables $\mu^{\frac{1}{k}}, \nu^{\frac{1}{k}}$, for the k above. Fix $\delta \in \mathrm{F}$ such that

$$\delta^k = \epsilon.$$

The maps

$$\xi : t(\mu^{\frac{1}{k}}, \nu^{\frac{1}{k}}) \mapsto t(\delta\mu^{\frac{1}{k}}, \nu^{\frac{1}{k}}) \text{ and } \zeta : t(\mu^{\frac{1}{k}}, \nu^{\frac{1}{k}}) \mapsto t(\mu^{\frac{1}{k}}, \delta\nu^{\frac{1}{k}}),$$

for $t(\mu^{\frac{1}{k}}, \nu^{\frac{1}{k}})$ Laurent series in the corresponding variables, obviously are automorphisms of K over F. In particular ξ maps μ to $\epsilon\mu$ and leaves ν fixed, and ζ maps ν to $\epsilon\nu$ and leaves μ fixed. Also note that the two automorphisms commute and both are of order ℓk.

Since $\mathbf{U}$ is F-definable, $\xi^m(e_\mu)$ is a $\mathbf{U}$-eigenvector with the eigenvalue $\epsilon^m \mu$, for any integer m.

By the properties of A-modules $\mathbf{V}e_\mu$ is an $\mathbf{U}$-eigenvector with the eigenvalue $\epsilon\mu$, so there is $\alpha \in \tilde{\mathrm{F}}$

$$(3) \qquad\qquad \mathbf{V}e_\mu = \alpha\xi(e_\mu).$$

But α is definable in terms of e_μ, $\xi(e_\mu)$ and C, so by elimination of quantifiers α is a rational function of the coordinates of the elements, hence $\alpha \in K$.

Since $\mathbf{V}$ is definable over F, we have for every automorphism γ of K,

$$\gamma(\mathbf{V}e) = \mathbf{V}\gamma(e).$$

So, (3) implies

$$\mathbf{V}\xi^i e_\mu = \xi^i(\alpha)\xi^{i+1}(e_\mu), \quad i = 0, 1, 2, \ldots$$

and, since

$$\mathbf{V}^{k\ell} e_\mu = \nu^{k\ell} e_\mu,$$

applying $\mathbf{V}$ to both sides of (3) $k\ell - 1$ times we get

$$(4) \qquad \prod_{i=0}^{k\ell-1} \xi^i(\alpha) = \nu^{k\ell}.$$

Now remember that

$$\alpha = a_0(\nu^{\frac{1}{k}}) \cdot \mu^{\frac{d}{k}} \cdot (1 + a_1(\nu^{\frac{1}{k}})\mu^{\frac{1}{k}} + a_2(\nu^{\frac{1}{k}})\mu^{\frac{2}{k}} + \cdots)$$

where $a_0(\nu^{\frac{1}{k}}), a_1(\nu^{\frac{1}{k}}), a_2(\nu^{\frac{1}{k}}) \ldots$ are Laurent series in $\nu^{\frac{1}{k}}$ and d an integer. Substituting this into (4) we get

$$\nu^{k\ell} = a_0(\nu^{\frac{1}{k}})^{k\ell} \delta^{\frac{k\ell(k\ell-1)}{2}} \mu^{d\ell} \cdot (1 + a_1'(\nu^{\frac{1}{k}})\mu^{\frac{1}{k}} + a_2'(\nu^{\frac{1}{k}})\mu^{\frac{2}{k}} + \cdots).$$

It follows that $d = 0$ and $a_0(\nu^{\frac{1}{k}}) = a_0 \cdot \nu$, for some constant $a_0 \in \mathrm{F}$. That is,

$$(5) \qquad \alpha = a_0 \cdot \nu \cdot (1 + a_1(\nu^{\frac{1}{k}})\mu^{\frac{1}{k}} + a_2(\nu^{\frac{1}{k}})\mu^{\frac{2}{k}} + \cdots).$$

Now we use the fact that $\zeta(e_\mu)$ is an $\mathbf{U}$-eigenvector with the same eigenvalue μ, so by the same argument as above there is $\beta \in K$ such that

$$(6) \qquad \zeta(e_\mu) = \beta e_\mu.$$

So,

$$\zeta^{i+1}(e_\mu) = \zeta^i(\beta)\zeta^i(e_\mu)$$

and taking into account that $\zeta^{k\ell} = 1$ we get

$$\prod_{i=0}^{k\ell-1} \zeta^i(\beta) = 1.$$

Again we analyse β as a Laurent series and represent it in the form

$$\beta = b_0(\mu^{\frac{1}{k}}) \cdot \nu^{\frac{d}{k}} \cdot (1 + b_1(\mu^{\frac{1}{k}})\nu^{\frac{1}{k}} + b_2(\mu^{\frac{1}{k}})\nu^{\frac{2}{k}} + \cdots)$$

where $b_0(\mu^{\frac{1}{k}}), b_1(\mu^{\frac{1}{k}}), b_2(\mu^{\frac{1}{k}}), \ldots$ are Laurent series of $\mu^{\frac{1}{k}}$ and d is an integer.

By an argument similar to the above using (7) we get

$$(7) \qquad \beta = b_0 \cdot (1 + b_1(\mu^{\frac{1}{k}})\nu^{\frac{1}{k}} + b_2(\mu^{\frac{1}{k}})\nu^{\frac{2}{k}} + \cdots).$$

for some $b_0 \in F$.

Finally we use the fact that ξ and ζ commute. Applying ζ to (3) we get

$$\mathbf{V}\zeta(e_\mu) = \zeta(\alpha)\zeta\xi(e_\mu) = \zeta(\alpha)\xi\zeta(e_\mu) = \xi(\beta)\zeta(\alpha)\xi(e_\mu).$$

On the other hand

$$\mathbf{V}\zeta(e_\mu) = \beta\mathbf{V}e_\mu = \beta\alpha\xi(e_\mu).$$

That is,

$$\frac{\alpha}{\zeta(\alpha)} = \frac{\xi(\beta)}{\beta}.$$

Substituting (5) and (7) and dividing on both sides we get the equality

$$\epsilon^{-1}(1 + a_1'(\nu^{\frac{1}{k}})\mu^{\frac{1}{k}} + a_2'(\nu^{\frac{1}{k}})\mu^{\frac{2}{k}} + \cdots) = 1 + b_1'(\mu^{\frac{1}{k}})\nu^{\frac{1}{k}} + b_2'(\mu^{\frac{1}{k}})\nu^{\frac{2}{k}} + \cdots$$

Comparing the constant terms on both sides we get the contradiction. This proves the proposition in the case $n = 2$.

To end the proof we just notice that the structure $\tilde{\mathbf{V}}(T_\epsilon^2)$ is definable in any of the other $\tilde{\mathbf{V}}(T_\epsilon^n)$, maybe with a different root of unity. This follows from the fact that the A-modules in all cases have similar description.□

Corollary *The structure $\tilde{\mathbf{V}}(U_\epsilon(\mathfrak{sl}_2))$ (Example 2.3.3) is not definable in an algebraically closed field.*

Indeed, consider

$$\mathbf{V}_0 = \{(a, b, c, d) \in \mathbf{V} : b \neq 0, \ c = 0\} \text{ and } \tilde{\mathbf{V}}_0 = \pi^{-1}(\mathbf{V}_0)$$

with the relations induced from $\tilde{\mathbf{V}}$.

Set $\mathbf{U} := K$, $\mathbf{V} = F$ and consider the reduct of the structure $\tilde{\mathbf{V}}_0$ which ignores the operators E and C. This structure is isomorphic to $\tilde{\mathbf{V}}(T_{\epsilon^2}^2)$ and is definable in $\mathbf{V}(U_\epsilon(\mathfrak{sl}_2))$, so the latter is not definable in an algebraically closed field. □

Remark Note that T_ϵ^2 here does not have any immediate connection to the non-classical Zariski curve T_N in [Z2]. So the Proposition does not "explain" the earlier examples, though an attentive reader could spot similarities in the proof of the Proposition and that of the non-algebraicity of T_N. A possible connection remains an open question.

3 Definable sets

3.1 Given variables $v_{1,1}, \ldots, v_{1,r_1}, \ldots, v_{s,1} \ldots, v_{s,r_s}$ of the sort $\tilde{V}$, variables $m_1, \ldots, m_s$ of the sort V, and variables $x = \{x_1, \ldots, x_p\}$ of the sort F, denote $A_0(e, m, t)$ the formula

$$\bigwedge_{i \leq s, \ j \leq N} E(e_i, m_i) \ \& \ P^A(\{t_{ikn\ell}\}_{k \leq d, \ \ell, n \leq N}; m_i) = 0$$

$$\& \ \bigwedge_{k \leq d, j \leq N, i \leq s} \mathbf{U}_k e_i(j) = \sum_{\ell \leq N} t_{ikj\ell} e_i(\ell).$$

Denote $A(e, m, t, z, v)$ the formula

$$A_0(e, m, t) \ \& \ \bigwedge_{i \leq s; \ j \leq r_i} v_{ij} = \sum_{\ell \leq N} z_{ij\ell} e_i(\ell).$$

The formula of the form

$$\exists e_1, \ldots, e_s \exists m_1, \ldots, m_s, \ \exists \{t_{ikjl} : k \leq d, \ i \leq s, \ j, \ell \leq N\} \subseteq \mathrm{F},$$

$$\exists \{z_{ijl} : i \leq s, \ j \leq r_i, \ \ell \leq N\} \subseteq \mathrm{F} \ \ A(e, m, t, z, v) \ \& \ R(m, t, x, z)$$

where R is a Boolean combination of Zariski closed predicates in the algebraic variety $\mathrm{V}^s \times \mathrm{F}^q$ over C, $q = |t| + |x| + |z|$ (constructible predicate over C) will be called **a core $\exists$-formula with kernel** $R(m, t, x, z)$ **over** C. The enumeration of variables v_{ij} will be referred to as the **partitioning enumeration**.

We also refer to this formula as $\exists eR$.

Comments (i) A core formula is determined by its kernel once the partition of variables (by enumeration) is fixed. The partition sets that $\pi(e_i(j)) = \pi(e_i(k))$, for every i, j, k, and fixes the components of the subformula $A(e, m, t, z, v)$.

(ii) The relation $A_0(e, m, t)$ defines the functions

$$e \mapsto (m, t),$$

that is given a canonical basis $\{e_i(1), \ldots, e_i(N)\}$ in M_{m_i} we can uniquely determine m_i and $t_{ikj\ell}$.

For the same reason $A(e, m, t, z, v)$ defines the functions

$$(e, v) \mapsto (m, t, z).$$

3.2 Lemma *Let* $a = \langle a_{1,1}, \ldots, a_{1,r_1}, \ldots, a_{s,1} \ldots, a_{s,r_s} \rangle \in \tilde{V} \times \cdots \times \tilde{V}$, $b = \langle b_1, \ldots, b_n \rangle \in \mathrm{F}^n$. *The complete type* $\mathrm{tp}(a, b)$ *of the tuple over* C *is determined by its subtype* $\mathrm{ctp}(a, b)$ *over* C *consisting of core $\exists$-formulas.*

Proof. We are going to prove that, given a', b' satisfying the same core type $\mathrm{ctp}(a, b)$ there is an automorphism of any $\aleph_0$-saturated model, $\alpha :$ $(a, b) \mapsto (a', b')$.

We assume that the enumeration of variables has been arranged so that $\pi(a_{ij}) = \pi(a_{kn})$ if and only if $i = k$. Denote $m_i = \pi(a_{ij})$.

Let e_i be bases of modules $\pi^{-1}(m_i)$, $i = 1, \ldots, s$, $j = 1, \ldots, N$, such that $\models A_0(e, m, t)$ for some $t = \{t_{ikj\ell}\}$ (see the notation in 3.1 and the assumption 2.1.3, in particular $e_i \in E_{m_i}$. By the assumption the corresponding systems span M_{m_i}, so there exist $c_{ij\ell}$ such that

$$\bigwedge_{i \leq s; \ j \leq r_i} a_{ij} = \sum_{\ell \leq N} c_{ij\ell} e_i(\ell),$$

and let $p = \{P_i : i \in \mathbb{N}\}$ be the complete algebraic type of (m, t, b, c).

The type $\mathrm{ctp}(a, b)$ contains core formulas with kernels P_i, $i = 1, 2, \ldots$ By assumptions and saturatedness we can find e' m', t' and c' satisfying the corresponding relations for (a', b'). In particular, the algebraic types of (m, t, b, c) and (m', t', b', c') over C coincide and $e'_i \in E_{m'_i}$. It follows that there is an automorphism $\alpha : F \to F$ over C such that $\alpha : (m, t, b, c) \mapsto (m', t', b', c')$.

Extend α to $\pi^{-1}(m_1) \cup \ldots \cup \pi^{-1}(m_s)$ by setting

$$(8) \qquad \alpha(\sum_j z_j e_i(j)) = \sum_j \alpha(z_j) e'_i(j)$$

for any $z_1, \ldots, z_N \in F$ and $i \in \{1, \ldots, s\}$. In particular $\alpha(a_{ij}) = a'_{ij}$ and, since $\alpha(\Gamma_{m_i}) = \Gamma_{m'_i}$, also $\alpha(E_{m_i}) = E_{m'_i}$.

Now, for each $m \in V \setminus \{m_1, \ldots, m_s\}$ we construct the extension of α, $\alpha_m^+ : \pi^{-1}(m) \to \pi^{-1}(m')$, for $m' = \alpha(m)$, as in 2.4. Use 2.1.2.1 to choose t_{ijk} satisfying P^A for m and find bases $e \in E_m$ and $e' \in E_{m'}$ with the $\mathbf{U}_i$'s represented by the matrices $\{t_{ijk} : k, j = 1, \ldots, N\}$ in $\pi^{-1}(m)$ and by $\{\alpha(t_{ijk}) : k, j = 1, \ldots, N\}$ in $\pi^{-1}(m')$. It follows that the map

$$\alpha_m^+ : \sum z_j e(j) \mapsto \sum \alpha(z_j) e'(j), \quad z_1, \ldots, z_N \in F$$

is an isomorphism of the A-modules

$$\alpha_m^+ : \pi^{-1}(m) \to \pi^{-1}(m').$$

Hence, the union

$$\alpha^+ = \bigcup_{m \in V} \alpha_m^+$$

is an automorphism of $\tilde{V}$. $\qquad\qquad\qquad\qquad\qquad\qquad\qquad\Box$

By the compactness theorem we immediately get from the lemma

Corollary *Every formula in* $\tilde{V}$ *with parameters in* $C \subseteq F$ *is equivalent to the disjunction of a finite collection of core formulas.*

3.3 We consider now a more general form of core formulas with parameters in both sorts $\tilde{V}$ and F.

The **general core formula** of variables $x = \{x_1, \ldots, x_p\}$ and $v = \{v_{ij} : i \leq s + u, j \leq r_i\}$ and parameters $C \subseteq F$, $\hat{e} \subseteq \tilde{V}$ will be of the form

$$\exists e_1, \ldots e_s \exists m_1, \ldots, m_s \ \exists \{t_{ikjl} : k \leq d, \ i \leq s, \ j,\ell \leq N\}$$
$$\exists \{z_{ijl} : i \leq s, \ j \leq r_i, \ \ell \leq N\} \ \exists \{y_{ijl} : i \leq u, \ j \leq r_{s+i}, \ \ell \leq N\}$$
$$A(e,m,t,z,v) \ \& \ B(\hat{e},y,v) \ \& \ R(m,t,x,y,z)$$

where $\hat{e} = (\hat{e}_{s+1}, \ldots, \hat{e}_{s+u})$ are names of fixed canonical bases of some modules $M_{\hat{m}_{s+1}}, \ldots, M_{\hat{m}_{s+u}}$ in $\tilde{V}$, m, t, z and A are the same as in 3.1, y is $\{y_{ijl} : i \leq u, \ j \leq r_{s+i}, \ \ell \leq N\}$, R is a Boolean combination of Zariski closed predicates in variables m, t, x, y, z and $B(\hat{e}, y, v)$ is the formula

$$\bigwedge_{i \leq u; \ j \leq r_{s+i}} v_{s+i,j} = \sum_{\ell \leq N} y_{ijl} \cdot \hat{e}_i(\ell).$$

As before we call R appearing in the general core formula the kernel of the formula and write $\exists e\, R$ for the general core formula with kernel R.

Remark Given the set in $\tilde{V}$ defined by a general core formula $\exists e\, R$ the values of parameters $\hat{m}_{s+1}, \ldots, \hat{m}_{s+u}$ are determined uniquely as $\pi(v_{ij})$ with $i = s + 1, \ldots, s + u$, $j \leq r_i$. Hence $\hat{e}_i$, $s < i \leq s + u$, are determined up to a linear transformation inside $M_{\hat{m}_i}$. So, choosing a different $\hat{e}' = \gamma\hat{e}$ one can still define the same set by using the formula $\exists e\, R'$ where $R'(m,t,x,y',z)$ is obtained from $R(m,t,x,y,z)$ by substituting $y'\gamma$ in place of y. In other words,
we may assume that two equivalent general core formulas have the same parameters $\hat{e}$.

Proposition *Every formula with parameters in* $\tilde{V}$ *is equivalent to the disjunction of a finite collection of general core formulas.*

Proof. By 3.1 it is enough to prove that there is such a form for the formula obtained from a core formula $\exists e\, R$ in variables $x = \{x_1, \ldots, x_p\}$

and $v = \{v_{ij} : i \leq s + u, j \leq r_i\}$ with parameters $C \subseteq F$,

(9) $\quad \exists e_1, \ldots e_s, \ldots e_{s+u} \exists m_1, \ldots, m_s, \ldots, m_{s+u}$

$$\exists \{t_{ikjl} : i \leq s + u, \ k \leq d, \ j, \ell \leq N\}$$
$$\exists \{z_{ijl} : i \leq s + u, \ j \leq r_i, \ \ell \leq N\}$$
$$A(e, m, t, z, v) \ \& \ R(m, t, x, z)$$

by substituting

$$v_{s+1,1} := a_{s+1,1}, \quad \cdots \quad v_{s+1,q_{s+1}} := a_{s+1,q_{s+1}},$$
$$\vdots \qquad \qquad \ddots \qquad \qquad \vdots$$
$$v_{s+u,1} := a_{s+u,1}, \quad \cdots \quad v_{s+u,q_{s+u}} := a_{s+u,q_{s+u}}$$

some $a_{ij} \in \tilde{V}$ and $1 \leq q_i \leq r_i$, $i = s + 1, \ldots, s + u$.

Notice that once the substitution $v_{s+i,1} := a_{s+i,1}$ occurred the value of m_{s+i} will be fixed as $m_{s+i} = \pi(a_{s+i})$. Denote this $\hat{m}_{s+i}$. Correspondingly there are finitely many possible values for $e_{s+i} \in E_{\hat{m}_{s+i}}$. Choosing any such canonical basis $\hat{e}_{s+i}$, the corresponding $t_{s+i,kjl}$ described in $A(e, m, t, z, v)$ will be fixed, denote the corresponding elements in F as $\hat{t}_{s+i,kjl}$. For the same reason we have the $z_{s+i,jl}$, for $j \leq q_{s+i}$, fixed as $\hat{z}_{s+i,jl}$ by $A(e, m, t, z, v)$.

So, $\exists e \, R^{v_I := a_I}$ is equivalent to

(10) $\quad \bigvee\limits_{\hat{e}_{s+1} \in E_{\hat{m}_{s+1}}, \ldots, \hat{e}_{s+u} \in E_{\hat{m}_{s+u}}} \exists e_1, \ldots e_s \exists m_1, \ldots, m_s$

$$\exists \{t_{ikjl} : i \leq s, \ k \leq d, \ j, \ell \leq N\}$$
$$\exists \{z_{ijl} : i \leq s, \ j \leq r_i, \ \ell \leq N\}$$
$$(A(e, m, t, z, v) \ \& \ R(m, t, x, z))^{v_I := a_I, m_I = \hat{m}_I, t_I = \hat{t}_I, z_I = \hat{z}_I}$$

Now we rename z_{ijl} with $s < i \leq s + u$ and $q_i < j \leq r_i$ as $y_{i-s, j-q_i, l}$. $R(m, t, x, z)^{v_I := a_I, m_I = \hat{m}_I, t_I = \hat{t}_I, z_I = \hat{z}_I}$ becomes then some constructible predicate in variables m, t, x, y, z and parameters C and $\hat{m}_I, \hat{t}_I, \hat{z}_I$.

We now want to reduce

$$A(e, m, t, z, v)^{v_I := a_I, m_I = \hat{m}_I, t_I = \hat{t}_I, z_I = \hat{z}_I}$$

to a suitable equivalent form. To this end we delete from the formula the conjuncts which are trivially true, namely $E(\hat{e}_i, \hat{m}_i)$ and the equalities of the form

$$P^A(\{\hat{t}_{ikn\ell}\}_{k \leq d, \; \ell,n \leq N}; m_i) = 0,$$

$$a_{ij} = \sum_l \hat{z}_{ijl}\hat{e}_i(l) \text{ and } \mathbf{U}_k\hat{e}_i(j) = \sum \hat{t}_{ikjl}\hat{e}_i(l)$$

for $i > s$. The only equations with indices $i > s$ remaining will have the form

$$v_{ij} = \sum_l y_{i-s,j,l}\hat{e}_i(l),$$

and the conjunction of all these will form our $B(\hat{e}, y, v)$ (we rename $\hat{e}_i$ as $\hat{e}_{i-s}$ in the final form). The remaining part of

$$A(e, m, t, z, v)^{v_I := a_I, m_I = \hat{m}_I, t_I = \hat{t}_I, z_I = \hat{z}_I}$$

will be exactly $A(e, m, t, z, v)$ where e, m, t, z, v are as in the definition of a general core formula. $\qquad \Box$

Remark We have also proved that the result $\exists e \, R^{v_I := a_I}$ of the substitution in a given core formula (9) with kernel $R(m, t, x, z)$ is equivalent to a disjunction of general core formulas each with the kernel

$$R(m, t, x, z)^{m_I = \hat{m}_I, t_I = \hat{t}_I, z_I = \hat{z}_I y_I},$$

where the substitution $z_I = \hat{z}_I y_I$ replaces z_{ijl} with $s < i \leq s+u$ by $\hat{z}_{ijl}$, for $j \leq q_i$, or by $y_{i-s,j-q_i,l}$, for $q_i < j \leq r_i$.

Corollary *Every formula in $\tilde{V}$ with parameters in $\tilde{V}$ is equivalent to the disjunction of a finite collection of general core formulas.*

3.4 We assume from now on the stronger assumption 2.1.4 and prove in this section that the core formulas in Corollaries of 3.2 and 3.3 can have a form more suitable for technical purposes.

Let Γ be the group in 2.1.4. Given a Zariski closed predicate $R :=$ $R(m, t, x, y, z)$ with m ranging in V^s and t, x, y, z tuples in F in accordance with the notation in 3.2, we define, for $\gamma = (\gamma_1, \ldots, \gamma_s) \in \Gamma^s$, the predicate $R^\gamma(m, t, x, y, z)$ of the same variables.

First we consider the case when R is irreducible. Set

$$V_R := \{m \in V^s : \exists t, x, y, z \, R(m, t, x, y, z)\},$$

the projection of R on V^s. Let $V_{R,\gamma}$ be the open subset of V_R equal to the domain of definition of the map (in variables m)

$$g(m, \gamma) := \langle g(m_1, \gamma_1), \ldots, g(m_s, \gamma_s) \rangle.$$

Set

$$\Gamma_R^s := \{\gamma \in \Gamma^s : V_{R,\gamma} \neq \emptyset\}.$$

This is a subgroup of Γ^s since $V_{R,\gamma}$ is a dense open subset when non-empty.

In case $\gamma \in \Gamma_R^s$ define R^γ to be the Zariski closure of the set

$$\{\langle m, t, x, y, z \rangle : \exists t', z' \; m \in V_{R,\gamma} \; \& \; t' = g(m,\gamma)^{-1} \cdot t \cdot g(m,\gamma) \; \& $$
$$\& \; z' = z \cdot g(m,\gamma) \; \& \; R(m, t', x, y, z')\}.$$

Remember that t is a collection of $N \times N$ matrices and z is a list of N-tuples, coordinates of elements of M_{m_i} in the corresponding canonical bases e. So in the definition above $z \cdot g(m,\gamma)$ corresponds to the coordinates of the same elements in bases $e' = g(m,\gamma) \cdot e$, and $g(m,\gamma)^{-1} \cdot t \cdot g(m,\gamma)$ is the result of the corresponding transformation of the matrices in t.

With an obvious abuse of notation we will often write $R(m, t^\gamma, x, y, z\gamma)$ for $R^\gamma(m, t, x, y, z)$, when $\gamma \in \Gamma_R^s$.

For a general Zariski closed R we first represent $R = R_1 \cup \cdots \cup R_k$ as the union of its irreducible components and then set

$$R^\gamma = R_1^\gamma \cup \cdots \cup R_k^\gamma,$$
$$\Gamma_R^s = \Gamma_{R_1}^s \cap \cdots \cap \Gamma_{R_k}^s.$$

Remarks (i) Obviously, $R^{\mathrm{id}} = R$, for id the unit element of Γ^s;

(ii) For $\gamma \in \Gamma_R^s$ the set $V_{R^\gamma} \cap V_R$ is a dense open subset of V_R equal to $V_R \cap V_{R,\gamma}$;

(iii) If R does not depend on t and z then $R^\gamma = R$ for every $\gamma \in \Gamma_R^s$;

(iv) Suppose $P \subseteq R$ is a Zariski closed relation. Then $P^\gamma \subseteq R^\gamma$ for every $\gamma \in \Gamma^s$ and $\Gamma_P^s \subseteq \Gamma_R^s$;

(v) Let $R^* = \bigcup_{\gamma \in \Gamma^s} R^\gamma$. Then

$$R^{*\gamma} = R^*$$

for every $\gamma \in \Gamma_R^s$.

We will say that R is Γ-invariant if $R^\gamma = R$ for every $\gamma \in \Gamma_R^s$.

3.5 Lemma 1. *We may assume that the kernels in core formulas in Corollary 3.2 are of the form $R(m, t, x, z) \; \& \; \neg S(m, t, x, z)$, where R, S are given by systems of equations and S is Γ-invariant.*

Proof. We go back to the proof of Lemma 3.2 and consider the deductively complete type p in the language of fields, $p = \{P_i\}$, with conjunctions of P_i appearing in the end as the kernels of core formulas. We may assume that each P_i is either a system of equations $R(m, t, x, z)$ in variables m, t, x, z or the negation $\neg S(m, t, x, z)$ of the system of equations S. We are going to prove that, for a given $\neg S \in p$ there is a system of equations $R \in p$, and a negation $\neg \bar{S} \in p$, with $\neg \bar{S}^\gamma = \neg \bar{S}$, for all $\gamma \in \Gamma^s_S$, such that $R \,\&\, \neg \bar{S} \models \neg S$. This implies that we can replace all P_i by $R \,\&\, \neg \bar{S}$ and would prove the Lemma.

If $\bigwedge_{\gamma \in \Gamma^s} \neg S^\gamma \in p$ then this formula, being equivalent to a negation $\neg \bar{S}$ of a system of equations, is invariant under Γ^s and satisfies $\neg \bar{S} \models \neg S$.

So we assume the opposite, $\bigwedge_{\gamma \in \Gamma^s} \neg S^\gamma \notin p$. Hence, for some nonempty proper subset $\Delta \subsetneq \Gamma^s$, with $1 \in \Delta$,

$$\neg T = \bigwedge_{\gamma \in \Delta} \neg S^\gamma \in p.$$

We assume Δ to be maximal with this property.

Obviously

$$\bigvee_{\gamma \in \Gamma^s} \neg T^\gamma \in p.$$

Denote

$$\mathrm{Stab}(\Delta) = \{\gamma \in \Gamma^s : \gamma \Delta = \Delta\}.$$

Since by maximality for any $\gamma \in \Gamma^s \setminus \mathrm{Stab}(\Delta)$ we have $\neg T^\gamma \,\&\, \neg T \notin p$, necessarily $T^\gamma \in p$ and so

$$\bigwedge_{\gamma \in \Gamma^s \setminus \mathrm{Stab}\Delta} T^\gamma \in p.$$

But

$$\bigvee_{\gamma \in \Gamma^s} \neg T^\gamma \,\&\, \bigwedge_{\gamma \in \Gamma^s \setminus \mathrm{Stab}\Delta} T^\gamma \models \bigvee_{\gamma \in \mathrm{Stab}\Delta} \neg T^\gamma$$

The latter is equivalent to $\neg T$, and $\neg T \models \neg S$. So we can take $\bigwedge_{\gamma \in \Gamma^s \setminus \mathrm{Stab}\Delta} T^\gamma$ for R and $\bigvee_{\gamma \in \Gamma^s} \neg T^\gamma$ for $\neg \bar{S}$. □

Lemma 2. *We may assume that in Corollary 3.2 the kernels of core formulas are of the form $R \& \neg S$ with R, S given by systems of polynomial equations and both are Γ-invariant.*

Proof. We use Lemma 1. Observe first that in general

Claim $\exists e, t, z \; A(e, m, t, v, z) \;\&\; P(m, t, z, x)$ is equivalent to $\exists e, t, z \; A(e, m, t, v, z) \;\&\; P^*(m, t, z, x)$, where

$$P^* = \bigvee_{\gamma \in \Gamma^s} P^\gamma.$$

Indeed, $\exists e \; P$ obviously implies $\exists e \; P^*$. To see the converse note that, given v and x, if for some e and $\gamma \in \Gamma^s$ we have

$$\models A(e, m, t, v, z) \;\&\; P(m, t^\gamma, z\gamma, x)$$

then, letting $e' = \gamma e$, we will have

$$\models A(e', m, t^\gamma, v, z\gamma)$$

and so,

$$\models A(e', m, t', v, z') \;\&\; P(m, t', z', x)$$

for $t' = t^\gamma$ and $z' = z\gamma$.

Applying the Claim to our $\exists e \; R \;\&\; \neg S$ we will get the equivalent formula $\exists e \; R^* \;\&\; \neg S$ since $S^* = S$. $\qquad\square$

Combining Lemma 2 with the Remark in 3.3 we get

Corollary *We may assume that in Corollary 3.3 the kernels of general core formulas are of the form $R \& \neg S$ with R, S given by systems of polynomial equations and both are Γ-invariant.*

Now we discuss general core formulas with Γ-invariant kernels.

Lemma 3. *Assuming that R_2 is Γ-invariant we have*
  (i) $\exists e \; (R_1 \;\&\; R_2) \equiv (\exists e \; R_1) \;\&\; (\exists e \; R_2)$;
  (ii) $\exists e \; \neg R_2 \equiv \neg \exists e \; R_2$.

Proof. (i) The left-hand-side obviously implies the formula on the right. Assume for converse that the right-hand-side is true. That is for given v, x and y there is e and e' such that

$$\models A(e, m, t, v, z) \;\&\; B(\hat{e}, y, v) \;\&\; R_1(m, t, x, y, z)$$

and

$$\models A(e', m, t', v, z') \;\&\; B(\hat{e}, y, v) \;\&\; R_2(m, t', x, y, z').$$

Since $e' = \gamma e$ for some $\gamma \in \Gamma^s$, we have

$$\models A(e, m, t, v, z) \;\&\; B(\hat{e}, y, v) \;\&\; R_2(m, t^\gamma, x, y, z\gamma).$$

But R_2 is Γ-invariant, hence we get

$$\models A(e,m,t,v,z) \ \& \ B(\hat{e},y,v) \ \& \ R_1(m,t,z,x) \ \& \ R_2(m,t,z,x),$$

as required.

(ii) We need only prove the implication from left to right. Assume that $\models A(e,m,t,v,z) \ \& \ B(\hat{e},y,v) \ \& \ \neg R_2(m,t,x,y,z)$. We need to check that for no e' it is possible $\models A(e',m,t',v,z') \ \& \ B(\hat{e},y,v) \ \& \ R_2(m,t',x,y,z')$. Indeed, as above by Γ-invariance the latter is equivalent to

$$\models A(e,m,t,v,z) \ \& \ B(\hat{e},y,v) \ \& \ R_2(m,t,x,y,z),$$

which would contradict the former. $\qquad\square$

Lemma 4 *Suppose* $\exists e R_1 \equiv \exists e R_2$, *both sides are general core formulas with the same partition of v-variables, u and $\hat{e}_1,\ldots,\hat{e}_u$ are same in both formulas, and R_1, R_2 are Γ-invariant. Then $R_1 \equiv R_2$.*

Proof. By Lemma 3

$$\exists e \ (R_1 \ \& \ \neg R_2) \equiv \exists e \ R_1 \ \& \ \exists e \ \neg R_2 \equiv \exists e \ R_1 \ \& \ \neg\exists e \ R_2$$

and so $R_1 \ \& \ \neg R_2$ is inconsistent, that is $\models R_1 \to R_2$. By symmetry $R_1 \equiv R_2$. $\qquad\square$

4 Zariski geometry

In this section we introduce on $\tilde{V}$ and its finite cartesian powers a topology which is naturally coming from the coordinate algebra A. To see that this is a Noetherian topology satisfying also the definition of a presmooth Zariski geometry (see [Z1] for this) we have to have more than just a quantifier elimination to existential formulas. To this end we carry out a more detailed analysis of general core formulas and their behavior under Boolean operations and projections.

4.1 We introduce the A-topology declaring **basic closed subsets** of $\tilde{V}^n \times F^p$ the subsets defined by general core formulas $\exists e R$ with kernels R given by Γ-invariant systems of polynomial equations with coefficients in F.

We also assume that R contains the equation $P^A(t,m) = 0$ (see 2.1.3), which is in the A-part of $\exists e R$.

We often denote $\hat{R}$ the closed set defined by the formula $\exists e R$.

The closed subsets of the topology are given by applying finite unions and arbitrary intersections to basic closed subsets.

Claim 1 Intersection of an infinite family of basic closed subsets of a Cartesian power of $\tilde{V}$ is equal to the intersection of its finite subfamily.

Indeed, since for a given set of variables there are finite number of ways to partition (enumerate) the variables as $\{v_{ij} : i \leq s, \; j \leq r_i\}$, we may assume that all core formulas defining the members of the family have the same partition of variables. Now by Lemma 2 and Lemma 3(i) of 3.5 the intersection of sets defined by $\exists e \, R_\alpha$, $\alpha \in I$, reduces to the intersection of Zariski closed sets defined by R_α, $\alpha \in I$, which obviously stabilises.

Using Koenig's Lemma we get

Claim 2 The A-topology is Noetherian.

Since for $s + u = 0$ a general core formula $\exists e \, R$ takes the form $R(x, y)$ the following is obvious.

Claim 3 The restriction of the A-topology to the sort F is the classical Zariski topology.

Claim 4 Any definable subset of a Cartesian power of $\tilde{V}$ is equal to the Boolean combination of closed subsets, that is, is **constructible**.

Indeed, by the Corollary in 3.3 it is sufficient to prove the statement for subsets defined by general core formulas. The Corollary in 3.5 together with Lemmas 3(ii) provide the rest.

We will also need a more detailed presentation of sets obtained by projecting closed sets onto coordinate subspaces, as well as fibers of these projections.

Lemma 1 *Let $\exists e \, R$ be the general core formula in the notation of 3.3 and $a \in \{1, \ldots, s + u\}$, $b \in \{1 \ldots, r_a\}$ some indices. Then the formula $\exists v_{ab} \exists e \, R$ is equivalent to a general core formula $\exists e' \, R'$ with the kernel R' equivalent to one of the following*

(i) $\exists y_{a-s,b1} \ldots y_{a-s,bN} R,$

(ii) $\exists z_{ab1} \ldots z_{abN} R,$ *or*

(iii) $\exists m_a \exists \{t_{akjl} : k \leq d, \; j, l \leq N\} \exists z_{ab1} \ldots z_{abN} R.$

Proof. (i) Suppose $s < a \leq s + u$. Since v_{ab} does not occur in

$A(e, m, t, z, v)$ and $R(m, t, x, y, z)$, the formula $\exists v_{ab} \exists e\, R$ is equivalent to

$$\exists e_1, \ldots e_s \exists \ldots$$
$$A(e, m, t, z, v) \;\&\; (\exists v_{ab} B(\hat{e}, y, v)) \;\&\; R(m, t, x, y, z),$$

with the quantifier prefix the same as in $\exists e\, R$. Looking at the form of $B(\hat{e}, y, v)$ one sees that $(\exists v_{ab} B(\hat{e}, y, v))$ is equivalent to some $B(\hat{e}, y', v')$ with $y' = y \setminus \{y_{a-s,b1} \ldots y_{a-s,bN}\}$ and $v' = v \setminus \{v_{ab}\}$. Now we can equivalently rewrite the formula as

$$\exists e_1, \ldots e_s \exists \ldots$$
$$A(e, m, t, z, v) \;\&\; B(\hat{e}, y', v') \;\&\; \exists y_{a-s,b1} \ldots y_{a-s,bN} R(m, t, x, y, z),$$

where $\exists y_{a-s,b1} \ldots y_{a-s,bN}$ moved from the quantifier prefix to the end of the formula. Of course, by quantifier elimination in algebraically closed fields, $\exists y_{a-s,b1} \ldots y_{a-s,bN} R$ is a constructible predicate.

(ii) and (iii). Suppose $a \leq s$. Then the formula $\exists v_{ab} \exists e\, R$ is equivalent to

$$\exists e_1, \ldots e_s \exists \ldots$$
$$(\exists v_{ab} A(e, m, t, z, v)) \;\&\; B(\hat{e}, y, v) \;\&\; R(m, t, x, y, z).$$

One can obviously eliminate the quantifier from $\exists v_{ab} A(e, m, t, z, v)$ by replacing v_{ab} everywhere in $A(e, m, t, z, v)$ by the term $\sum_{\ell \leq N} z_{ab\ell} e_a(\ell)$. This makes the conjunct $v_{ab} = \sum_{\ell \leq N} z_{ab\ell} e_a(\ell)$ in $A(e, m, t, z, v)$ a tautology and after removing it we get a formula without v_{ab} which in the case $r_a \neq 1$ is again of the form $A(e, m, t, z', v')$, where $v' = v \setminus \{v_{ab}\}$, $z' = z \setminus \{z_{ab1} \ldots z_{abN}\}$. This gives us a general core formula of the form (ii) for $\exists v_{ab} \exists e\, R$.

In case $r_a = 1$ the conjunct $v_{ab} = \sum_{\ell \leq N} z_{ab\ell} e_a(\ell)$ is the only one that uses the variables e_a. By eliminating this conjunct we made other subformulas containing e_a redundant. So we eliminate

$$\bigwedge_{j \leq N} E(e_a, m_a) \;\&\; P^A(\{t_{akn\ell}\}_{k \leq d,\; \ell, n \leq N}; m_a) = 0 \;\&$$

$$\&\; \bigwedge_{k \leq d, j \leq N} \mathbf{U}_k e_a(j) = \sum_{\ell \leq N} t_{akj\ell} e_a(\ell)$$

from $A(e, m, t, z, v)$ as well (notice that by our assumptions, $P^A(\{t_{akn\ell}\}_{k \leq d,\; \ell, n \leq N}; m_a) = 0$ is also copied in R). The resulting formula is again of the form $A(e', m', t', z', v')$, with $v' = v \setminus \{v_{ab}\}$,

$m' = m \setminus \{m_a\}$, $z' = z \setminus \{z_{ab1} \ldots z_{abN}\}$ and $t' = t \setminus \{t_{akjl} : k \leq d, \ j, \ell \leq N\}$. Now we may push the quantifiers $\exists m_a$, $\exists \{z_{ab1} \ldots z_{abN}\}$ and $\exists \{t_{akj\ell} : k \leq d, \ j, \ell \leq N\}$ to the end of the formula and get the general core formula of the form (iii). $\qquad \square$

Lemma 2 *Suppose $\exists e \, R$ does not contain x, free variables of the sort F. Let $\exists e' R'$ be the general core formula equivalent to $\exists v_{ab} \exists e \, R$ as in the above Lemma. More precisely $R' = \exists u R(u, w)$, some u depending on the case. Let $v' = v \setminus \{v_a b\}$ and $\hat{v}'$ is a tuple in $\tilde{V}$ satisfying $\exists e' R'$. Then $(\exists e R)^{v' := \hat{v}'}$ has kernel of the form $R(u, \hat{w})$, for some $\hat{w}$.*

Proof. Follow the analysis in the proof of Proposition in 3.3. In case (i) the substitution

$$v' := \hat{v}'$$

fixes the whole of e, m, t, z and $y \setminus \{y_{a-s,b,1} \ldots, y_{a-s,b,N}\}$, so the kernel is $R^{m=\hat{m}, t=\hat{t}, z=\hat{z}, y'=\hat{y}'}$. In other words, in this case we satisfied the requirement of the Lemma with $u = (y_{a-s,b,1} \ldots, y_{a-s,b,N})$ and $w = (m, t, z, y')$.

In case (ii) again $\hat{v}'$ fixes the whole of e, m, t, y and $z \setminus \{z_{ab1} \ldots z_{abN}\}$. In case (iii) $\hat{v}'$ fixes $e \setminus e_a$ $m \setminus m_a$, $t \setminus t_a$, y and $z \setminus \{z_{ab1} \ldots z_{abN}\}$. $\qquad \square$

4.2 For further purposes we need a more detailed understanding of intersections of closed sets.

Let $\{v_1, \ldots, v_n\}$ be a linear reenumeration of variables $\{v_{ij} : i \leq s, j \leq r_i\}$, $n = r_1 + \cdots + r_s$, of sort $\tilde{V}$ in a general core formula $\exists e R$ (the variables $\{v_{ij} : s < i \leq s + u, j \leq r_i\}$ remain unchanged). We write $k \sim_R k'$ for $k, k' \in \{1, \ldots, n\}$ if k and k' correspond to some (i, j) and (i, j') in the old enumeration. This is an equivalence relation. We denote I_R the subset $\{1, \ldots, n\}$ corresponding to $\{(i, 1) : i = 1, \ldots, s\}$ in the partitioning enumeration, the set of representatives of $\sim_R$-classes.

We use the abbreviation e_i for $\{e_i(1), \ldots, e_i(N)\}$, t_i for $\{t_{ikj\ell} : k \leq d; \ j, l \leq N\}$ and z_i for $\{z_{ij\ell} : j \leq r_i; \ \ell \leq N\}$, $i = 1, \ldots, n$, along with other obvious abbreviations. In particular, $\mathbf{U} e_i = t_i e_i$ stands for

$$\bigwedge_{k \leq d, j \leq N, i \leq s} \mathbf{U}_k e_i(j) = \sum_{\ell \leq N} t_{ikj\ell} e_i(\ell).$$

We rewrite equivalently the general core formula $\exists e\, R$ of 3.3 as $\tilde{R}(v,x)$:

$$\exists e_1, \ldots e_n\, \exists t_1, \ldots, t_n,\, \exists z_1, \ldots, z_n\, y\, \exists m_1, \ldots m_n$$

$$\bigwedge_{i \le n} E(e_i, m_i)\ \&\ \mathbf{U}e_i = t_i e_i\ \&\ v_i = z_i e_i\ \&\ \bigwedge_{i \sim_R j} e_i = e_j\ \&\ B(\hat{e}, y, v)$$

$$\&\ R(m, t, z, x, y)\ \&\ \bigwedge_{i \sim_R j} m_i = m_j\ \&\ t_i = t_j.$$

This is not a core formula because of the component $\bigwedge_{i \sim_R j} e_i = e_j$.

Remark 1 In R only m_i, t_i with $i \in I_R$ as well as z_i, $i \le n$, y and x occur explicitly. Let $d_R = \dim R$, the dimension of the variety defined by R in the space given by these variables. Obviously we may assume that R depends on all variables m_i, t_i, z_i, $i \le n$, y and x. Then in the bigger ambient space we still have

$$d_R = \dim[R\ \&\ \bigwedge_{i \sim j} m_i = m_j\ \&\ t_i = t_j].$$

Let $\exists e\, S$ be another general core formula of the same variables with possibly different partitioning enumeration $\{v_{ij} : i \le s' + u, j \le r'_i\}$. We assume that variables $\{v_{ij} : s' < i \le s' + u, j \le r'_i\}$ and parameters $\hat{e}_{s'+1}, \ldots, \hat{e}_{s'+u}$ are the same in both $\exists e\, R$ and $\exists e\, S$.

We re-enumerate the variables $\{v_{ij} : i \le s', j \le r'_i\}$ linearly as $v_1, \ldots, v_n$. We have the corresponding equivalence relation $\sim_S$ on $\{1, \ldots, n\}$ and a set of its representatives I_S. As above $\exists e S$ can be equivalently rewritten as the formula $\tilde{S}(v, x)$:

$$\exists e'_1, \ldots e'_n\, \exists t'_1, \ldots, t'_n,\, \exists z'_1, \ldots, z'_n\, y\, \exists m_1, \ldots m_n$$

$$\bigwedge_{i \le n} E(e'_i, m_i)\ \&\ \mathbf{U}e'_i = t'_i e'_i\ \&\ v_i = z'_i e'_i\ \&\ \bigwedge_{i \sim_S j} e'_i = e'_j\ \&\ B(\hat{e}, y, v)$$

$$\&\ S(m', t', x, y, z')\ \&\ \bigwedge_{i \sim_S j} m_i = m_j\ \&\ t'_i = t'_j.$$

Lemma *The formula* $\tilde{R}(v, x)\ \&\ \tilde{S}(v, x)$ *is equivalent to the formula*

$\tilde{T}(v, x)$:

$$\exists e_1, \ldots e_n \; \exists t_1, \ldots, t_n, \; \exists z_1, \ldots, z_n \; y \; \exists m_1, \ldots m_n$$

$$\bigwedge_{i \le n} E(e_i, m_i) \; \& \; \mathbf{U}e_i = t_i e_i \; \& \; v_i = z_i e_i \; \& \bigwedge_{i \sim_{RS} j} e_i = e_j \; \& \; B(\hat{e}, y, v)$$

$$\& \; R(m, t, x, y, z) \; \& \; S(m, t, x, y, z) \; \& \bigwedge_{i \sim_{RS} j} m_i = m_j \; \& \; t_i = t_j,$$

where $\sim_{RS}$ is the transitive closure of the composition of the two equivalence relations $\sim_R$ and $\sim_S$.

Proof. The implication $\tilde{T}(v, x) \to \tilde{R}(v, x) \; \& \; \tilde{S}(v, x)$ is obvious.

For converse suppose $\tilde{R}(v, x) \; \& \; \tilde{S}(v, x)$ holds. This implies the existence of $e_i, e_i', t_i, t_i', z_i, z_i', m_i \; (i = 1, \ldots, n)$ and y which satisfy

$$\bigwedge_{i \le n} E(e_i, m_i) \; \& \; \mathbf{U}e_i = t_i e_i \; \& \; v_i = z_i e_i \; \& \bigwedge_{i \sim_R j} e_i = e_j \; \& \; B(\hat{e}, y, v)$$

$$\& \; R(m, t, x, y, z) \; \& \bigwedge_{i \sim_R j} m_i = m_j \; \& \; t_i = t_j$$

and

$$\bigwedge_{i \le n} E(e_i', m_i) \; \& \; \mathbf{U}e_i' = t_i' e_i' \; \& \; v_i = z_i' e_i' \; \& \bigwedge_{i \sim_S j} e_i' = e_j' \; \& \; B(\hat{e}, y, v)$$

$$\& \; S(m, t', x, y, z') \; \& \bigwedge_{i \sim_S j} m_i = m_j \; \& \; t_i' = t_j'.$$

m_i must be the same in both formulas since $m_i = \pi(v_i)$. It follows from the assumption 2.1.4 that for some $\gamma_i \in \Gamma_m$, $e_i = \gamma_i e_i'$. Since R is Γ-invariant we can exchange, for $i \in I_R$, e_i by $\gamma_i e_i$, t_i by $t_i^{\gamma_i}$ and z_i by $z_i \gamma_i$ without changing the validity of R and so may assume that $\gamma_i = 1$ and $e_i = e_i'$ for $i \in I_R$.

By symmetry we can reduce to the situation that also $e_j = e_j'$ for $j \in I_S$.

Claim We can choose $e_i = e_i'$ for all $i \le n$.

Proof. By induction on n. We have already $e_i = e_i'$ for all $i \in I_R \cup I_S$, so we assume that $I_R \cup I_S \subseteq \{1, \ldots, n-1\}$ and we can choose $e_i = e_i'$ for all $i \le n - 1$. We have $e_n = e_\ell = e_\ell'$ for some $\ell \in I_R$, $\ell \sim_R n$, and $e_n' = e_k' = e_k$, for some $k \in I_S$, $k \sim_S n$. From the equivalences it follows that $m_i = m_n = m_k$, i.e. the modules coincide. So, $e_n = \gamma e_n'$ for some $\gamma \in \Gamma_m$.

Let $J_k = \{i \le n : e_i = e_k\}$, $J_k' = \{i \le n : e_i' = e_k\}$. Note that $n \notin J_k$.

Apply the substitution $e_i \mapsto \gamma e_i$ and $e'_j \mapsto \gamma e'_j$ for all $i \in J_k$, $j \in J'_k$, leaving e_i and e'_j for $i \notin J_k$, $j \notin J'_k$ unchanged in

$$\bigwedge_{i \leq n} E(e_i, m_i) \ \& \ \mathbf{U} e_i = t_i e_i \ \& \ v_i = z_i e_i \ \& \bigwedge_{i \sim_R j} e_i = e_j$$

and

$$\bigwedge_{i \leq n} E(e'_i, m_i) \ \& \ \mathbf{U} e'_i = t'_i e'_i \ \& \ v_i = z'_i e'_i \ \& \bigwedge_{i \sim_S j} e'_i = e'_j.$$

This induces the corresponding transformation of t, t', z, z' which, by Γ-invariance does not change the validity of $R(m, t, x, y, z)$ and of $S(m, t', x, y, z')$.

This preserves all the existing equalities and gives $e'_n = e_n$ for the new value of e'_n. Claim proved.

This brings us to the situation with $e_i = e'_i$, $t_i = t'_i$ and $z_i = z'_i$ in the formulas above. Thus

$$\bigwedge_{i \leq n} E(e_i, m_i) \ \& \ \mathbf{U} e_i = t_i e_i \ \& \ v_i = z_i e_i \ \& \bigwedge_{i \sim_{RS} j} e_i = e_j$$

$$\& \ R(m, t, z, x) \ \& \ S(m, t, z, x) \ \& \bigwedge_{i \sim_{RS} j} m_i = m_j \ \& \ t_i = t_j,$$

hold. This proves the converse implication. □

Corollary 1 *The intersection of two basic closed sets given by general core formulas $\exists e\, R$ and $\exists e\, S$ with arbitrary partitioning enumerations and the same parameters $\hat{e}_{s+1}, \ldots, \hat{e}_{s+u}$ is a basic closed set given by a core formula $\exists e\, T$, with T equivalent to*

$$R(m, t, x, y, z) \ \& \ S(m, t, x, y, z) \ \& \bigwedge_{i \sim_{RS} j} m_i = m_j \ \& \ t_i = t_j.$$

Indeed, $\tilde{T}(v, x, y)$ can be transformed into a core formula by the following process.

Let $I_{RS} \subseteq \{1, \ldots, n\}$ be a set of representatives of $\sim_{RS}$-classes. Assuming $I_{RS} \subseteq \{1, \ldots, u\}$ re-enumerate $v_1, \ldots, v_n$ as $v_{ij} : i \leq u, j \leq r_i\}$, v_{i1} is the v_i in the linear enumeration and indices (ij) correspond to indices equivalent to i by $\sim_{RS}$.

Using the equalities $e_i = e_j$, $m_i = m_j$ and $t_i = t_j$, for $i \sim_{RS} j$, we delete e_j, m_j and t_j, with $j > u$, everywhere from $\tilde{T}$ along with the subformulas stating the equalities.

We re-enumerate z-variables in accordance with enumeration v_{ij}, so that now the formula $\tilde{T}$ says now that $v_{ij} = z_{ij}e_i$ for every $i \leq u$ and $j \leq r_i$.

After that $\tilde{T}$ transforms to

$$\exists e_1, \ldots, e_u \, \exists t_1, \ldots, t_u \, \exists m_1, \ldots, m_u \, \exists z_{11}, \ldots, z_{ur_u}$$
$$\bigwedge_{i \leq u} E(e_i, m_i) \ \& \ \mathbf{U}e_i = t_i e_i \ \& \ \bigwedge_{j \leq r_u} v_{ij} = z_{ij}e_i \ \& \ B(\hat{e}, y, v)$$
$$\& \ R'(m, t, x, y, z) \ \& \ S'(m, t, x, y, z),$$

where R' and S' obtained by substituting t_i and m_i instead of t_j and m_j, for $j \sim_{RS} i$, $j > u$. This is a general core formula.

Remark 2 By the Remark in 3.3 we can always assume that parameters $\hat{e}$ in both formulas are the same.

Combining Corollary 1 with (i) and (ii) of the definition 4.1 we get.

Corollary 2 *Every closed set in the A-topology is equal to the union of a finite family of closed sets each of the form $\hat{P}$, for P a Zariski closed predicate.*

Corollary 3 *Given a basic closed set $\mathbf{P} \subseteq \tilde{V}^n \times F^p$, there is a core formula $\exists e \, P$ defining $\mathbf{P}$ with the finest partition of the v-variables. That is, for every $\exists e \, R$ defining the same set the partition $\sim_P$ is refining $\sim_R$.*

Fixing a choice of parameters $\hat{e}$ (one of the finitely many), the Zariski closed relation P above is determined uniquely by the set $\mathbf{P}$. Any other choice $\hat{e}'$ of parameters for $\mathbf{P}$ determines a Zariski closed relation P' obtained from P by a linear transformation in variables y.

Indeed, take for $\exists e \, P$ the formula obtained by taking the conjunction of all possible representations $\exists e \, R$ of $\mathbf{P}$, R Zariski closed, using Corollary 1.

Lemma 4 in 3.5 implies the uniqueness of P. $\qquad \square$

We will say that the algebraic constructible set $P(F)$ for P and $\mathbf{P}$ as above is **associated** with $\mathbf{P}$. If $P(F)$ is Zariski closed we call $P(F)$ **the variety associated** with the closed set $\mathbf{P}$.

4.3 From now on when we write a basic closed set in the form $\hat{P}$ (equivalently, use the core formula $\exists e\, P)$ the kernel P is canonical, that is is uniquely determined by the set $\hat{P}$.

We define

$$\dim \hat{P} := \dim P(\mathrm{F}),$$

where dim on the right is the dimension of the algebraic variety.

For a constructible set S we define

$$\dim S := \dim \bar{S}, \text{ where } \bar{S} \text{ is the closure of } S.$$

Suppose $v = v_1^\frown v_2$, $|v_1| = n_1$, $|v_2| = n_2$, $x = x_1^\frown x_2$, $|x_1| = k_1$, $|x_2| = k_2$ and let

$$\mathrm{pr} : \tilde{\mathrm{V}}^{n_1+n_2} \times \mathrm{F}^{k_1+k_2} \to \tilde{\mathrm{V}}^{n_1} \times \mathrm{F}^{k_1}$$

be the projection $\mathrm{pr} : v_1^\frown v_2^\frown x_1^\frown x_2 \mapsto v_1^\frown x_1$.

Proposition *Let $S \subseteq \tilde{\mathrm{V}}^{n_1+n_2} \times \mathrm{F}^{p_1+p_2}$ be a closed set. Then*

(i) $\mathrm{pr}(S)$ is a constructible set;

(ii) for each $a \in \mathrm{pr}(S)$, the set $S \cap \mathrm{pr}^{-1}(a)$ is closed;

(iii) for each nonnegative integer ℓ the set

$$\{a \in \mathrm{pr}(S) : \ \dim S \cap \mathrm{pr}^{-1}(a) \geq \ell\}$$

is constructible. If $\ell > \min_{a \in \mathrm{pr}(S)} \dim S \cap \mathrm{pr}^{-1}(a)$ then the set is contained in a proper subset closed in $\mathrm{pr}(S)$.

(iv) assuming S is irreducible, we have

$$\dim S = \dim \mathrm{pr}(S) + \min_{a \in \mathrm{pr}\, S} \dim S \cap \mathrm{pr}^{-1}(a).$$

(v) for any two irreducible $S_1, S_2 \subseteq \tilde{\mathrm{V}}^n \times \mathrm{F}^p$, for every irreducible component S_0 of $S_1 \cap S_2$,

$$\dim S_0 \geq \dim S_1 + \dim S_2 - \dim \tilde{\mathrm{V}}^n \times \mathrm{F}^p.$$

Proof. (i) Follows from Claim 4 in 4.1.

(ii) Just notice that

$$S \cap \mathrm{pr}^{-1}(a) = S \cap \{v_1^\frown x_1 = a\}$$

and notice that $\{v_1^\frown x_1 = a\}$ is a basic closed set.

(iii) By our definition of dimension and the two Lemmas in 4.1 this is equivalent to the same statement for S an affine algebraic variety. This is a well-known theorem for algebraic varieties used as an axiom (FC) for Zariski geometries in [Z1].

(iv) First observe

Claim If S is irreducible then $S = \hat{P}$ with the associated variety P of the form

$$\bigcup_{\gamma \in \Gamma^s} R^\gamma, \quad R \text{ irreducible.}$$

Indeed, by definition S is a union of sets of the form $\hat{P}$. Since it is irreducible there is just one such in the union. Let $R(\mathrm{F})$ be an irreducible component of the variety $P(\mathrm{F})$. By Γ-invariance $R^\gamma(\mathrm{F})$ is also a component of P, for all $\gamma \in \Gamma_P$. By irreduciblity of S the union of all R^γ is equal to P. Claim proved.

Again as in (iii), by 4.1, (iv) is equivalent to

$$\dim P = \dim \mathrm{pr}\, P + \min_{b \in \mathrm{pr}\, P} \dim P \cap \mathrm{pr}^{-1}(b)$$

for an appropriate projection pr. But this is the known *addition formula* for algebraic varieties and more generally Zariski structures, see [Z1].

(v) First observe that by the Claim above irreducible S_i ($i = 0, 1, 2$) have to be of the form $\hat{P}_i$, for P_i of the form $\bigvee_{\gamma \in \Gamma^s} R_i^\gamma$, R_i irreducible.

The rest follows from Corollary 1 of 4.2 (see also Remark 1 in the same section). In the present notation we get by the Corollary that the kernel in $\hat{P}_1 \cap \hat{P}_2$ corresponds to the intersection of two algebraic subvarieties of dimensions $\dim P_1$ and $\dim P_2$ in the ambient affine space of dimension $\dim \tilde{V}^n + \dim \mathrm{F}^p$. By the Dimension Theorem for affine spaces we get the required inequality.   □

Theorem *For any algebra* A *satisfying the assumptions 2.1(1–4) the structure* $\tilde{V}$ *is a Zariski geometry, satisfying the presmoothness condition provided the affine algebraic variety* V *is smooth.*

Proof. The Proposition above and the topological subsection 4.1 prove all the assumptions defining Zariski geometries, see [Z1].   □

References

[BG]   K. Brown and K. Goodearl, *Lectures on Algebraic Quantum Groups*, Birkhäuser, 2002.

[C]   A. Connes, *Noncommutative geometry*, Academic Press 1994.

[H]   E. Hrushovski, Groupoids, imaginaries and internal covers, arXiv:math.LO/0603413.

[HZ]   E. Hrushovski and B. Zilber, Zariski Geometries, *Journal of Amer. Math. Soc.* 9(1996), 1–56.

[Man]   Yu. Manin, *Topics in Noncommutative geometry*, Princeton Univ. Press, Princeton, 1991.

[Maj]   Sh. Majid, *A quantum groups primer*, Lond. Math. Soc. Lecture Notes Ser. 292, Cambridge University Press, 2002.

[P]   B. Poizat, L'égalité au cube, *J. Symb. Logic* 66,(2001), 1647–1676.

[RVW]   N. Reshetikhin, A. Voronov, A. Weinstein, Semiquantum geometry, http://arxiv.org/abs/q-alg/9606007.

[Sk]   Z. Skoda, Noncommutative localization in noncommutative geometry, *Lond. Math. Soc. Lect. Note Ser.* 330 (2006), 220-313.

[Z1]   B. Zilber, *Lecture notes on Zariski structures*, 1994-2004, http://www.maths.ox.ac.uk/~zilber.

[Z2]   B. Zilber, Non-commutative geometry and new stable structures, Newton Institute preprints series NI05048-MAA, 2005.

Model theory guidance in number theory?

Ivan Fesenko

University of Nottingham

This note mentions several areas of number theory and related parts of mathematics where different aspects of model theory can potentially offer important new insights. The situations listed below are very well known to number theorists, but probably not so well to model theorists. I include just a short presentation of each of the examples, together with references to the literature. In some of them one can feel important similarities between two mathematical theories, which are still not formalized and well understood. A model theoretical analysis may provide a valuable help.

The main reason to hope for such developments involving model theory, for example as a bridge between two currently separated areas in mathematics, is that for many of the situations listed below it is natural to anticipate existence of certain common structures remaining invisible at the current level of knowledge. Model theoretical analysis could help to reveal some of those structures. In some of the situations one seeks a more algebraic construction lying behind analytical objects. And it is well known that model theory (e.g. parts such as nonstandard mathematics, geometric stability theory) provides a sort of algebraization of analytical constructions.

It is also appropriate to recall that Poizat compares the inclusion model theory – mathematical logic with the inclusion arithmetic – mathematics.

This note is an extended version of a talk given at a conference in spring 2005 inside the INI programme Model Theory and Applications to Algebra and Analysis. The reader is advised to read this note together with [Fe1].

1 Commutative – noncommutative

The endomorphism ring of a saturated model of a commutative group may become much more noncommutative than the endomorphism ring of the original object. For example, the endomorphism ring of the saturated group of nonstandard integers $^*\mathbb{Z}$ is a large noncommutative ring, see [Fe1, sect. 1].

Nonstandard commutative theories may sometimes be related to (parts of) noncommutative theories at the classical level, see [Fe1, sect.4]. This leads to many open questions about nonstandard commutative interpretation of various objects in representation theory and noncommutative geometry, and its application. See in particular [Fe1, sect. 6–10]. One challenge is to develop and apply a model theoretic point of view to the real multiplication programme [Fe1, sect.9]; for some of recent developments see [T1,T2,T3].

In particular, [T1] introduces the fundamental group of a locally internal space, a space in which every point has an internal neighbourhood, or in a different setting, a locally defined neighbourhood. It is interesting to study this and potentially other fundamental groups, defined using model theory, from the point of view of their possible relation to motivic fundamental groups.

Generally speaking, objects which are called motivic (fundamental group, cohomology theory, zeta function) seem to be a natural object of analysis for model theorists.

2 Connecting different characteristics and different p-adic worlds

Many observed but unexplained analogies between theories in characteristic zero and those in positive characteristic, or those in a geometric situation, are well known. The first mathematician who emphasized the importance of such analogies and their use in number theory and algebra was Kronecker.

2a. If one can express various constructions related to the archimedean valuation in a form symmetric to the form for nonarchimedean valuations, this would have many important consequences. See the book [H1] on many aspects of this in relation to zeta functions. Some first attempts to provide a nonstandard interpretation of some of its ideas are contained in [C].

2b. The anticipated underlying symmetry between primes reveals itself in Arakelov geometry (see, e.g. [SABK], [Mo]) only to a very partial degree. A model theoretical analysis of its main concepts would be very useful.

2c. It is well known that an analogue of the Hurwitz formula in characteristic zero in the number field case implies the ABC conjecture, see [Sm]. Could model theory be helpful in getting further insights for such a formula or inequality?

2d. A related activity is the algebra of and algebraic geometry over the "field of one element $\mathbb{F}_1$" and also so called absolute derivations, see, e.g. [K1], [K2], [So], [H2], [Dt], [KOW]. They do cry out for a model theoretic input.

2e. It was an observation of Tate and Buium that a "non-additive derivative" $\frac{da}{dp}$ with respect to prime p, a in the completion of the maximal unramified extension of $\mathbb{Q}_p$, could be defined as $\frac{\varphi(a)-a^p}{p}$ where φ is the Frobenius operator. This "non-additive derivative" was used in [Bu]. It would be interesting to have a model theoretical analysis of this "derivative" with respect to p. The work [BMS] provides a possible answer in this direction.

2f. The expression $\frac{\varphi(a)-a^p}{p}$ is involved in the definition of a p-adic logarithm function and plays a very important role in explicit formulas for the (wild) Hilbert symbol, see [FV, p.259], which themselves are a more elaborate version of much simpler formulas for the tame symbol and formulas for pairings involving differential forms for Riemann surfaces. A model theoretic insight into the unified structure of explicit formulas, which play a fundamental role in arithmetic geometry, would be of great importance for modern number theory and arithmetic geometry.

3 Representation and deformation theories

3a. Morita equivalence is not compatible with the standart part map, see [Fe1, sect.6]. To what extent can this be used for applications of model theory to representation theory?

3b. Kazhdan's principle in representation theory of reductive groups over local fields says that the theory in positive characteristic zero is often the "limit" of theories in characteristic zero when the ramification index tends to infinity, see e.g. [DKV]. Model theory in the ramified case is quite difficult, but still it is very interesting to get a model theoretical insight into this observation.

3c. Deformation theory often plays an important role, in algebraic geometry, representation theory, group theory, mathematical physics (see e.g. [G], [BGGS], [St]) but we are lacking a reasonable conceptual understanding of its general structure. Could model theoretical analysis make any progress possible?

4 Higher dimensional objects in arithmetic

Local arithmetic in dimension one and in higher dimensions works with iterated inductive and projective limits of very simple objects, finite abelian p-groups, endowed with several additional structures.

4a. In p-adic representation theory the central role is played by Fontaine's rings (see, e.g. [Be]). They are still waiting for their best definition and a model theoretic analysis may provide it.

4b. Two dimensional local fields, topology on its additive and multiplicative group, their arithmetic, and two dimensional class field theory (see [FK]) are much more difficult than the one dimensional theory. On the other hand, one can view a two dimensional local field as a subquotient of a saturated model of a one dimensional local field (see [Fe2, sect.13]), and hence one can ask for a model theoretical approach to higher local fields and their properties.

4c. Questions on a model theoretical interpretation of translation invariant measure and integration on higher dimensional local fields were asked in [Fe2,Fe3]. The answer has not required a long wait: it is contained in the recent work [HK]. In particular [HK] and its further extension unifies the translation invariant integration with the so called motivic integration. This work can be viewed as part of more general systematic use of model theory in algebraic geometry, representation theory and number theory, based on the philosophy of Hrushovski and Kazhdan that model theory allows one to naturally extend the formalism of Grothendieck's approach to the case of algebraic geometry over fields with additional structures, like henselian fields.

5 Quantum physics, and of similarities between it and number theory

5a. Hyperdiscrete (saturation of discrete) constructions in model theory correspond well to the way physicists argue in quantum physics. The combination of both discrete and continuous properties in hyperdiscrete objects is extremely promising for applications in mathematical physics: hyperdiscrete objects are ideally suited to describe the familiar type of wave–particle behaviour in physics through images of nonstandard objects under the standard part map.

5b. For a first model theoretic insight into the noncommutative structures of quantum physics, from the point of view of stability theory, see the recent work of Zilber [Z1,Z2].

5c. Divergent integrals ubiquitous in field theories can be naturally viewed via associating to them a nonstandard complex unlimited number, and various renormalization procedures could have enlightening nonstandard interpretations. In particular, nonrigorous physical constructions could be given mathematically sound justification. It is very surprising that almost nothing has been done in this direction. For first steps in nonstandard interpretations of aspects of quantum physics theories see e.g. [Y], [YO].

5d. Much of what is known about quantum field theory comes from perturbation theory and applications of Feynman diagrams for calculation of scattering amplitudes. The Feynman path integral is extremely difficult to give a mathematically sound theory. In particular, the value of the integral has a rigorous mathematical meaning as a hypercomplex number, manipulations with which do produce standard complex numbers seen in the recipes of Feynman. Recall that Wiener measure (often used in mathematical approaches to the path integral) can be viewed as the Loeb measure associated to a hyper random discrete walk. On the other hand, the translation invariant $\mathbb{R}((X))$-valued measure on two dimensional local fields [Fe2, Fe3] can be viewed as induced from a hyper Haar measure. The additive group of two dimensional local fields, an arithmetic loop space on which that measure is defined, is reasonably close to the loop space on which one calculates the Feynman integral. Could model theory provide a better understanding of the Feynman integral?

5e. Vafa in [Va] indicates that number theory remains the most important part of mathematics with which quantum physics has not had any essential interrelation. Moreover he suggests that quantum mechanics would be reformulated in this century using number theory. I hope that the future understanding of relations of quantum physics with number theory should involve model theory as an interpreteer and friendly guide.

Of course many other areas have not been mentioned in this short note. For some of the most important (interrelations number theory – complex analysis, number theory – dynamical systems, number theory – algebraic topology) see [Vo], [Dn1], [Dn2], [Ma], [Fu].

References

[Be]   L. Berger, An introduction to the theory of p-adic representations, in *Geometric Aspects of Dwork Theory*, 255–292, Walter de Gruyter, Berlin, 2004, available from www.ihes.fr/~lberger/

[BGGS] Ph. Bonneau, M. Gerstenhaber, A. Giaquinto, D. Sternheimer, Quantum groups and deformation quantization: explicit approaches and implicit aspects, *J. Math. Phys.* 45 (2004), no. 10, 3703–3741.

[BMS] L. Bélair, A. Macintyre, T. Scanlon, Model theory of Frobenius on Witt vectors, 2002, available from math.berkeley.edu/~scanlon/papers/papers.html

[Bu]   A. Buium, Differential characters of abelian varieties over p-adic fields, *Invent. Math.* 122 (1995), 309–340.

[C]   B. Clare, Properties of the hyper Riemann zeta-function, preprint, Nottingham 2004.

[DKV] P. Deligne, D. Kazhdan, M.-F. Vignéras, *Représentations des groupes réductifs sur un corps local*, Travaux en Cours, Hermann, Paris, 1984.

[Dn1] C. Deninger, Number theory and dynamical systems on foliated spaces, *Jahresber. DMV* 103 (2001), no. 3, 79–100.

[Dn2] C. Deninger, A note on arithmetic topology and dynamical systems, Algebraic number theory and algebraic geometry, 99–114, *Contemp. Math.*, 300, AMS 2002.

[Dt]   A. Deitmar, Schemes over $\mathbb{F}_1$, in *Number fields and function fields— two parallel worlds*, 87–100, Progr. Math., 239, Birkhäuser Boston, Boston, MA, 2005.

[Fe1] I. Fesenko, Several nonstandard remarks, in *Representation theory, dynamical systems, and asymptotic combinatorics*, 37–49, Amer. Math. Soc. Transl. Ser. 2, 217, Amer. Math. Soc., Providence, RI, 2006, available from www.maths.nott.ac.uk/personal/ibf/rem.pdf

[Fe2] I. Fesenko, Analysis on arithmetic schemes. I, in *Kazuya Kato's fiftieth birthday*, Docum. Math. Extra Volume (2003) 261–284, available from www.maths.nott.ac.uk/personal/ibf/ao1.pdf

[Fe3] I. Fesenko, Measure, integration and elements of harmonic analysis on generalized loop spaces, Proceed. St. Petersburg Math. Soc., Vostokov Festschrift, vol. 12 (2005), 179–199, available from www.maths.nott.ac.uk/personal/ibf/aoh.ps

[FK]   I.B. Fesenko, M. Kurihara (eds.), *Invitation to higher local fields*

Geometry and Topology Monographs vol 3, Warwick 2000, available from www.maths.nott.ac.uk/personal/ibf/mp.html

[FV]   I.B. Fesenko, S.V. Vostokov, *Local fields and their extensions*, 2nd ed., AMS 2002, available from www.maths.nott.ac.uk/personal/ibf/book/book.html

[Fu]   K. Fujiwara, Algebraic number theory and low dimensional topology, Nagoya Univ. preprint 2005.

[G]   F. Gouvêa, Deformations of Galois representations, in *Arithmetic algebraic geometry (Park City, UT, 1999)*, 233–406, Amer. Math. Soc., Providence, RI, 2001.

[H1]   Sh. Haran, *The mysteries of the real prime*, Clarendon Press, Oxford 2001.

[H2]   Sh. Haran, Arithmetic as geometry I. The language of non-additive geometry, preprint, Technion 2005.

[HK]   E. Hrushovski, D. Kazhdan, Integration in valued fields, available from arxiv.org/abs/math.AG/0510133.

[K1]   N. Kurokawa, Absolute Frobenius operators, *Proc. Japan Acad. Ser. A Math. Sci.* 80 (2004), no. 9, 175–179.

[K2]   N. Kurokawa, Zeta functions over $\mathbf{F}_1$, *Proc. Japan Acad. Ser. A Math. Sci.* 81, no. 10 (2005), 180–184.

[KOW]   N. Kurokawa, H. Ochiai, M. Wakayama, Absolute derivations and zeta functions, in *Kazuya Kato's fiftieth birthday*, Doc. Math. (2003) Extra Vol., 565–584

[Ma]   Yu.I. Manin, Lectures on zeta functions and motives (according to Deninger and Kurokawa), Columbia University Number Theory Seminar (New York, 1992), *Astérisque* No. 228, (1995), 4, 121–163.

[Mo]   A. Moriwaki, Diophantine geometry viewed from Arakelov geometry, *Sugaku Expositions* 17 (2004), no. 2, 219–234.

[SABK]   C. Soulé, D. Abramovich, J. F. Burnol, J. K. Kramer, *Lectures on Arakelov Geometry*, Cambridge Univ. Press 1995.

[Sm]   A. L. Smirnov, Hurwitz inequalities for number fields, *Algebra i Analiz* 4 (1992), no. 2, 186–209; Engl. transl. in *St. Petersburg Math. J.* 4 (1993), 357–375.

[So]   C. Soulé, Les variétés sur le corps à un élément, *Mosc. Math. J.* 4 (2004), no. 1, 217–244

[St]   D. Sternheimer, Deformation theory: a powerful tool in physics modelling. In *Poisson geometry, deformation quantisation and group representations*, 325–354, LMS Lecture Note Ser., 323, Cambridge Univ. Press, Cambridge, 2005.

[T1]   L. Taylor, A hyperreal manifold with real multiplication, preprint, Nottingham 2005.

[T2]   L. Taylor, Pseudo-analytic torus, preprint, Nottingham 2005.

[T3]   L. Taylor, The double since function and theta functions on the noncommutative torus, preprint, Nottingham 2005.

[Va]   C. Vafa, On the future of mathematics/physics interaction, in *Mathematics: frontiers and perspectives*, 321–328, AMS 2000.

[Vo]   P. Vojta, Nevanlinna theory and Diophantine approximation, in *Several complex variables (Berkeley, CA, 1995–1996)*, 535–564, Math. Sci. Res. Inst. Publ., 37, Cambridge Univ. Press, Cambridge, 1999.

[Y]   H. Yamashita, Nonstandard methods in quantum field theory I: a hyperfinite formalism of scalar fields, *Intern. J. Th. Phys.* 41(2002),

511–527.

[YO]   H. Yamashita, M. Ozawa, Nonstandard representations of the canon-
         ical commutation relations, *Rev. Math. Phys.*, 12(2000), 1407–1427.

[Z1]   B. Zilber, Pseudo-analytic structures, quantum tori and
         non-commutative geometry, preprint 2005, available from
         `www.maths.oxford.ac.uk/~zilber`

[Z2]   B. Zilber, Non-commutative geometry and new stable structures,
         preprint 2005, available from `www.maths.oxford.ac.uk/~zilber`